NATO ASI Series

Advanced Science Institutes Series

A series presenting the results of activities sponsored by the NATO Science Committee, which aims at the dissemination of advanced scientific and technological knowledge, with a view to strengthening links between scientific communities.

The Series is published by an international board of publishers in conjunction with the NATO Scientific Affairs Division

A	Life Sciences	Plenum Publishing Corporation
B	Physics	London and New York
C	Mathematical and Physical Sciences	Kluwer Academic Publishers
D	Behavioural and Social Sciences	Dordrecht, Boston and London
E	Applied Sciences	
F	Computer and Systems Sciences	Springer-Verlag
G	Ecological Sciences	Berlin Heidelberg New York
H	Cell Biology	London Paris Tokyo Hong Kong
I	Global Environmental Change	Barcelona Budapest

PARTNERSHIP SUB-SERIES

1. Disarmament Technologies	Kluwer Academic Publishers
2. Environment	Springer-Verlag/Kluwer Acad. Publishers
3. High Technology	Kluwer Academic Publishers
4. Science and Technology Policy	Kluwer Academic Publishers
5. Computer Networking	Kluwer Academic Publishers

The Partnership Sub-Series incorporates activities undertaken in collaboration with NATO's Cooperation Partners, the countries of the CIS and Central and Eastern Europe, in Priority Areas of concern to those countries.

NATO-PCO DATABASE

The electronic index to the NATO ASI Series provides full bibliographical references (with keywords and/or abstracts) to about 50000 contributions from international scientists published in all sections of the NATO ASI Series. Access to the NATO-PCO DATABASE compiled by the NATO Publication Coordination Office is possible in two ways:

- via online FILE 128 (NATO-PCO DATABASE) hosted by ESRIN, Via Galileo Galilei, I-00044 Frascati, Italy.

- via CD-ROM "NATO Science & Technology Disk" with user-friendly retrieval software in English, French and German (© WTV GmbH and DATAWARE Technologies Inc. 1992).

The CD-ROM can be ordered through any member of the Board of Publishers or through NATO-PCO, Overijse, Belgium.

Series I: Global Environmental Change, Vol. 53

Springer
Berlin
Heidelberg
New York
Barcelona
Budapest
Hong Kong
London
Milan
Paris
Santa Clara
Singapore
Tokyo

Atmospheric Ozone Dynamics

Observations in the Mediterranean Region

Edited by

Costas Varotsos
University of Athens
Department of Physics, Division of Applied Physics
Laboratory of Meteorology
Panepistimioupolis, Build PHYS.V
Athens, GR 157 84, Greece

With 139 Figures (12 Colour Plates)

Springer
Published in cooperation with NATO Scientific Affairs Division

Proceedings of the NATO Advanced Research Workshop "Ground Level and Satellite Ozone Observations. Changes in the Mediterranean Region", held in Athens, Greece, October 31 – November 4, 1995

Cataloging-in-Publication Data applied for

Die Deutsche Bibliothek - CIP-Einheitsaufnahme

Atmospheric ozone dynamics : observations in the Mediterranean region ; [proceedings of the NATO Advanced Research Workshop "Ground Level and Satellite Ozone Observations. Changes in the Mediterranean Region", held in Athens, Greece, October 31 - November 11, 1995] / ed. by Costas Varotsos. Publ. in cooperation with NATO Scientific Affairs Division. - Berlin ; Heidelberg ; New York ; Barcelona ; Budapest ; Hong Kong ; London ; Milan ; Paris ; Santa Clara ; Singapore ; Tokyo : Springer, 1997
(NATO ASI series : Ser. I, Global environmental change ; Vol. 53)
ISBN-13: 978-3-642-64558-7 eISBN-13: 978-3-642-60797-4
DOI:10.1007/978-3-642-60797-4

Softcover reprint of the hardcover 1st edition 1997

Typesetting: Camera ready by the authors/editors
Printed on acid-free paper
SPIN: 10525492 31/3137 - 5 4 3 2 1 0

Preface

An increasing annoyance because of the anthropogenically induced ozone depletion and relevant increase of the surface ultraviolet (UV) radiation (including erythemal part of the UV spectrum) has resulted in the growth of interest to total ozone (TOZ) surface and satellite observations, surface UV measurements (with an emphasis on UV-B) and the interpretation of observation results to assess potential impacts of the UV radiation enhancement on man and biosphere. The significance of this phenomenon is different in various countries. Special attention has been paid, for instance, to ozone depletion and UV radiation increase at high latitudes (this problem has also been discussed during the Workshop). It is equally clear, however, that low and mid-latitude environmental dynamics requires carefull monitoring and assessment of environmental trends as well. Such a conclusion is especially true for the Mediterranean Region where numerous resort places are located and subject to excessive levels of Solar Ultraviolet Radiation. This is why the NATO Advanced Research Workshop "Ground-Level and Satellite Ozone Observations: Changes in the Mediterranean Region" was organized and took place in Athens (Greece) during the time period 31 October - 4 November 1995. It should be pointed out, however, that presentations discussed during the Workshop went far beyond the regional scale of consideration. This has been very important, because it opened an opportunity to analyse the Mediterranean environmental situation in the broader context of the European global changes. An important advantage of the Workshop's programme has also been the consideration of not only problems of TOZ and UV-B variability but also tropospheric ozone dynamics as well as those causes (chemical and photochemical reactions coupled with air transport) which are responsible for changes of the ozone concentration in the troposphere.

Total and tropospheric ozone variability takes place due to impacts of a number of factors which contributions have not been reliably assessed as yet. A critical issue is a relationship between natural and anthropogenically induced changes. It is important that impacts of various factors have been discussed (chemical and photochemical processes, meteorological conditions, impacts of stratospheric aerosols, solar activity etc.). In a substantial part of the papers tropospheric pollution has been considered in the context of ozone changes. Apart from the analysis of observation data some key problems of numerical modelling have been discussed, such as tropospheric chemical composition changes, Monte Carlo radiative transfer calculations, and others. The subject of a number of papers has been the development of remote sensing techniques to retrieve minor optically active components of the atmosphere.

Thus, the programme of the Workshop has been characterised by a system's approach to the solution of the problems considered which has been the basis of the successful accomplishment of the Workshop's aims. It has been equally important that many outstanding experts presented their contributions. The Workshop more than justified the expectations. The support from NATO and local sources (Training Centre of National Bank S.A. and others) has helped to reach a high level of the organisation. The contribution of the collaborators of the University of Athens and other Greek institutions to the success of the Workshop was outstanding.

The editor is grateful to the NATO Scientific Affairs Division, the European Environment Agency, the Greek Ministry of Development, the Greek Ministry of Environment, Physical Planning and Public Works, the Greek Ministry of National Economy, the Greek Ministry of Culture, the University of Athens, the Hellenic Airline Association, the Hellenic Opticians Association, the Hellenic Civil Aviation Authority, the PHARMA-CARE, the EKO Lubricant and Refineries ABEE, the S. A. Vairamoglou S.A. and the Ventouris Lines Companies, the Hellenic Foreign Trade Board the National Mortgage Bank of Greece, for their sponsorship.

Finally, this Workshop is dedicated to the memory of the scientists Patrick Aimedieu and Kjell Henriksen, who passed away.

The Organising Committee of this NATO ARW

Contents

Name and address of each author/editor

Name	Title	Address
Aguirre de Cancer I.	Dr.	Departamento Física de Materiales, C-IV Universidad Autónoma de Madrid, Cantoblanco, 28049-Madrid, Spain
Bock H. G.	Dr.	Interdisziplinäres Zentrum für Wissenschaftliches Rechnen, Universität Heidelberg, D-69120 Heidelberg, Germany
Borrell Peter	Dr.	EUROTRAC ISS, Fraunhofer Institute (IFU), Garmisch-Partenkirchen, Germany
Builtjes Peter	Dr.	EUROTRAC ISS, Fraunhofer Institute (IFU), Garmisch-Partenkirchen, Germany
Cartalis Costas	Dr.	University of Athens, Department of Physics, Division of Applied Physics, Laboratory of Meteorology, Panepistimioupolis, Build. PHYS V, Athens, GR 157 84, Greece
Cracknell Arthur P.	Prof.	Department of Applied Physics and Electronic and Mechanical Engineering, University of Dundee, DUNDEE DD1 4HN, Scotland, UK
Cussó F.	Dr.	Departamento Física de Materiales, C-IV Universidad Autónoma de Madrid, Cantoblanco, 28049-Madrid, Spain
Dalyuk I.rina	Dr.	Main Geophysical Observatory, Karbyshev Str., 7, St. Petersburg 194021, Russia
Degórska Malgorata	Dr.	Institute of Geophysics, Polish Academy of Sciences, 01-452 Warsaw, Ks. Janusza 64, Poland
de la Morena Benito	Dr.	Departamento de Teledetección y Aeronomía, Instituto Nacional de Técnica Aeroespacial, C/ Ajalvir km 4, 28850 Torrejón de Ardoz,

Hov Øystein	Dr.	EUROTRAC ISS, Fraunhofer Institute (IFU), Garmisch-Partenkirchen, Germany
Jakobs Hermann J.	Dr.	University of Cologne, Institute for Geophysics and Meteorology, EURAD, Aachener Str. 201-209, 50931 Cologne, Germany
Jaque F.	Dr.	Departamento Física de Materiales, C-IV Universidad Autónoma de Madrid, Cantoblanco, 28049-Madrid, Spain
Kallos George	Asst. Prof.	University of Athens, Department of Physics, Division of Applied Physics, Laboratory of Meteorology, Panepistimioupolis, Build. PHYS V, Athens, GR 157 84, Greece
Köhler Ulf	Dr.	Deutscher Wetterdienst, Meteorological Observatory Hohenpeissenberg, Albin-Schwaiger-Weg 10, 82383 Hohenpeissenberg, Germany
Kondratyev Kirill Ya.	Prof.	Russian Academy of Science, Research Centre of Ecological Safety, 18 Korpusnaya St., 197110, St. Petersburg, Russia
Kotroni Vassiliki	Dr.	University of Athens, Department of Physics, Division of Applied Physics, Laboratory of Meteorology, Panepistimioupolis, Build. PHYS V, Athens, GR 157 84, Greece
Lagouvardos Kostas	Dr.	University of Athens, Department of Physics, Division of Applied Physics, Laboratory of Meteorology, Panepistimioupolis, Build. PHYS V, Athens, GR 157 84, Greece
Lifante G.	Dr.	Departamento Física de Materiales, C-IV Universidad Autónoma de Madrid, Cantoblanco, 28049-Madrid, Spain
Marquard Lutz	Dr.	Institut für Umweltphysik, Universität Heidelberg, D-69120 Heidelberg, Germany

EUROTRAC: Applications to Photo-oxidants and the future project

Peter Borrell[1], Peter Builtjes, Peringe Grennfelt, Øystein Hov, Roel van Aalst, David Fowler, Gérard Mégie, Nicolas Moussiopoulos, Peter Warneck, Andreas Volz-Thomas, Richard Wayne

[1] EUROTRAC ISS, Fraunhofer Institute (IFU), Garmisch-Partenkirchen, Germany

Abstract. The conclusions of the EUROTRAC Application Project concerning photo-oxidants in Europe are presented, together with a brief account of the proposed project, EUROTRAC-2, which it is hoped will follow when the present project finishes at the end of 1995.

Keywords. photo-oxidants, air pollution, troposphere, ozone, NO_X, VOC, PAN, trans-boundary problems, atmospheric modelling, source-receptor relationships, environmental policy, EUROTRAC, EUREKA

1 Introduction

EUROTRAC, the European co-ordinated research project within the EUREKA initiative studying the transport and chemical transformation of pollutants in the troposphere, is now nearing the end of the first phase. The project, which comprises some 250 research groups organised into 14 subprojects, has achieved a remarkable scientific success since its start in 1988, bringing together international groups of scientists to work on problems directly related to the transport and chemical transformation of trace substances in the troposphere [1,2].

To ensure that full use is made of the results to support environmental policy development, the Application Project (AP) was formed to assimilate the scientific results from EUROTRAC and present them in a condensed form so that they are suitable for use by those responsible for environmental planning and management in Europe.

One of the three themes addressed by the AP was "Photo-oxidants in Europe; in the free troposphere; in rural and urban atmospheres". This paper presents a preliminary account of the major conclusions concerning photo-oxidants in Europe.

EUROTRAC itself finishes at the end of 1995, and the paper also presents a brief description of the plans for a future project, EUROTRAC-2.

2 The Application Project Report: the source of the conclusions in this paper

The EUROTRAC Application Project, of which the authors are members, has been working for nearly two years to extract and summarise the policy-related scientific information obtained within EUROTRAC. The following paragraphs are taken

NATO ASI Series, Vol. I 53
Atmospheric Ozone Dynamics
Observations in the Mediterranean Region
Edited by Costas Varotsos

from the Executive Summary [3] of the AP final report. No attempt is made here to present the scientific results themselves, or the arguments upon which the conclusions are based. These are to be found in the report of the project, which will appear as volume 10 of the final scientific report of EUROTRAC [4].

3. Photo-oxidants in EUROPE

3.1 Policy applications of EUROTRAC scientific results

The application of effects-based control strategies by European governments, in which the maximum environmental benefit is being sought for the investment in control technology, places great demands on our knowledge and understanding of the links between sources, deposition and effects. The research accomplished in EUROTRAC provides substantial scientific support for the negotiations in the second generation of abatement strategy protocols under the UN-ECE Convention on the Long Range Transport of Air Pollution (LRTAP), in particular the second revised NO_x protocol and the revisions of the recently signed sulphur protocol, and for work within the European Environmental Agency, the EU Framework Directive for Ozone and in the development of national strategies. EUROTRAC models and measurements are also used by WMO/UNEP and IPCC for assessment of the current ozone budget and its sensitivity to changes in precursor concentrations.

3.2 The concentration of photo-oxidants in Europe is strongly influenced by photochemical production from man made precursors that are emitted within the region

From experimental studies it is concluded that the natural background of ozone over Europe at the turn of the century, within the atmospheric boundary layer, was about 10 to 15 ppb at ground level and 20 to 30 ppb one to two kilometres above the ground. Today, the concentration of ozone near the sea surface is 30 to 35 ppb before air masses move into Europe from the west. On a seasonal basis, photochemical processes over western and central Europe add about 30 to 40 % to this background in summer, and subtract about 10% in winter.

Within Europe very high concentrations of more than 100 ppb are observed during photochemical episodes under unfavourable meteorological conditions, i.e. high solar radiation combined with stagnant air or circulating wind systems.

In the free troposphere, that is from the top of the atmospheric boundary layer (1 to 2 km above the ground) to the tropopause which constitutes the boundary with the stratosphere (10 to 12 km above the ground), the background concentration before the air masses pass over Europe is higher than in the atmospheric boundary layer, being about 40 to 50 ppb in winter and autumn and 50 to 70 ppb in spring and summer.

The concentration of free tropospheric ozone over Europe is influenced not only by European emissions but also by North American and Japanese emissions.

3.3 Tropospheric ozone in the northern hemisphere has increased since the 1950s

The concentration of ozone in the troposphere north of 20°N has increased since the beginning of modern measurements. This increase was larger at northern mid-latitudes than in the tropics, and larger over Europe and Japan than over North America. Comparison with historical data suggests that ozone in the troposphere over Europe has doubled since the turn of the century and that most of the increase has occurred since the 1950s. Measurements of nitrate in ice cores from Alpine glaciers provide strong circumstantial evidence for man-made emissions being responsible for the observed ozone trend.

3.4 Tropospheric ozone increase slowed down in the 1980s

Long term observations show that the increase of ozone in the free troposphere was smaller in the eighties than in the seventies. The average ozone concentrations in the boundary layer near the ground have even decreased at some locations, for example at Garmisch-Partenkirchen in Germany and at Delft, a polluted site in the Netherlands. The concentration of peroxyacetylnitrate (PAN), a photo-oxidant like ozone, increased by a factor of three at Delft in the 1970s and stabilised in the 1980s.

3.5 Weekday/weekend differences in the emissions of ozone precursors

The understanding of the temporal and spatial resolution of the emissions of NO_X and VOC has improved considerably. Results show for example that emissions are approximately 30% lower at weekends than during the week. Such variations provide a regular "natural experiment" for examining the effects of the short term reductions of emissions. The effects of the reductions on the photo-oxidant concentrations in this case appear to be rather small.

3.6 Should NO_x or VOC be controlled, or both?

Recent model simulations have suggested that the effective abatement of elevated ozone concentrations in Europe require the reduction of the emissions of both NO_X and VOC, with more emphasis being put on VOCs, especially in north-west and central Europe. However some field experiments in EUROTRAC have identified possible shortcomings in the models that are presently used for quantifying ozone/VOC and ozone/NO_X relationships. These experiments emphasise the greater importance of NO_X emissions in controlling the photochemical ozone balance. The reasons are:

- There is an indication, based on ambient measurements, that biogenic VOC emissions, which cannot be subjected to abatement, form a base level of VOCs which is higher than that assumed previously.

- The photochemistry in urban plumes seems to proceed faster than is assumed in models. The results suggest that the oxidation of VOCs leads to more peroxy radicals and, hence, more ozone over a shorter time than predicted by the photochemical schemes currently used in atmospheric models, and to a faster removal of NO_X, the catalyst in ozone formation.

Based on today's knowledge, the following picture emerges:

- An effective way to reduce ozone concentrations on urban and suburban scales appears to be to reduce VOC emissions. However, NO_x reductions are required to reduce the concentrations of other oxidants such as NO_2 and PAN.

- Both NO_x and VOC reductions are required in order to reduce ozone levels on a European scale. NO_x reductions are essential to reduce ozone concentrations on a global scale.

- The emission reductions need to be substantial (40 to 60 %) to obtain noticeable reductions in ozone concentrations.

3.7 Are photo-oxidants a home made or trans-boundary problem?

The pre-industrial concentration of ozone of about 10 to 15 ppb at ground level resulted from the approximate balance between the transfer of ozone from the stratosphere to the troposphere, the destruction in the troposphere by photochemical reactions and by deposition to the ground. The 15 to 20 ppb difference between today's ozone levels, near the sea surface before air masses pass over Europe, and the pre-industrial ozone concentration is probably due to photochemical formation from precursors such as VOC and NO_x, emitted in other parts of the northern hemisphere, in particular North America. Reduction in the background tropospheric level will require agreement on a hemispheric scale. In addition, precursors emitted from biomass burning have a large impact on ozone concentrations in the tropics and in the southern hemisphere.

The very high ozone, NO2 and PAN concentrations that are observed in some urban and suburban areas (photochemical smog) are due to photochemical production in the atmospheric boundary layer from precursors that are mostly emitted within the area.

On a continental scale, the enhanced photo-oxidant concentrations observed are a consequence of both in-situ chemistry and transport from regions with higher emissions. Studies in EUROTRAC have greatly added to our understanding of the relevant processes for quantifying ozone/precursor relationships and, hence, provide a better basis for determining how reductions in precursor emissions in one region would reduce the photo-oxidant levels in regions downwind, especially in moderately populated and rural areas (scales >50 km). However, source-receptor relationships are still difficult to assign because the chemistry is non-linear and there are large differences in the residence times of photo-oxidants and precursors in the atmospheric boundary layer close to the ground, compared with the residence times in the free troposphere.

While the highest photo-oxidant levels can be counteracted by local pollution control measures, abatement of enhanced ozone formation on a European scale requires a co-ordinated abatement strategy.

3.8 Practical applications of photo-oxidant models on all scales
Strategies to abate photochemical air pollution at any relevant scale may be assessed with models developed within EUROTRAC.

On the local scale, a zooming model (EZM) has already been successfully utilised to optimise the air pollution abatement strategy for Athens, to support the decisions taken with regard to traffic regulations in Barcelona during the 1992 Olympics and to interpret the observations during measuring campaigns in the Upper Rhine Valley.

Models have also been developed to describe the distribution of ozone in the global troposphere. The results show that the concentrations of ozone throughout large parts of the northern hemisphere have been substantially increased by anthropogenic emissions of nitrogen oxides, VOCs and carbon monoxide. Since ozone is a greenhouse gas, these elevated levels could be making an appreciable contribution to global warming..

The EURAD model as well as global models have been used to calculate the influence of aircraft emissions on upper tropospheric ozone levels.

4. EUROTRAC-2: the proposed project

EUROTRAC was judged a success both by the scientists involved and also by the supporting governments. The scientists benefited from their inclusion in the subprojects, which had well-defined scientific aims and provided a framework of helpful co-ordination which facilitated the work. The governments recognised that the research groups which they funded were working towards the solution of an inter-disciplinary, trans-boundary environmental problem. It was agreed by both the scientific community and the participating governments that an effort should be made to establish a new project to follow EUROTRAC when it finishes at the end of 1995. A project description was developed and an application made to EUREKA for the admission of EUROTRAC-2 as a new project. A decision is expected early in 1996. The following sections are extracts from the project description[5].

4.1 EUROTRAC-2: a summary of the proposal
The proposed project, EUROTRAC-2, will contribute directly to the further development of strategies for reducing the anthropogenic contribution to the abundance of photo-oxidants and acidic substances over Europe, by improving the quantitative understanding of the factors determining the formation, transport, chemical transformation, deposition and impact of photo-oxidants and acidic substances in the troposphere. It will thus assist in alleviating the growing environmental problems which stem from these species

EUROTRAC-2 will provide an integrated scientific evaluation of photo-oxidants and acidifying substances in the troposphere over Europe. Within the project advanced techniques linking numerical atmospheric models and observational data will be developed and used to provide an unprecedented capability for understanding the complex interactions between the sources, the concentrations and deposition of trace substances appropriate to environmental impact. The measurements and modelling will cover Europe as a whole but additional attention will be paid to eastern and southern Europe, including Russia, which for geographical, meteorological and social reasons, offer particular environmental and scientific challenges at the present time. The project will then cover the whole cycle of atmospheric pollution within a European context from emission to deposition and so establish the required source-receptor relationships.

The results obtained in EUROTRAC-2 will provide a common scientific understanding upon which effective environmental abatement and control strategies in Europe can be based. The proposed project will build upon the strengths of the first phase of EUROTRAC, harnessing the research resources of the participating European countries to create viable active research groups to address inter-disciplinary and international trans-boundary problems.

The project, like its predecessor, will be within the EUREKA initiative. It is also intended that it should be closely linked with other ongoing European air-quality activities, such as EMEP, the EEA topic centre for air quality, the EC (DG-XI) group on the ozone directive and the atmospheric chemistry environmental program of the EC (DG-XII). It is further expected that the project will be integrated with the appropriate IGBP/IGAC activities.

4.2 EUROTRAC-2: the objectives

The overall objective of EUROTRAC-2 (*The Transport and Chemical Transformation of Environmentally Relevant Trace Constituents in the Troposphere over Europe; Second Phase*) is

> " to support the further development of abatement strategies within Europe by providing an improved scientific basis for the quantification of source-receptor relationships for photo-oxidants and acidifying substances".

The project will provide an integrated scientific evaluation of photo-oxidants and acidifying substances in the troposphere over Europe, using the extended knowledge base and improved methodologies gained in the first phase of EUROTRAC. In particular advanced techniques in linking numerical atmospheric models and observational data will be used to provide an unprecedented capability for understanding the complex interactions between the sources, transformations and deposition which determine the concentrations of trace substances appropriate to environmental impact. The work will thus provide scientific support for future measures needed to control photo-oxidants and the deposition of acidifying substances and nutrients.

The specific objectives are

1. *Quantification of atmospheric interactions*

   To quantify the anthropogenic and natural contributions of relevant emissions and atmospheric processes to the abundance and the long term changes of photo-oxidants and acidifying substances in the planetary boundary layer and free troposphere.

2. *Evaluation of feedback mechanisms*

   To evaluate the consequences of feed-back mechanisms, for example: the feed-back between the concentrations of tropospheric photo-oxidants and biogenic emissions; the feed-back between the concentrations of photo-oxidants and those of climatically relevant atmospheric constituents; and the feed-back between the changing intensity of ultraviolet radiation and photo-oxidant production.

3. *Contribution to the formulation of abatement strategies and future air quality*

   To contribute to the formulation and improvement of strategies for reducing the anthropogenic contribution to the abundance of photo-oxidants and acidifying substances and to the prediction of future air quality on shorter and longer time scales.

The project description identifies seven sets of tasks which should be undertaken in order to obtain the scientific understanding and experience, necessary to provide the desired source-receptor relationships.

If accepted as a EUREKA project and fully supported by the participating governments, the new project will last for six years and should help to provide the firm scientific basis required for future environmental policy development in Europe.

References

1. EUROTRAC ISS, *The EUROTRAC Annual Report for 1993*, parts 1 to 9, Garmisch-Partenkirchen, 1994.
2. Borrell, P.M., Borrell, P., Cvitas, T., & Seiler, W. *(eds.), Transport and Transformation of Pollutants in the Troposphere*, Proc. .EUROTRAC Symp. '94, SPB Academic Publishing, Hague, 1994.
3. Borrell, P., Builtjes, P., Grennfelt, P., Hov, Ø., van Aalst, R., Fowler, D., Mégie, G., Moussiopoulos, N., Warneck, P., Volz-Thomas, A. & Wayne R. Photo-oxidants, Acidification and Tools; *Policy Applications of EUROTRAC Results; the Executive Summary*, EUROTRAC Newsletter 16, 16-22, 1995.
4. Borrell, P., Builtjes, P., Grennfelt, P., Hov, Ø., van Aalst, R., Fowler, D., Mégie, G., Moussiopoulos, N., Warneck, P., Volz-Thomas, A. & Wayne R. *Photo-oxidants, Acidification and Tools; Policy Applications of EUROTRAC Results*, Springer Verlag, Heidelberg, 1996 (in press).
5. EUROTRAC IEC, SSC & ISS, *EUROTRAC Phase 2*, EUROTRAC Newsletter 15, 11-19, 1995.

The spatial variation of ozone depletion in Europe

A. P. Cracknell[1], C. P. D. McArthur[1] and C. Varotsos[2]

[1] Department of Applied Physics and Electronic and Mechanical Engineering University of Dundee, DUNDEE DD1 4HN, Scotland, UK

[2] Department of Applied Physics, University of Athens, 33 Ippokratous Str., 106 80, Athens, Greece

Abstract. We have previously used data from the TOMS instrument, which was flown on Nimbus-7, to study total ozone column amounts for a number of different locations in Europe, including Athens, Thessaloniki, Crete, Dundee and St. Petersburg. In each case we have studied the monthly average ozone amount over a period of a decade or more. There is no doubt that there has been an overall depletion during this period. However, there are very considerable differences among the behaviour of the monthly averages. For some months and for some locations there is a quite large depletion, while for other months and locations there are quite small depletions or even no significant depletion at all. We have recently extended the network of locations at which we have studied these monthly averages to include London, Berlin, Mallorca and Naples. The results will be presented in this paper and compared with the results obtained for other locations previously.

Keywords. TOMS, Nimbus-7, ozone depletion

1 Introduction

Measurements of the total ozone content in the atmospheric column are made with ground-based instruments, namely Dobson spectrophotometers and Brewer spectrophotometers, and with the satellite-flown Total Ozone Monitoring Spectrometer (TOMS) which has been flown on Nimbus-7, Meteor-3 and two of the spacecraft in the NOAA series of polar-orbiting meteorological satellites. The ground-based instruments give rather direct measurements but only at a restricted network of points. The TOMS instruments do not give such direct measurements and they need to be calibrated against the ground-based instruments, but they do have the advantage of giving frequent global coverage.

One can see published diagrams in the literature illustrating ozone depletion on a very coarse global scale and derived from either ground-based data or TOMS data, see figure 14 of WMO and UNEP, 1995, for example. However, such figures have their limitations by presenting seasonal data over a very large area; the principal limitations are:

NATO ASI Series, Vol. I 53
Atmospheric Ozone Dynamics
Observations in the Mediterranean Region
Edited by Costas Varotsos

(a) This is a temporal average. If we looked at monthly means we would see substantially different patterns for different seasons.
(b) This represents a very broad spatial average and hides a large amount of local variation in detail.

We have previously used data from the TOMS (Total Ozone Monitoring Spectrometer) flown on Nimbus-7 to study the depletion of the total ozone amount at particular locations over the period 1979-1992. These locations include Athens (Varotsos and Cracknell 1993) Crete and Thessaloniki (Cracknell *et al.* 1994), Dundee (Cracknell and Varotsos 1994) and St. Petersburg (Kondratyev *et al.* 1994).

The TOMS data is supplied by NASA as a global dataset on a CD-ROM, along with a computer program to run on a PC to extract the total ozone amount at a location and over a time period that the user is free to specify. This program will generate total ozone amounts in the vertical column at the specified location and on a day-by-day basis within the period specified. The algorithms used by NASA in this program are ultimately validated by data from ground-based measurements, principally obtained using Dobson spectrophotometers. We have used TOMS version 6 data in this work.

In our previous work on Athens, for example, we have used the NASA CD-ROM to generate the daily ozone values for Athens and then used these values to generate monthly means (Varotsos and Cracknell 1993), Figure 1.1 shows the monthly means at Athens. One can notice two things: (a) There is a very clear oscillation with a period of one year. The values are high, *ca* 380 D.U. in the early part of the year (February, March and April), thereafter they drop in May, June and July and then remain fairly steady at about 300 - 320 D.U. until December, rising again in January to about 360 D.U; (b) There is a downward trend as well, but if one just looked at the curve without the straight line one might not find it very convincing. Taken over the whole period, the depletion comes out over the period from January 1979 to January 1992 at (6 ± 2) %, or a monthly depletion rate of circa 0.135 D.U. per month.

In the paper already mentioned (Varotsos and Cracknell 1993) we also plotted the data for each month separately. Each month shows a depletion, but the value varies, see Figure 1.2. One can see that the magnitude of the depletion (as a percentage) is largest in the early part of the year, which is when the absolute values themselves are largest. Notice that the depletion varies between 1 and 2% in the later part of the year but is in the range 7-11% in the early part of the year. This means that, since the absolute values of the ozone concentration are largest in the early part of the year, if we looked at absolute depletions, rather than the percentages plotted in Figure 1.2, the difference between different times of the year would be more striking.

Two questions naturally arise from a consideration of Figure 1.2:
(a) The first question concerns the extent to which these results depend on the fact that we have chosen the rather arbitrary periods of calendar months over which to perform our averaging. The lengths vary slightly (28 days → 31 days) but this is probably not too important. The starting date, 1 January, is arbitrary and bears no simple relation to the dates of the solstices and equinoxes (which come at about 24th or 25th of a month). It would be much more logical to take months of equal length

(e.g. lunar months) which began, or were centred, at the solstices/equinoxes. To our knowledge the effect of changing the definition of the month over which averaging is performed has not been studied. It would make an interesting exercise. However, given the size of the errors, we would not anticipate this effect to be of any very great consequence.

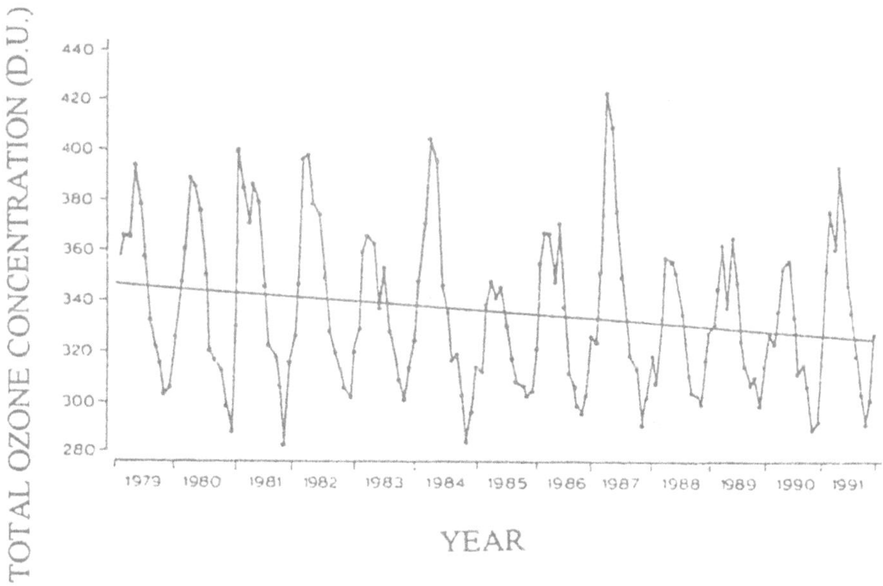

Fig 1.1. Monthly means of total ozone at Athens deduced from Nimbus-7 TOMS data (Varotsos and Cracknell 1993).

(b) The second question, and the one that we have been studying in the work reported in this paper, concerns the extent to which the behaviour shown in Figure 1.2 changes as a function of space, if one moves a distance of 50 km, or 250 km or 500 km or more, for example. As a start to approaching this question, we note that in various papers results in the same format as Figure 1.2 already exist for a quite widely separated set of locations in Europe, namely Athens, Crete, Thessaloniki, Dundee and St. Petersburg. It is worth recalling the results of that work before considering a more systematic approach to this problem.

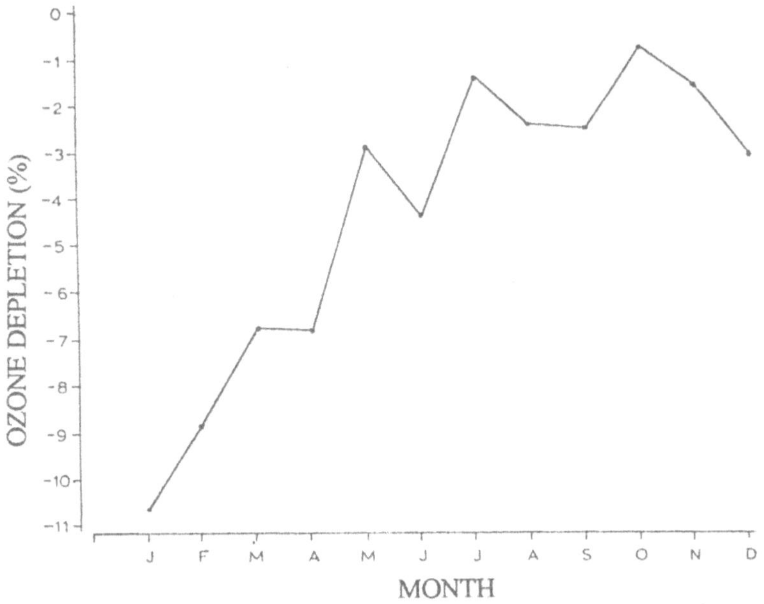

Fig. 1.2. The trend of TOMS total ozone versus month of the year for Athens during the period 1979-1991 (after Varotsos and Cracknell 1993).

2 Previous results

2.1 Trends in the total ozone content

We show, in Figures 2.1 and 2.2, the mean monthly ozone depletions for Dundee and St.Petersburg, respectively, that are analogous to the results shown in Figure 1.2 for Athens. (Note that the results for St. Petersburg are given per decade whereas the results for Athens and Dundee are for the whole period of observation of about 13 years.) It is perhaps worth stressing at this point that very good consistency has been found between these TOMS results and the ground data for those two locations (Athens and Thessaloniki) for which ground data from Dobson or Brewer spectrophotometers were available (Cracknell *et al.* 1994). The depletion calculated for each of these locations is given in Table 2.1.

When one comes to examine the details of the results in Figures 1.2, 2.1 and 2.2, it is immediately apparent that many of the differences between the mean monthly depletions at different locations are not insignificant. We demonstrate this in Figure 2.3 with the data for Athens and Dundee; in this Figure we have plotted the mean monthly depletion for Athens (•), along with the error bars for each of these values, and we have plotted the mean monthly depletion for Dundee(o). To avoid confusing the diagram we have not included the error bars for the Dundee data but they are comparable with those for the Athens data. From this Figure we see that for some months, January, April, July and August, there is clearly no significant difference between the depletion rates for Athens and Dundee. For some months, February, May and December, the differences are quite clearly significant, whereas for the remaining months, March, June, September, October and November, the situation is not clear cut either way. Looked at in another way, the Dundee data in Figure 2.1 shows no significant evidence of any depletion in the monthly means for February and June.

We therefore set out to study this variability of ozone depletion and, hopefully, to try to relate it to the known meteorology of the area of western Europe. The results that will be described in the next section represent only a first step in that work.

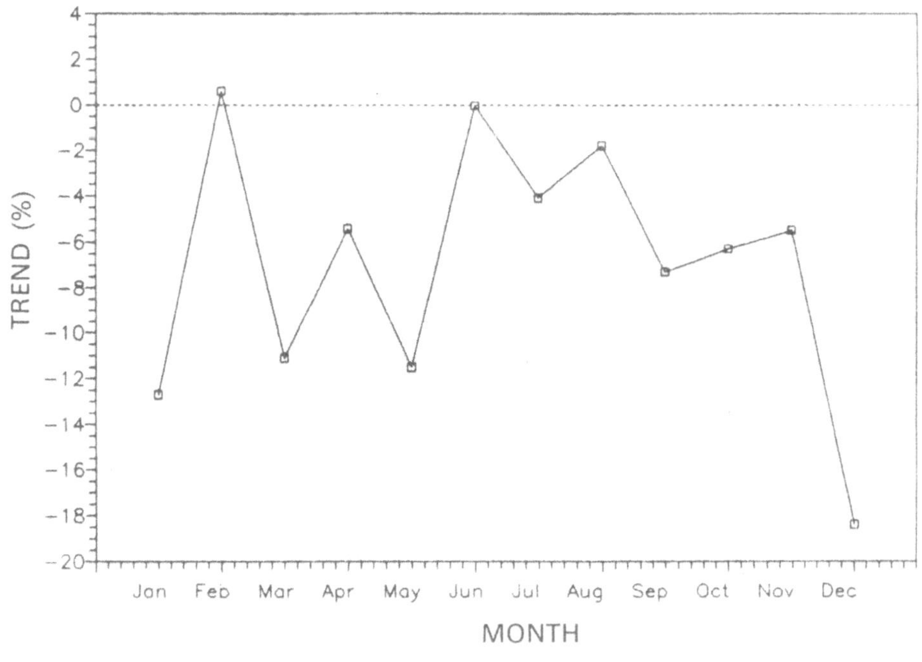

Fig. 2.1. The trend of TOMS total ozone versus month of the year for Dundee during the period 1979-1992 (Cracknell and Varotsos 1994).

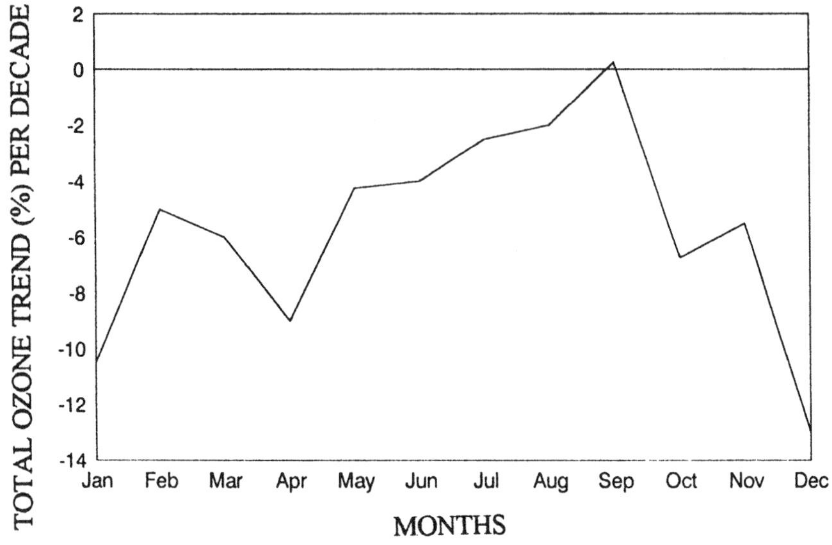

Fig. 2.2. The trend of TOMS total ozone per decade versus month of the year for St. Petersburg during the period 1978-1992 (Kondratyev *et al.* 1994).

Table 2.1. Annual depletion of total ozone (Figures in brackets denote °W).

	Location		Depletion (% pa)
	N	**E(W)**	
Athens	38	24	0.5 ± 0.1
Crete	35	25.3	0.5 ± 0.1
Thessaloniki	40.5	22.9	0.5 ± 0.1
Dundee	56.5	(3)	0.6 ± 0.1
St. Petersburg	60	30	0.8 ± 0.3

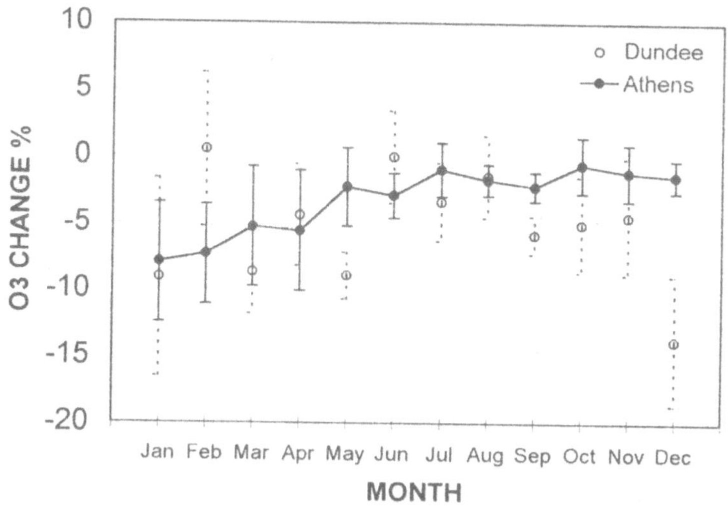

Fig. 2.3. The mean monthly depletion of total ozone for Athens(•), along with the error bar for each of these values, and for Dundee (o).

2.2 Significance of the total ozone trends

It is of great importance to examine the statistical significance of the trend in the total ozone content derived in the latter paragraph. This becomes more interesting for locations characterized by large interannual ozone variability which is a crucial component for the statistical significance of the total ozone trend.

Recently Chandra and Varotsos (1995) suggested that: a) There is a large longitudinal variation of the monthly trend in total ozone over the northern mid-latitudes ranging from 1.5 to 8.5 per cent per year with a large standard error, b) The trend in total ozone content over the Mediterranean area varies in a similar way with the zonal average total ozone trend over mid-latitudes. Also, the trend of the total ozone over Athens, Greece, is representative of the whole Mediterranean region and so it is representative of the zonal average total ozone trend over the northern mid-latitudes, and c) The interannual variability of the amplitude of the annual wave in the total ozone amount over the Mediterranean region compares extremely well with the interannual variability of the total ozone amount over this region.

Taking into account the aforementioned results it is evident that the Athens location is the most representative for the detection of the statistical significance of the trend.

The data subjected to statistical analysis consist of: TOMS total ozone measurements for the period extending from January 1979 to December 1991 and total ozone measurements performed at the University of Athens using a Dobson spectrophotometer covering the period from January 1992 to May 1994.

The two data sets were combined into a single data set assuming that they are samples originating from the same population. The validity of this assumption was tested using the Kolmogorov-Smirnov and the Mann-Whitney statistical tests, for the homogeneity of two samples, but was found to be statisticallly not acceptable at 95 per cent confidence level. The same tests were also applied using the Dobson data set combined with progressively reduced TOMS data sets. Using a set of TOMS data covering the period from January 1980 to December 1991, combined with the complete Dobson data set, the hypothesis of homogeneity could not be rejected at 95 per cent confidence level according to the Kolmogorov-Smirnov test, while the Mann-Whitney test indicated acceptable homogeneity only when we considered TOMS data for the period from January 1981 to December 1991, the second sample always being the complete Dobson data set. The assumption of normality in the Mann-Whitney test is justified on the basis of the samples' sizes in all cases.

The fact that the observed lack of homogeneity in our data set is a result of the first one or two years of data, enables us to consider the whole data set in our analysis, since the effect of these data on the results is expected to be minimal.

The linear correlation coefficient, R, and the slope, B, are tested for being equal to zero, using a two-tailed t-test in the mean monthly total ozone values. The true level of significance is calculated for each test.

The data were also tested for the existence of trend using the Kendall-Mann rank statistic. The test was applied separately for each month as well as for all the data together.

From the analysis it is concluded that for all cases, both R and B are negative indicating a negative trend of the total ozone against time.

It is also concluded from this analysis that for January, February, March, August, November, December and for the all months case, the hypothesis of R being equal to zero can be rejected at a level of confidence better than 95 per cent.

These remarks lead us to the conclusion that there is a statistically significant negative correlation between total ozone and time, for the above mentioned months, at a level of confidence better than 95 per cent.

The results of the Kendall-Mann rank test are in close agreement with those of the linear regression, discussed above. The test index u(n) value exceeds -1.96 which is the lower limit for a 95 per cent confidence interval, thus indicating the existence of a negative trend for January, February, March, June, August, November, December and for the all months case.

3 More recent calculations

In the work which we have described so far we have looked at monthly averages of ozone concentrations over a period of the order of a decade and we have seen substantial differences between, for example, the results for Athens and for Dundee. In Athens the depletion is quite large in the early months of the year but it becomes quite small later on, particularly in July and October. In Dundee, on the other hand, the small depletions are early in the year, in February and June. We have also looked at other locations in Greece (Crete and Thessaloniki), for which the results are not reproduced here, and also at St. Petersburg, see Figure 2.2. The results are all different in terms of the pattern of depletion on a month by month basis. As a start towards a more systematic study we have determined the mean monthly depletion of ozone for a further set of well-distributed sites in western Europe, namely for the cases of Mallorca, Spain, (39.45°N, 3.43°W), Naples, Italy, (40.50°N, 14.15°E), Berlin, Germany, (52.32°N, 13.25°E) and London, England, (51.30°N, 0.10°W) from January 1979 - January 1989. The depletion rates of the mean monthly total ozone concentrations over this ten-year period show strong seasonal variations. Similarities in such trends between both the aforementioned locations and that of those found in previous research work are observed, while a general depletion rate exists of approximately 7-8% per decade. The results are shown in Figures 3.1-3.4 for, respectively, Mallorca, Naples, Berlin and London.

4 Conclusions

It is really very difficult to see any obvious pattern of behaviour or to draw any simple or obvious conclusions. More work is clearly needed to study the TOMS data on a finer network of points and to include a study of the general features of the stratospheric transport and stratospheric chemistry of the area in relation both to the processes of ozone formation and destruction and to ozone transport. The situation is further complicated by the fact that NASA is about to produce a new version (version 7) of their TOMS processing algorithm/program. It would be unwise to go on using the old version. However, if we change to the new version we shall need to go back and re-do the calculations for all the locations that we have already studied. That is going to involve quite a lot of work.

As far as the statistical significance of the trend, one can conclude that there is a significant negative trend of the total ozone with time, which started during the mid-eighties.

The rate, per year, at which this depletion is taking place scales as 0.92 per cent for January, 0.71 per cent for February, 0.67 per cent for March, 0.36 per cent for August, 0.46 per cent for November and 0.48 per cent for December. The case of all months has an intermediate rate of 0.51 per cent per year, as expected, due to the influence of the non-significant months.

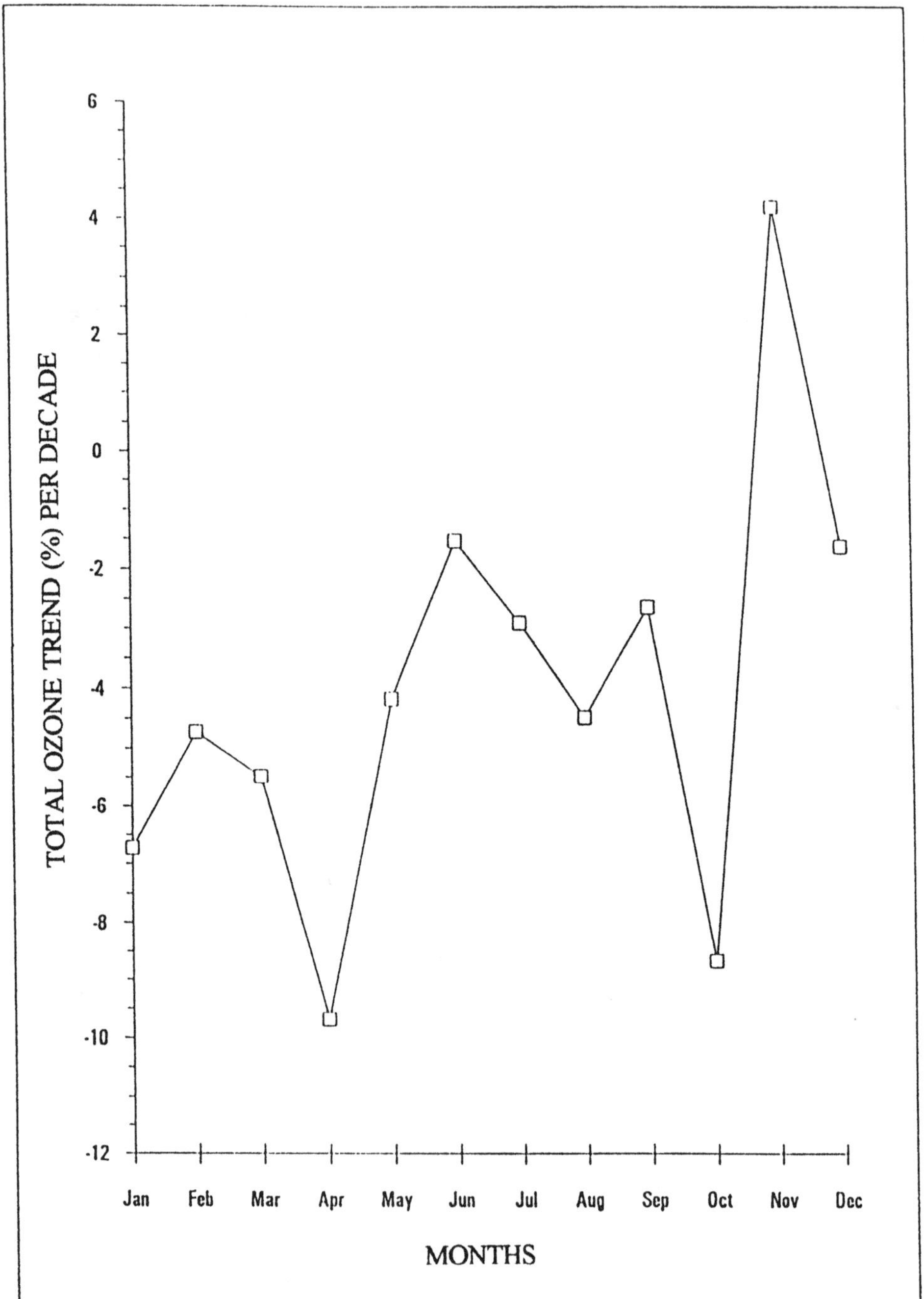

Fig. 3.1. The trend of TOMS total ozone versus month of the year for Mallorca during the period 1979-1992.

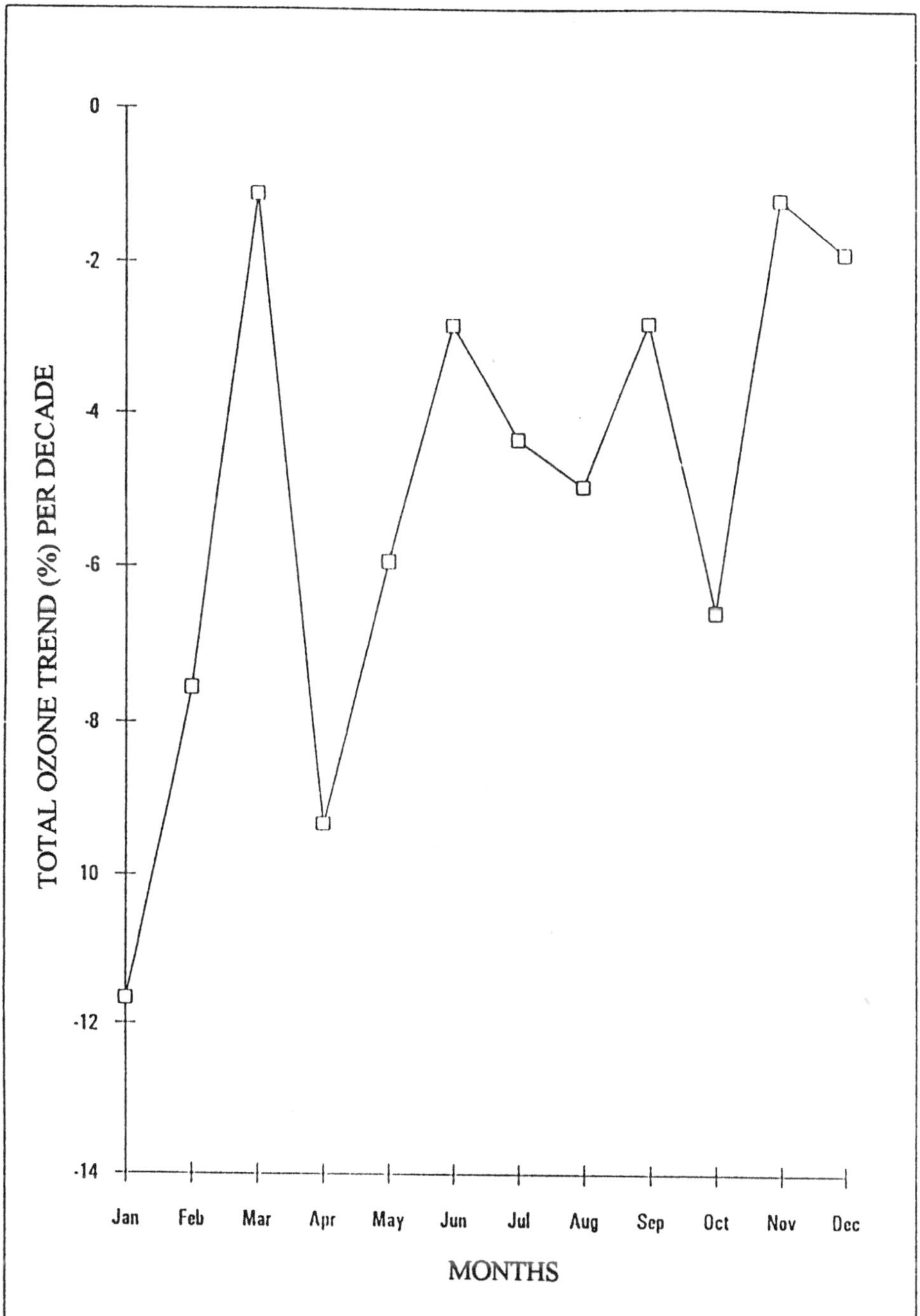

Fig. 3.2. The trend of TOMS total ozone versus month of the year for Naples during the period 1979-1992.

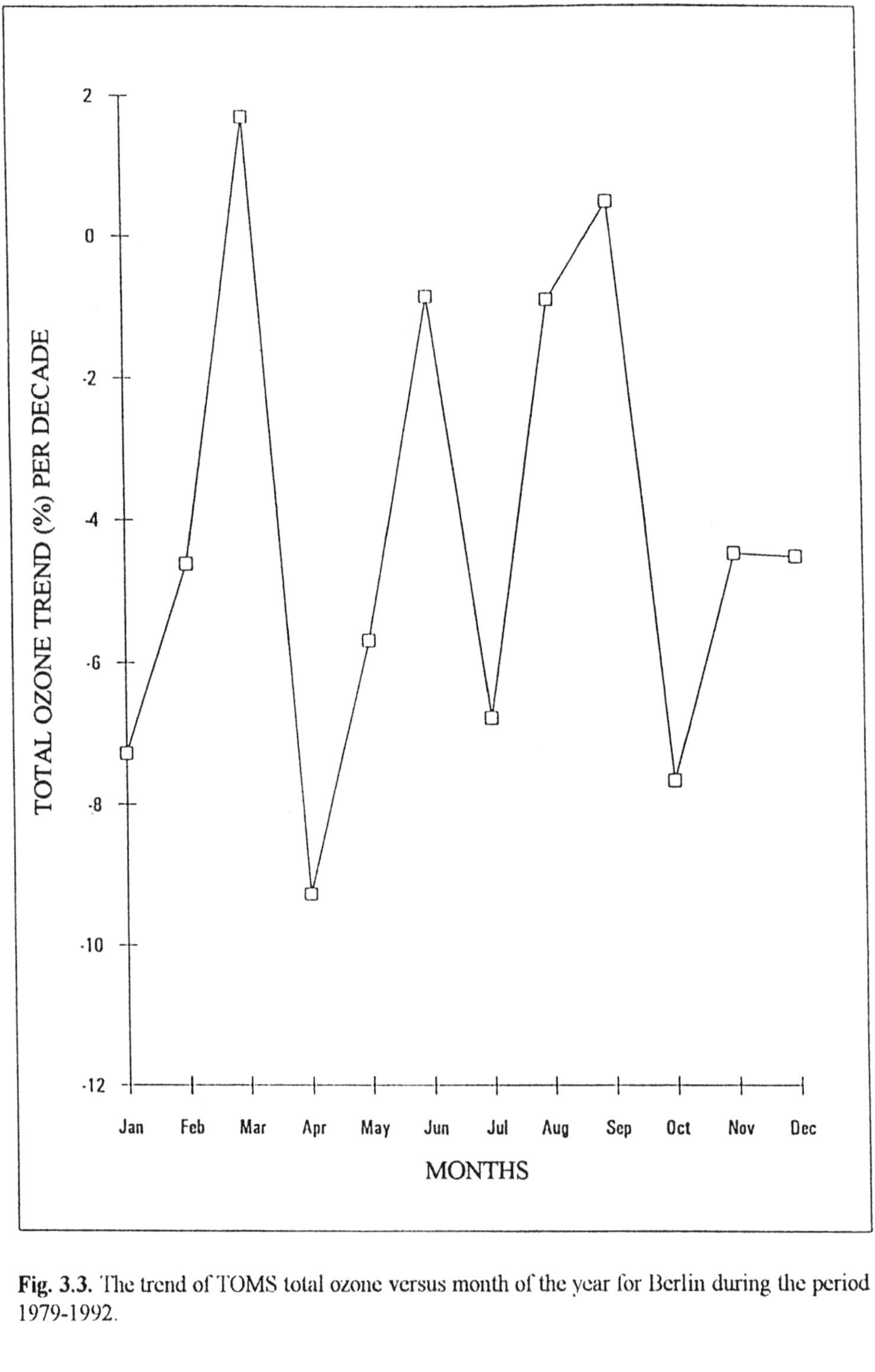

Fig. 3.3. The trend of TOMS total ozone versus month of the year for Berlin during the period 1979-1992.

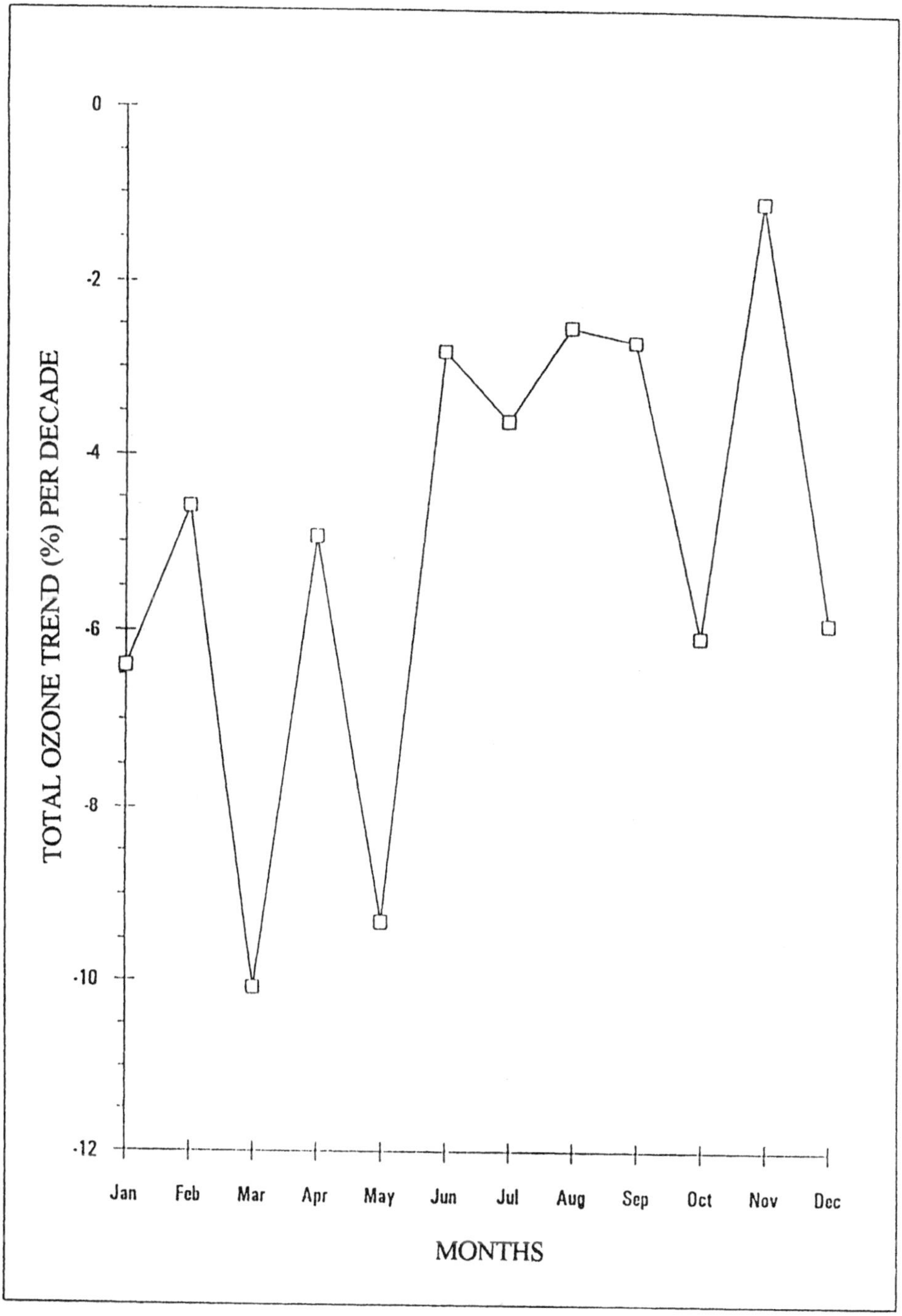

Fig. 3.4. The trend of TOMS total ozone versus month of the year for London during the period 1979-1992.

References

WMO and UNEP, 1995, *The changing ozone layer*, by Bojkov R.D..

Chandra, S., and Varotsos, C.A., 1995, Recent trends of the total column ozone: implications for the Mediterranean region. *International Journal of Remote Sensing*, **16,** 1765-1769.

Cracknell, A.P., and Varotsos, C.A., 1994, Ozone depletion over Scotland as derived from Nimbus-7 TOMS measurements. *International Journal of Remote Sensing*, **15,** 2659-2668.

Cracknell, A.P., Varotsos, C.A., and Asimakopoulos, D.N., 1994, On the total ozone depletion over Greece derived from satellite-flown and ground-based instruments. *International Journal of Remote Sensing*, **15,** 3285-3293.

Kondratyev, K.Ya., Varotsos, C.A., and Cracknell, A.P., 1994, Total ozone amount trend at St. Petersburg as deduced from Nimbus-7 TOMS observations. *International Journal of Remote Sensing*, **15,** 2669-2677.

Varotsos, C.A., and Cracknell, A.P., 1993, Ozone depletion over Greece as deduced from Nimbus-7 TOMS measurements. *International Journal of Remote Sensing*, **14,** 2053-2059.

Statistical Analyses of Total Ozone Spatial Variability in the Mediterranean

I. Dalyuk, O. Pokrovsky

Main Geophysical Observatory , Karbyshev str.,7, St.Petersburg, Russia, 194021
Main Geophysical Observatory , Karbyshev str.,7, St.Petersburg, Russia, 194021

Abstract. Our aim is to present EOF method extended to temporal scale and to demonstrate its opportunities in reproduction of combined temporal and spatial TO variability. The method of biorthogonal decomposition was proposed in order to reach this aim. Numerical algorithm of its realization is discussed as well as formulas for representation of generalised EOF. Main application issues are: the dividing of the analyzed area onto a number of statistically homogeneous sub regions with respect to spatial variability and outlines for optimization of surface ground-based ozonometric networks. The analyses of TO spatial variability spectrum reveals main exremal zones for Europe and the Mediterranean.

Keywords. total ozone, intricent variability, biorthogonal decomposition, homogeneous sub regions, optimization, ozonometric network

1. Approach

There are a number of papers devoted to application of the empirical orthogonal functions (EOF) technique for the analysis of total ozone (TO) spatial fields. Most of them demonstrated the high efficiency of TO fields representation by means of the EOF expansion.Here the biorthogonal expansion technique was used (Rao C.R.,1967) as the background for analyzing the spatial and temporal structure of TO fields. It provides the possibility of an adequate TO variations spectral representation, i.e. the sufficient approximation rate of spatial and temporal TO variations apart as well as of combined spatial+temporal fields. The generalized empirical orthogonal functions were used as the basic functions. Let **X** be the input data matrix in deviations from the multi-annual means defined over certain space area and certain period of time. Then **V**, the set of EOF, can be constructed so that $\sum_x = V \Lambda V^T$, where $\sum_x$ is the covariance matrix. Thereupon the temporal coefficient can be obtained by C = **X V**, and the approximation of the spatial TO variations is $\hat{X} = CV^T$ (T-transposition sign). Next. for every vector C^j from the set **C** we can apply the orthogonal expansion technique once more and get a set of conjugated functions U^j, so that $\sum_{C^j} = U^j M U^{j^T}$, where **M** is the matrix containing eigen values of $\sum_{C^j}$ as diagonal elements, U^j is the matrix constructed from

NATO ASI Series, Vol. I 53
Atmospheric Ozone Dynamics
Observations in the Mediterranean Region
Edited by Costas Varotsos

eigen vectors of $\sum_{C^j}$. Thus we obtain a number of function sets describing temporal variability for each spatial harmonic. Now the representation of combined temporal+spatial structure of the initial field can be written as follows:

$$\hat{X}_{ml} = \sum_{k=1}^{k1} V_{lk} \sum_{k'=1}^{k2} C^{k}_{mk'} \cdot U^{k}_{mk'}$$

where $\hat{X}_{ml}$ is the estimate for TO variation for fixed time and space coordinate (m,l-indexes),

k1 - number of spatial functions,

k2 - number of conjugated vectors,

$_{lk}$ - cutset of **V**-matrix

$C^{k}_{mk'}$ - set of coefficients ,

$^{k}_{mk'}$ - cutset of conjugated **U**-basis.

The technique applied gives the following benefits: it allows to obtain 1) the set of linear unbiased estimates for the TO spatial variations as well as for the TO temporal evolution; 2) The optimal convergence with respect to variances; 3) Consequently, the best approximation of combined spatial+temporal TO structure.

2. Spatial Structure of Total Ozone Variations

Thus, we have the set of functions **V** describing the spatial structure and for each spatial function we have the set of temporal harmonics defining the TO variations spectrum.

The structure of TO natural variability fields (Fig. 1) is, possibly, well known: the values increase poleward from 30 DU to 68 DU for the first period (February-March-April) and from 14 DU to 36 DU for the second period (September-October-December). In the first case the zonal character of variability is revealed. In the second case the "trace' of North-Atlantic anomaly can be seen. (ocean-atmosphere interaction). The intricent variability in the Mediterranean compared to Europe is rather small. It varies from 30 DU to 42 DU for spring and from 14 to 22 DU for autumn.

The variations for two 5-year-long periods (1982-1986 and 1987-1991) for the Mediterranean can be compared. The zonal character prevails, but the contour lines have the tendency to bend northward in the region between the Balkan and Apennine Peninsulas., curving the Ionic sea. Other pecularities can be indicated while analyzing TO intricent variability fields.

Next, Fig. 2 presents the first two TO EOF for Europe for two mentioned above periods (spectrum of TO space oscillations). The first functions for both time periods reflect the zonal character of TO distribution: increasing northward and locating the anomaly over the Scandinavian Peninsula. The second two functions reveal two extremal zones :1) in the North Atlantic (for autumn season it is shifted to the south-west); 2)in the region of 45°-60° of Eastern longitude and 56°-63°Northern latitude.. The zones of extremal values

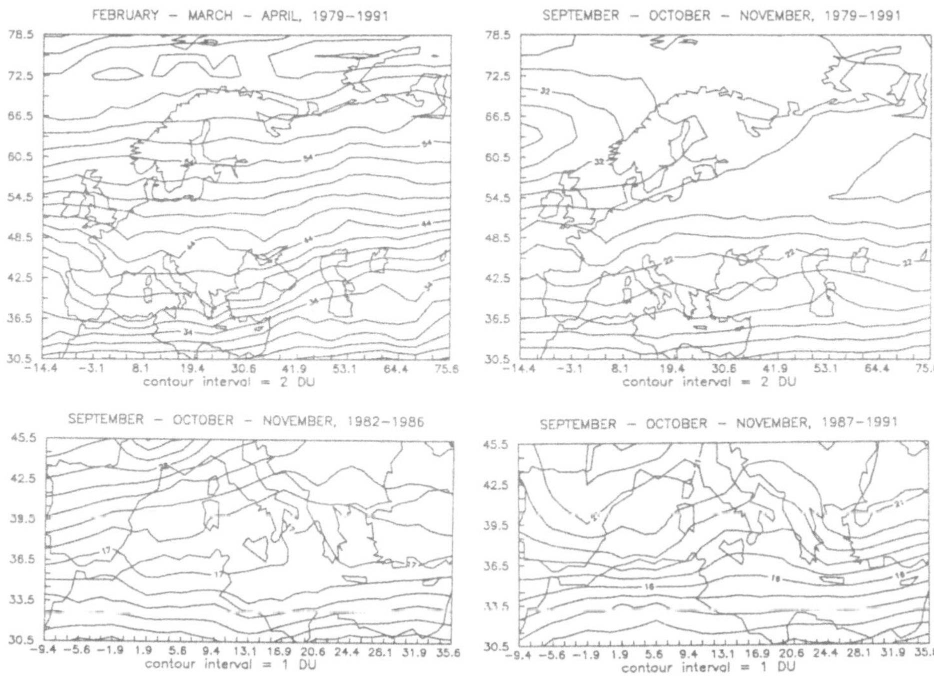

Fig.1. Total ozone natural variability over Europe for two seasons and over the Mediterranean for two five-year-long periods.

have opposite signs, i.e. they are anti correlated. The size of anomaly regions for spring season are larger then for autumn.

So, the EOF technique allows to present the spatial TO fields as a combination of slow oscillation components and reveals the main features of TO variations distribution.As far as the Mediterranean region as part of Europe is concerned it can be noted that only the second function for autumn and the third function for spring provide the sign-altering lines crossing this area.

Let's consider the spatial structure of the Mediterranean region itself (Fig.3) The following main extremal zones can be revealed : 1) the first functions locate the region over the Northern part of the Apennine Peninsula ,for the second time period it is shifted to the north-west.; 2) The second two functions reveal two anti correlated zones, the first one is over the Pyrenees, the second extends from Asia Minor to the Balkan Peninsula. The processes for the first period have larger spatial scales , about thousands km, for the second period scales are smaller. The zones of anomalies are shifted to north-west as in the case of the first function.

3. Regionalization

EOFs usually have rather definite structure (Pokrovsky and Piatigorsky, 1995). The first function (correspondent to maximal eigen value or maximal variance) has a constant sign, other functions have the sign-altering lines. The

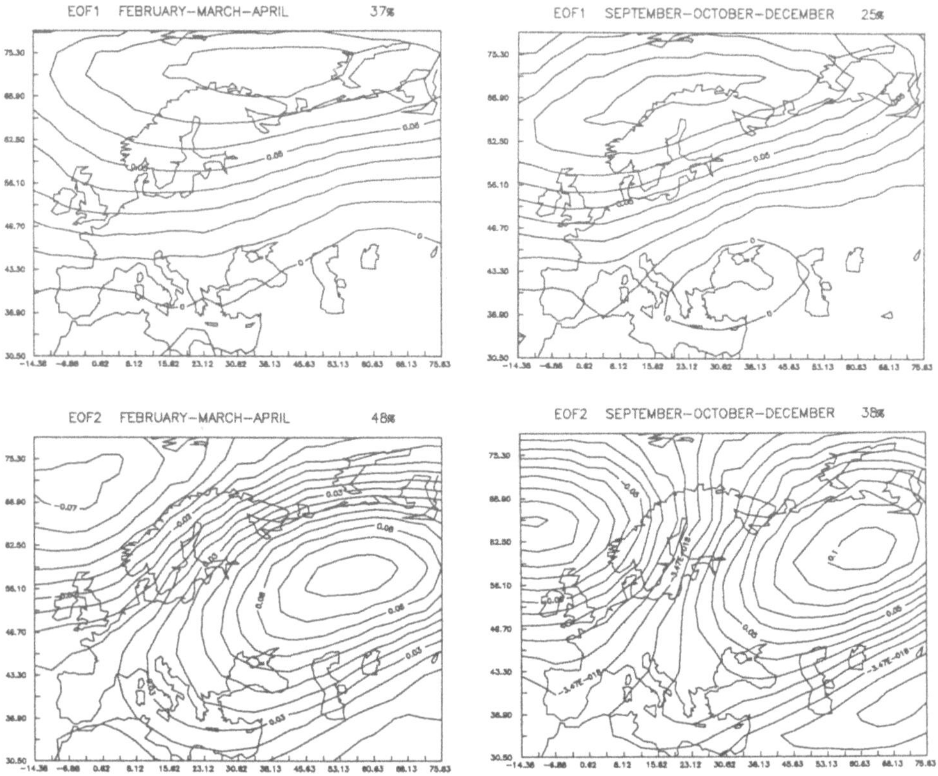

Fig.2. Total ozone empirical orthogonal functions; spatial field expansion; 1979-1991 (the value of relative cumulative variance is in the right up corner).

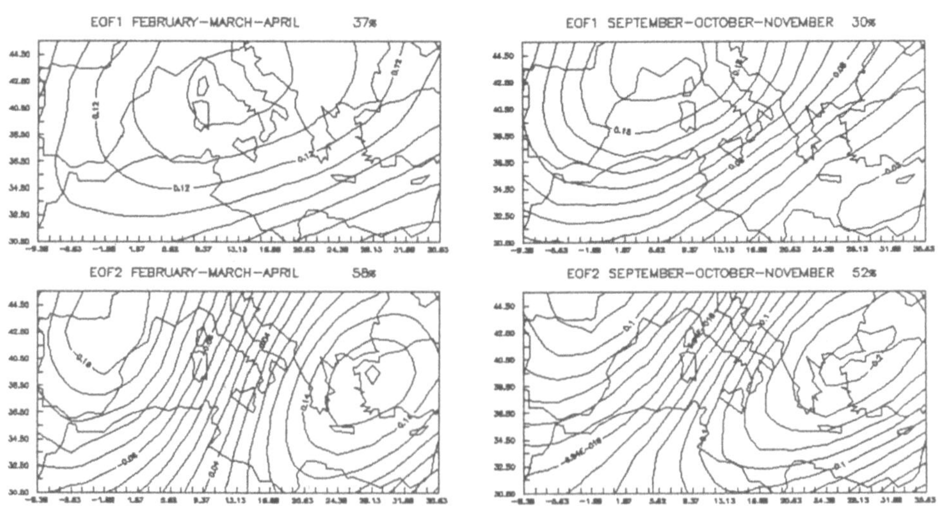

Fig.3. Total ozone empirical orthogonal functions; spatial field expansion for the Mediterranean; 1979-1991 .

latter ones can be considered to be boundaries of a statistically homogeneous regions. The combination of these lines forms the sub regions (we call this procedure 'regionalization'). The example of it is presented for the Mediterranean. Five functions were used for two seasons (relative cumulative variance 80% and 76%) to construct this subdivision (Fig 4). As a result we obtain nine homogeneous (up to a certain extent) sub regions for the first period and eleven sub regions for the second. In both cases the area is divided mostly in zonal and meridional directions. The numbers define the number of function associated with the curve. As far as the stability of this subdivision is concerned: some division lines have rather stable character: They appear for different samples (different year-long periods) calculated. For example, the line 3 crossing the Balkan Peninsular. Other boundaries can have some 'degree of freedom'.

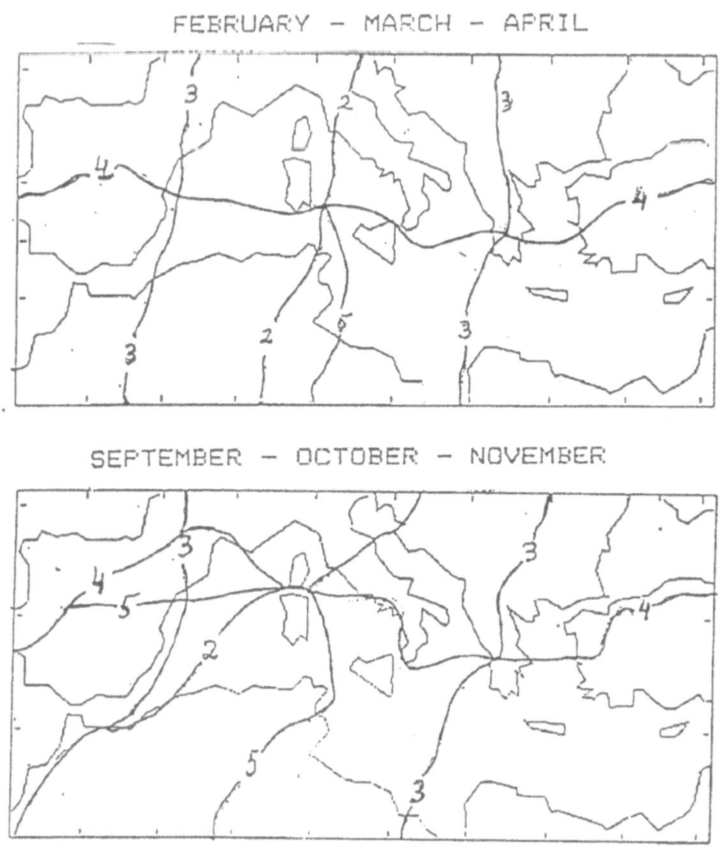

Fig.4. The subdivision of the Mediterranean area (regionlization).

Applications. The TO fields could be presented with the minimal number of spectral functions inside this regions. The local photo-chemical models too better to use in this way selected areas. The optimization of an observational network can be carried out (Pokrovsky, 1984). In this case the *a priori* information from the observational network is needed. Then the functional describing the measure of informational content for each site could be derived. The sites could be ranged according to the set of values, characterising the efficiency of each one .

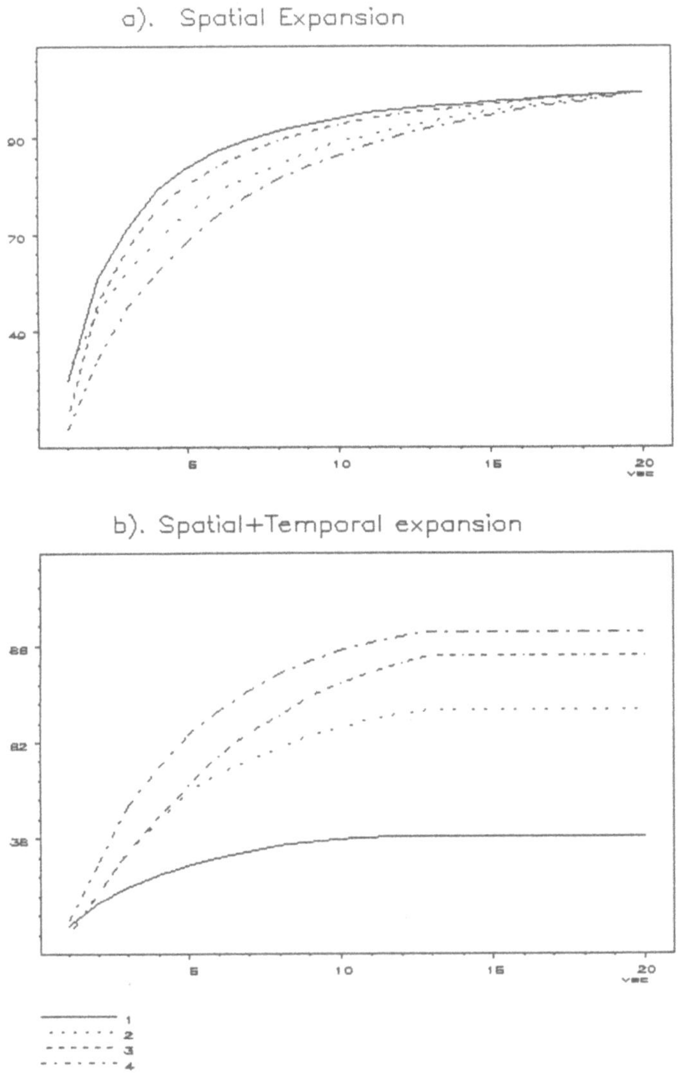

Fig.5. Relative cumulative variance: a) the spatial expansion only , 1 - the Mediterranean, 2 - Europe, February-March-April; 3 - the Mediterranean, 4 - Europe, September-October-November; b) spatial+temporal expansion: 1 - with 1 spatial function, 2,3,4 - 3,5,7 spatial functions

4. Relations Between Temporal and Spatial Oscillations and rates of Convergence

So, each spatial harmonic is connected with a set of temporal functions. First three temporal functions connected with the first spatial harmonic have the characteristic time scales of about 5-7 days.(the space scales of the spatial harmonic were about thousand of kilometres). For the second and the third harmonic the frequency of oscillations increases and gives 2-3 days in the latter case.

Fig. 5 demonstrates rates of approximation for four selected variants. The fastest convergence has place in the case of the Mediterranean, February-March-April., as it must be. The explained variance is about 86 % for 7 eigen functions. But these functions don't describe the temporal structure.Below the relative cumulative variance for combined spatial+temporal expansion is presented. The numbers on the curves detect the number of spatial harmonics utilized. The rate of convergence for the global TO field expansion is much more lower. Seven spatial functions together with conjugated temporal functions explains only 70% of variance. It must be noted that the same number of spatial functions taken without their temporal components explains only 23% of variance.

4. Conclusion

The statistical technique applied to analysis of global (combined spatial+temporal) TO fields gives following results: 1) spectral expansions demonstrate sufficient rate of convergence and can be used for statistical modelling of TO variations for the aims of forecasting of TO spatial fields.; 2) the ideology of 'regionalization' allows to construct the optimal from the statistical point of view observational network, furthermore, the scenario of observational network extension can be considered. 3) Spatial and temporal structure of the TO variations and relationships between them can be examined .

REFERENCES

1. Pokrovsky O.M., 1984., Optimization of the meteorological remote sensing. - Leningrad, Hydrometeoizdat, 264 p. (in Russian).
2. Pokrovsky O.M. and Piatigorsky A.G., 1995. The use of empirical orthogonal functions for regionalization of meteorological fields.-St.Petersburg, Meteorology and Hydrology, 1995, N6(in Russian).
3. Rao C.R.,1967, Linear statistical methods and its applications.- N.Y.,547p.

Total Ozone Winter-Spring Deficiencies in the Northern Mid-Latitudes

Małgorzata Degórska and Bonawentura Rajewska-Więch

Institute of Geophysics, Polish Academy of Sciences
01-452 Warsaw, Ks. Janusza 64, Poland

Abstract. Results of observational and modeling studies have confirmed that polar and mid-latitude ozone losses are linked to halogen chemistry. However, dynamics appears to play an important role through, e.g., the transport of chemically perturbed air towards the lower latitudes. An association has been known between the interannual variation of total ozone at northern mid-latitudes and the behaviour of the stratospheric polar vortex, depending on the characteristics of mid-winter stratospheric warming. For example, at Belsk (52^0N, 21^0E) during the 32-year measurement period, the occurrence of major warmings has always (with the only one exception) been connected with the winter-spring total ozone means larger than the long-term average. The strongest ozone depletions have occurred during the winters with the minor stratospheric warmings. In particular, this concerns the record low total ozone during winter-spring 1991/1992 and 1992/1993 over Europe and during winter-spring 1994/1995 over Siberia.

Keywords. Total ozone, polar vortex, statospheric warming

1. Introduction

Among dynamic factors influencing the ozone distribution on hemispheric scale, one of the most important seems to be that connected with the behaviour of the stratospheric polar vortex. Phenomena of major mid-winter warmings, temporarily leading to a change of westerly to easterly stratospheric polar circulation, with a breakdown of the polar vortex and strong planetary wave activity, supply ozone to the middle and high latitude regions from the equatorial regions. Contrary to this, the minor warmings occur without a breakdown of the polar vortex but only with its displacing towards the lower latitudes. We suggest that they play a significant role in producing the strongest ozone deficiencies in Northern Hemisphere by transporting the potentially ozone-poor air to mid-latitudes.

It seems useful to examine this suggestion basing on the data from long-term operating stations. Our attention has been focused on the ozone data from Belsk, Poland. This station has a continuous record of sufficient duration (32 years) to establish a reliable long-term mean values. Additionally, the reliability of the Belsk station data has grown (WMO, 1989; NOAA Techn. Rep. NESDIS 60, 1992) after their retrospective reevaluation (Dziewulska-Łosiowa et al., 1983; Degórska and Rajewska-Więch, 1988; 1991).

NATO ASI Series, Vol. I 53
Atmospheric Ozone Dynamics
Observations in the Mediterranean Region
Edited by Costas Varotsos

2. Results

In Fig.1 one can evidently see that there is a close association between the characteristics of the mid-winter stratospheric warmings and total ozone variability at Belsk. During the 32-year measurement period, the occurrence of major warmings has always (with the only one exception) been connected with the winter-spring total ozone means larger than the long-term average. The strongest ozone depletions have occurred during the winters with the minor stratospheric warmings.

Fig.1 shows, like many other recently published papers (e.g. WMO, 1991; 1995; Stolarski et al., 1992), that the total ozone above the northern middle latitude has been declining in the past two decades. Our interest is focused on the winter-spring ozone variability of last five years. Figures 2a, 2b, 2c, 2d, and 2e present the time series of the total ozone daily means at Belsk in winter-spring seasons; the solid line depicts the long-term (1964-1994) average value, broken lines indicate the $\pm 1\sigma$ limits. Generally, the ozone level during 1990/1991 was close to the long-term average values (Fig.2a) and a major mid-winter warming occurs with the breakdown of polar vortex (Naujokat et al., 1991).

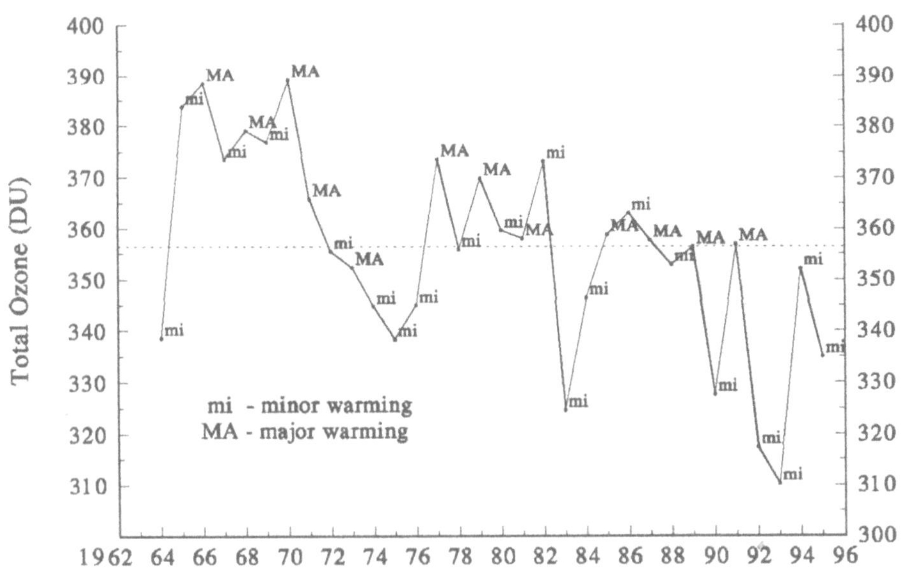

Fig.1 Seasonal means of total ozone (December to March) during 1964-1995. Characteristics of the mid-winter warmings from Naujokat et al., (1994)

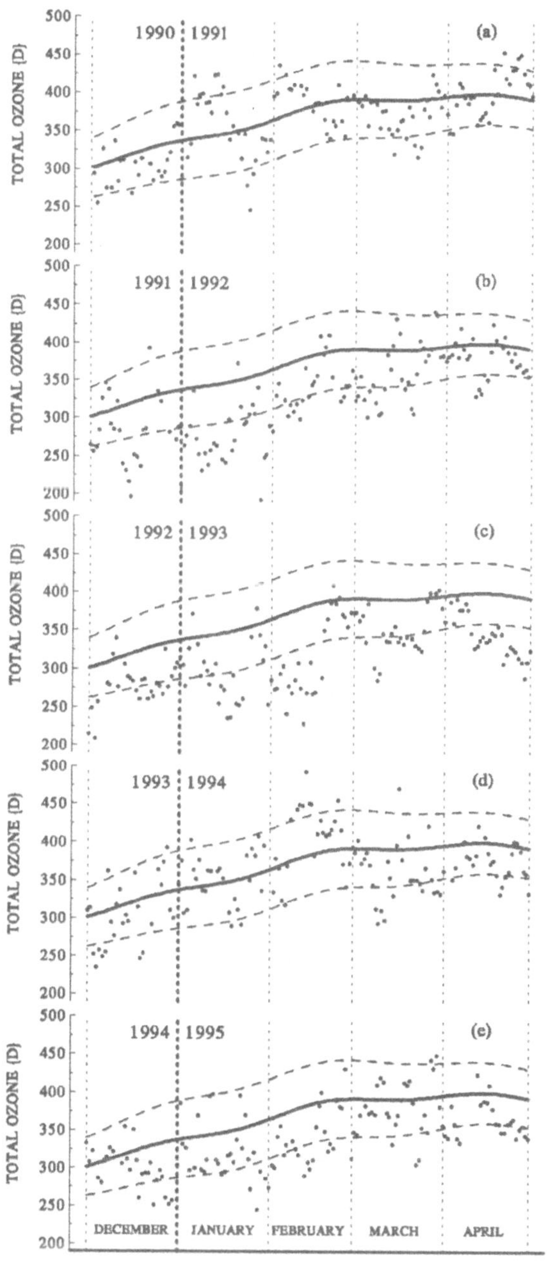

Fig.2. Time series of total ozone daily means at Belsk in winter-spring seasons of 1991-1995

Labitzke and Van Loon (1988, 1992) showed that the occurrence of major warmings was well anticorrelated with the solar activity (11-year cycle) during the easterly phase of the quasi-biennial oscillation of equatorial winds in the lower stratosphere (QBO). Thus, in concurrence with their findings, in 1992 and 1993 minor warmings occurred in January and February, respectively (Naujokat et al., 1992; 1993), not leading to a breakdown of the polar vortex. Contrary to 1991, during these winters the polar vortex was not much affected by the stratospheric warming, but only displaced towards northern Europe in January 1992 and one month later in 1993. It is worth to note that in January 1992 the record low values of total ozone over Europe and Atlantic ocean were observed by satellite and at various ground stations, as for example Belsk (Fig.2b). Potential vorticity and 30 hPa temperature analysis indicates (Naujokat et al., 1992) that Europe and great part of Atlantic ocean were located inside the polar vortex in January 1992, after an extremely strong minor warming over the polar regions. A similar situation was observed in February 1993. A strong minor warming occurred, also not leading to a break-down of the vortex, but with the cold air being shifted towards Europe. Noteworthy, the lowest values of total ozone over Europe in 1993 were observed in February (Fig.2c).

In the winter-spring season of 1993/94, the total ozone over Belsk was close to its long-term average (Fig.1). We can see (Fig.2d) that for 1993/94, the January and February daily mean ozone values were mainly above the long-term average. Only in March, total ozone at Belsk was about 6% below the long-term average. Larger negative ozone and temperature anomalies were noted over the European Arctic where the cold vortex was shifted (Naujokat et al., 1994).

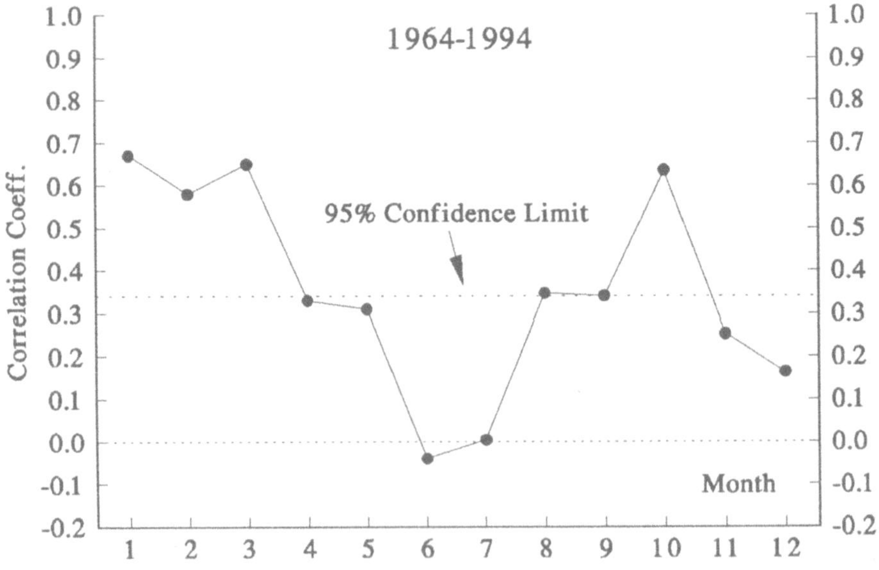

Fig.3. Correlation between the total ozone monthly means and the 30-hPa temperature over Belsk

In the winter-spring season 1994-1995 the ozone deficiencies over the middle and high latitudes were observed, mainly over Siberia, where the minimum of temperature in the stratosphere and the new record low ozone occurred (Bojkov et al., 1995). A strong minor warming occured in February, when the polar vortex was displaced over Siberia but did not break (Naujokat et al., 1995).

The positive correlation between the total ozone and the stratospheric temperature has been established (e.g., Tung and Yang, 1988). Figure 3 indicates that total ozone and the 30-hPa temperature over Belsk are significantly correlated during January, February, March and in October. The month-to-month changes in the ozone-temperature correlation may suggest the significant role of large-scale transport of ozone connected with the polar vortex build-up, displacement or breakdown.

3. Conclusions

The behaviour of the polar vortex (characterized by a considerable interannual variability) has strong implication to the northern hemisphere total ozone distribution. The occurrence of major warmings is connected with relatively high total ozone winter-spring average. The strongest ozone depletions have occurred during the winters with the minor stratospheric warmings, especially when the polar vortex is displaced towards the ozone measuring station. This implicates the strong longitudinal dependence of the winter-spring ozone deficiencies as well as their good correlation with the stratospheric temperature.

REFERENCES

Bojkov R.D., Fioletov V.E., Balis D.B., Zerefos C.S., Kadygrova T.V.,Shalamjansky A.M., 1995, Further ozone decline during northern hemisphere winter-spring of 1994-1995 and the new record low ozone over Siberia, submitted to Geophys. Res. Lett.

Degórska M., Rajewska-Więch B., 1988, The re-evaluated ozone profiles obtained from Umkehr measurements taken at Belsk, Poland, in the years 1963-1981, Pubs. Inst. Geophys. Pol. Acad. Sc., D-30 (220), 27-76.

Degórska M., Rajewska-Więch B., 1991, A retrospective reevaluation of total ozone and Umkher profile data from Belsk, Poland, Publs. Inst. Geophys. Pol. Acad. Sc., D-36 (246), 71-76.

Dziewulska-Łosiowa A., Degórska M., Rajewska-Więch, 1983, The normalized total ozone data record, Publs. Inst. Geophys. Pol. Acad. Sc., D-18 (169), 23-73.

Labitzke K., van Loon H., 1988, Associations between the 11-year solar cycle, the QBO, and the atmosphere. Part I: the troposphere and stratosphere in the Northern Hemisphere in winter, J. Atmos. Terr. Phys., 50, 197-206.

Labitzke K., van Loon H.,1992, On the association between the QBO and the extratropical stratosphere, J. Atmos. Terr. Phys., 54, 1453-1463.

Naujokat B., Petzoldt K., Labitzke K., Lenschow R., Rajewski B., Wiesner M., Wohlfart R.C., 1991, 1992, 1993, 1994, 1995, Beilage zur Berliner Wetterkarte, Berlin, Germany.

NOAA Technical Report NESDIS 60, 1992, International Dobson Data Workshop Summary Report, Washington, D.C., USA.

Stolarski R., Bojkov R., Bishop L., Zerefos C., Staehelin J., Zawodny J., 1992, Measured trends in stratospheric ozone, Science, 256, 343-349.
Tung K., and H. Yang, 1988, Dynamic variability of column ozone, J. Geophys. Res., 93, 11,123-11,128.
World Meteorological Organization. 1989, Report of the International Ozone Trends Panel 1988, WMO Global Ozone Research and Monitoring Project, Report No. 18.
World Meteorological Organization, 1991, Scientific Assessment of Ozone Depletion: 1991, WMO Global Ozone Research and Monitoring Project, Report No.25.
World Meteorological Organization, 1995, Scientific Assessment of Ozone Depletion: 1994, WMO Global Ozone Research and Monitoring Project, Report No. 37.

Regional Modelling of Tropospheric Ozone Distribution and Budgets

Adolf Ebel, M. Memmesheimer and H.J. Jakobs

EURAD Project, University of Cologne, Aachener Str. 201–209, 50931 Cologne, Germany

Abstract. Air quality modelling has become an indispensable tool for research and application. This paper puts emphasis on the development and use of comprehensive regional model systems. A specific version, namely the European Air Pollution Dispersion Model (EURAD) was applied to the simulation of photo-oxidants over Europe during a photo–smog episode. The results are discussed as an example of model application. Budgets of air pollutants (ozone, NO_x, SO_2) of smaller domains are calculated and discussed with respect to processes controlling them. It is shown that the ozone budget significantly depends on vertical transport into lower tropospheric layers during photo–smog events. The need of model evaluation is stressed.

Keywords. Ozone, numerical modelling, atmospheric chemistry, transport, troposphere, air pollution, budgets

1 Introduction

Regional air quality models have been developed for application to a great variety of scientific and practical problems. Though by far not yet perfect with regard to their scientific content and computational performance, they have become an indispensable tool for air pollution research. Due to different applicational needs and computational constraints (which were a seriously limiting factor for numerical model design in the past and still are an important aspect for advanced model strategies despite eminent recent growth of computer capacity) a large variety of chemical transport models originated during the last two or three decades. The history of these models is reviewed in an illuminating article by Peters et al. (1995). In this study we will make use of results generated with a specific advanced regional air pollution dispersion model as described in the following section. It has been designed with the aim of application to European air quality issues. This, of course, includes the treatment of the Mediterranean region though the main emphasis of recent studies used to be upon the most heavily polluted areas of central Europe. Other models with their focus on central European conditions will briefly be mentioned in Section 2.

NATO ASI Series, Vol. I 53
Atmospheric Ozone Dynamics
Observations in the Mediterranean Region
Edited by Costas Varotsos

Europe exhibits a broad range of peculiar features regarding air pollution and its impact on ecological and socio-economic systems. Differences in climate and land type as well as land use pose largely contrasting problems of cause and effects of air pollution. Whereas discussions on acid deposition and forest damage presently seem to occur with considerably reduced intensity when compared to the treatment of these topics about a decade ago in central Europe, sulfur emission, oxidation and deposition are still high priority problems in northern Europe, particularly Scandinavia, and, of course, in parts of Europe where large fossil fuel consumption without modern desulfurization technology is still in use.

Apparently, the major problem of central and south Europe regarding air quality is presently the formation of ozone and other photo-oxidants and their impact on local and regional scales. This study will preferrably deal with this aspect of air pollution. As regards photo-smog it is evident that a high degree of complexity of the problem is inferred by the interaction of small- to large-scale processes in space and time. An example of the spatial small-scale aspect is the appearance of "hot spots" of air pollution which is typical for southern Europe. With regard to temporal scales the impact of meteorological processes is important. Photo-smog events are of episodic character and exhibit a broad spectrum of temporal variability.

Another issue with similar spatial and temporal characteristics like photo-smog and related to this phenomenon is that of aerosol and haze formation and transport. This problem is not or only marginally treated in most air quality models though it is well recognized and attempts are made to improve existing models in this respect (e.g. Ziegenbein et al., 1994; Ackermann et al., 1995; Binkowsky and Shankar, 1995). Another class of regional problems is the impact of air traffic emissions on the composition of the tropopause region. Usually, air traffic emissions are treated from the global point of view focusing on the possible influence on climate change. Yet, it has been shown that regional processes cannot be neglected when assessing the role of aircraft exhaust for the composition of the atmospheric layers around the troposphere (Petry et al., 1994).

The great variety of questions to be tackled by air pollution calculations and deviating applicational needs have lead to the design of quite different types of air pollution models. Hierarchies of more or less consistent models have been established to broaden the range of application of these numerical tools to scientific and practical questions. Clearly, there is a strong need to design consistent comprehensive systems for reliable air pollution dispersion simulations as discussed by Hansen et al. (1994).

This study exploits results which were gained in the framework of a joint European activity aiming at regional air quality simulation with a small number of selected model systems. The project European Modelling of Atmospheric Constituents (EUMAC, a part of the European environmental research project EUROTRAC, 1987-1995) followed the idea of consistent comprehensive modelling. There exist, of course, other models of similar type and character, and studies have been initiated to compare the ones used in EUMAC with a limited number of competing models, e.g. by the project "Athenian Photo-Chemical Smog Intercomparison of Simulations" (APSIS, Moussiopoulos et al., 1994) and a special activity focusing on continental-scale models (Hass et al., 1994, 1996). Yet, it is beyond the scope of this paper to complement or expand these studies. Here results from a specific model are used to demonstrate the potential of regional modelling. This is mainly done on the European, i.e. continental, scale with some emphasis on application to Mediterranean problems.

2 The regional model

The air quality model system used for this study is EURAD (European Air Pollution Dispersion Model System). It is an episodic Eulerian gridpoint model system and consists of three main parts, namely a mesoscale meteorological model (MM5), a chemical transport model (EURAD-CTM) and an emission model (EEM). Descriptions of the EURAD model system can be found in several recent publications (e.g. Jakobs et al., 1995; Feldmann et al., 1996; Hass et al., 1996) so that the model description can be restricted to its main characteristics in this paper.

Chemical and transport calculations are carried out in the EURAD-CTM (CTM: Chemical Transport Model; Hass, 1991) which was derived from the Regional Acid Deposition Model (RADM; Chang et al., 1987). 63 species are treated in 158 reactions using the RADM2 chemical gas phase mechanism (Stockwell et al., 1990). It contains highly parameterized calculations of cloud effects (Walcek and Taylor, 1986; Mölders et al., 1994; Chang et al., 1987). CTM uses the method of operator splitting and employs the advection algorithm of Bott (1989a,b) to integrate the continuity equations for the concentrations C_i of the species i ($i = 1-63$):

$$\frac{\partial C_i}{\partial t} = -\nabla(\mathbf{u}C_i) + \nabla(K\nabla C_i) + f_i + W_i + E_i \tag{1}$$

($\mathbf{u}$: wind vector; K eddy diffusion tensor; f_i: production and loss due to dry chemistry; W_i: cloud effects including wet chemistry, tracer redistribution and wash- and rain-out; E_i: emission from elevated sources). The boundary conditions are defined by fluxes and concentrations, respectively. In particular, the lower boundary condition is

$$F_{Z_i} = F_{D_i} + F_{E_i} \tag{2}$$

where F_{Z_i} is the total vertical flux at the earth's surface, F_{D_i} the flux due to dry deposition and F_{E_i} the emission at or close to the surface. The mesoscale meteorological model MM5 drives the EURAD-CTM by providing the necessary meteorological fields calculated off-line. MM5 can be run in a hydrostatic and non-hydrostatic mode. Yet, what is equally important for its application in the EURAD system is its nesting ability (Grell et al., 1994; Jakobs et al., 1995).

Initial and boundary conditions are generated using ECMWF analyses (horizontal resolution approximately 1.68 degrees; global data sets every 6 hours). The nudging technique is employed for lower layer temperature and horizontal wind (Stauffer and Seaman, 1990).

The EURAD model system needs at least emission data for 19 species, namely $NO, NO_2, CO, SO_2, SO_4, NH_3$ and various non-methane volatile organic compounds (VOCs). Gridded hourly values have to be provided. They are generated by the EURAD Emission Model (EEM) designed by Memmesheimer et al. (1991). In its most straightforward version the EEM calculates emission input employing the EMEP emission inventory and adding biogenic VOCs using an emission parameterization by Lübkert and Schöpp (1989).

The horizontal resolution of the model can be chosen in a rather broad range, say between 10 and 100 km, depending on the size and type of the domain to which it is applied. For simulations covering larger parts of Europe 50–80 km are typical. In nested applications the gridsize is scaled down by a factor of 1/3 in two steps, i.e. the ratio of gridpoint distances is 9:3:1. The terrain following σ-coordinate is used for the vertical. One has

$$\sigma = (p - p_o)/(p_s - p_o) \tag{3}$$

where p, p_o and p_s are the pressure of a given model level, at the upper boundary and at the surface, respectively. For standard applications 16 levels between the surface and upper boundary are usually chosen with p_o = 100 hPa, i.e. about 16 km. The vertical resolution changes with height (σ = 1, .99, .98, .96, .93, .89, .84, .78, .7, ...1,0). The thickness of the lowest layer corresponds to about 75 m, that of the highest one to about 4 km. The horizontal grid is the Arakawa-C grid (Arakawa and Lamb, 1977). An operator splitting technique is used for integrating equ. (1) (Hass, 1991). The basic time step is currently 300 sec in the EURAD-CTM. Transport is calculated with 1/2 of the basic time step whereas chemistry time steps depend on specific reaction conditions.

3 Simulation of episodes

Comprehensive complex air quality model systems like EURAD are most beneficially applied to the simulation of short episodes with particular pollution phenomena like winter and summer smog. They enable rather detailed analyses of temporal and spatial changes of pollutant concentrations as well as studies of processes controlling transport and chemical transformations. The EURAD model system is computationally demanding (about 7 hours CPU on a CRAY-YMP90 for a full calculation of a 2-day episode). Various measures can be taken to reduce required computer time and storage. For instance, the number of gridpoints could be reduced or a simpler chemical mechanism be employed. For some applications it may be sufficient to treat only the boundary layer and use larger time steps (long. term trend calculations; Builtjes, 1992). Though such simplifications can have their benefits they certainly lead to a loss of information content of the models. Considering scientific needs for the implementation of more physics and chemistry, the demand of increasing the reliability of models for the policy task of environmental assessment and the development of abatement strategies and, last but not least, the rapid growth of computational power it seems worthwile and promising to proceed on the way to more advanced comprehensive air quality modelling. Following this philosophy, an example of regional model application is presented in this section. A period containing an episode with high photo-oxidant concentrations in July and August 1990 was chosen.

3.1 Photo-smog calculations

The event selected as an example was characterized by high ozone concentrations in central Europe between about August 3 and 5, 1990. A special feature making the episode particularly attractive is its termination by a well defined coldfront. This front slowly moved through western and central Europe from northwest to southeast generating interesting features of ozone redistribution during its passage (Ebel et al., 1995a). Fig. 1 shows a flat pressure distribution over most of Europe on August 4, 1990. This distribution is typical for the larger part of the episode and implies weak transport of air. Also, only little cloud coverage was observed. Alltogether, favourable conditions for photo-oxidant formation existed. The coldfront can be identified by strong temperature gradients extending from northern Spain to Denmark. A noteworthy feature is the strong southwesterly flow over Scandinavia generated by the low pressure system over the Northatlantic with strong convergence over north and central Germany, i.e. in a region with increased ozone production and concentrations. Obviously, transport of air enriched with manmade ozone towards northeast (Baltic Sea and Scandinavia) can be expected. This is an important factor for southern Scandinavian ozone budgets as will be demonstrated in Section 4.

Emission estimates of this simulation are based on EMEP inventories exhibited in Fig. 2a and b for NO_x and anthropogenic VOCs, respectively. Using the EURAD Emission Model, finer resolved hourly values are calculated from such inventories taking into account population density in Europe. Presently, the emission input is improved by applying more sophisticated GENEMIS estimates (Friedrich, 1993).

The front moving southeastward gradually erodes the polluted air over central Europe as can be seen in Fig. 3 exhibiting the horizontal ozone distribution in the lowest model layer, i.e. near the surface. It also shows enhanced concentrations over the Baltic Sea caused by prefrontal southwesterly winds. Furthermore, polluted air is transported into the Mediterranean region through southeast France. What is also evident is the formation of hot spots in the southern and southwestern parts of Europe. Evidently, maxima of ozone concentration are displaced with respect to spots of highest emissions, for instance, Athens. The reason is transport of precursors from their sources before ozone can be generated and

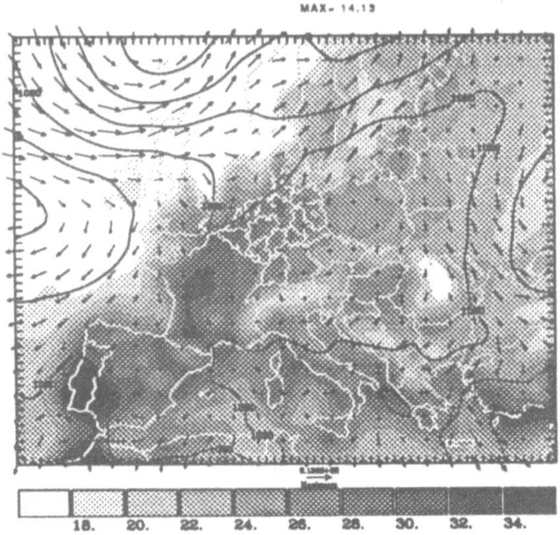

Figure 1: Temperature (in °C, scale at bottom), sea level pressure (contour lines, in hPa) and horizontal wind (arrows, scale at bottom) for Europe on August 4, 1990, 12 GMT. Results calculated for the lowest model layer are displayed.

advection of ozone itself. The model calculates vertical tracer distributions and thus allows to study higher level effects in the boundary layer and the free troposphere. As an example a vertical cross section along the front is shown in Fig. 4 for August 5, 1990, 14 GMT. The well known elevated ozone maximum in the upper part of the boundary layer and a prefrontal region with enhanced convective transport of ozone rich air are easily identifiable.

Peroxyacetyl nitrate (PAN) as another important photo–oxidant shows a spatial distribution (Fig. 5) in general similar to ozone but with some different details. They can be explained by the dependence of PAN on temperature leading to a longer lifetime at higher colder levels than in the warmer surface layer. For instance, the extended Baltic tongue of PAN can be interpreted as a result of long–range transport at upper colder levels and downward mixing. The less pronounced values over the Gulf of Genova seem to indicate losses of PAN due to higher temperature. Though the simulated PAN distribution looks reasonable there are not enough observations for its evaluation. In this respect the situation is better but not ideal regarding ozone, sulfur compounds and NO_x. A comparison of ozone calculations and observations is shown in Fig. 6 for an area where photo–smog formation occurred. The modelled time series resemble the observed ones rather well though a tendency to underprediction of maxima and overprediction of minima is obvious. Evaluation studies for the EURAD model system were carried out with different approaches and for various episodes (e.g. Hass et al., 1993; Feldmann et al., 1994; Ebel et al., 1995b). They have been continued and will remain a permanent task with the aim of model improvement and development.

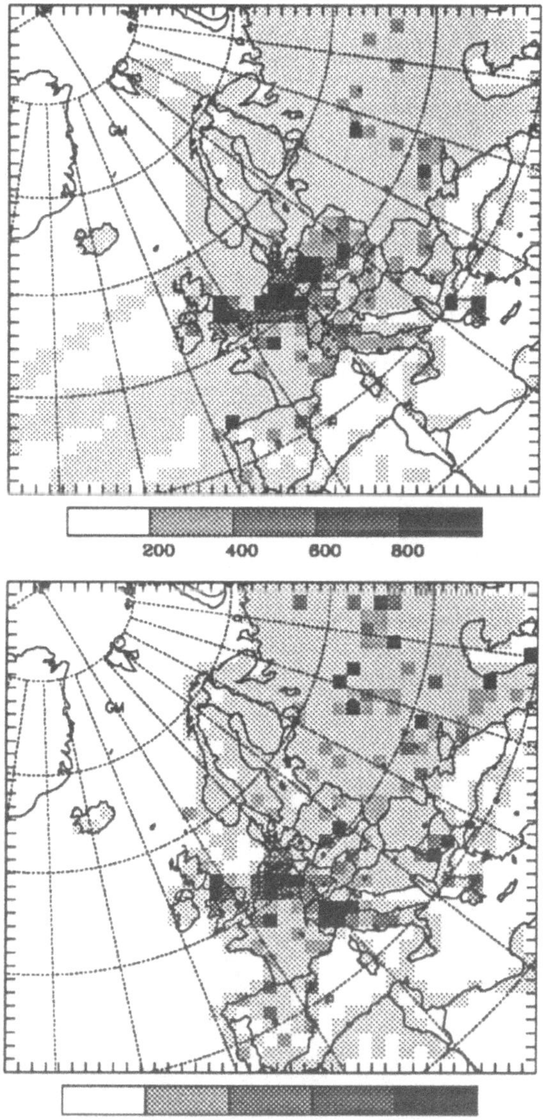

Figure 2: Emissions of NO_x (upper panel) and VOCs (lower panel) after the EMEP inventory for 1990. In units of kg/month per square (150 x 150 km^2). Mass of NO_x is given as mass of NO_2; VOC as VOC.

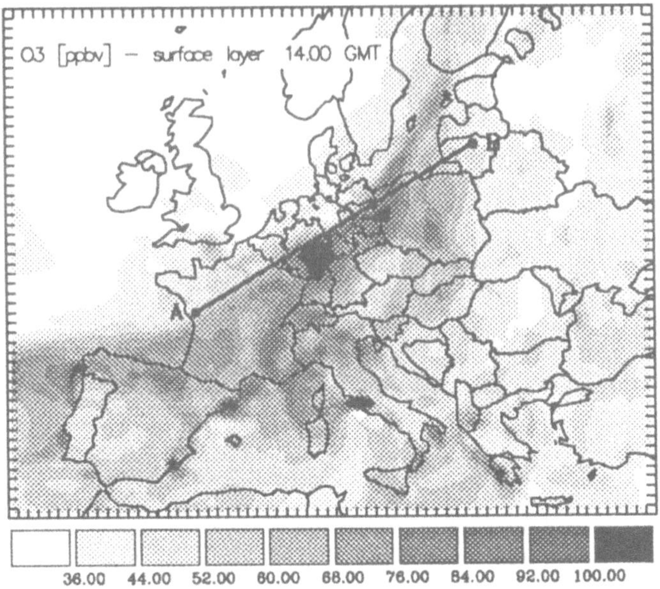

Figure 3: Simulated ozone concentration (in ppbv) over Europe, August 5, 1990, 14 GMT in the lowest model layer. Line A–B marks the position of the cross section shown in Fig. 4.

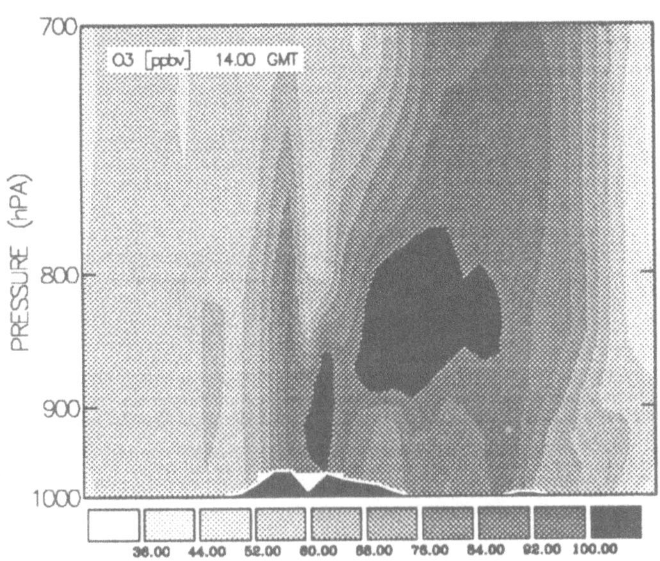

Figure 4: Vertical cross section along line A–B in Fig. 3 representing the calculated ozone distribution between about 1000 and 700 hPa close to a cold front, August 5, 1990, 14 GMT.

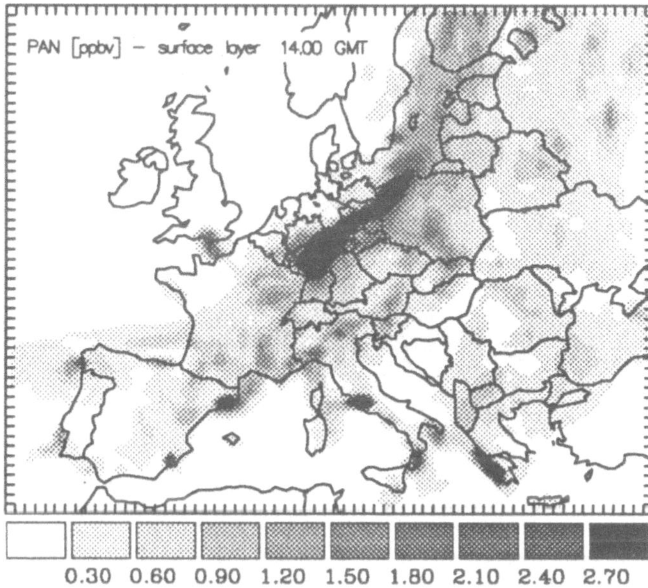

Figure 5: As Fig. 3, but for PAN (in ppbv).

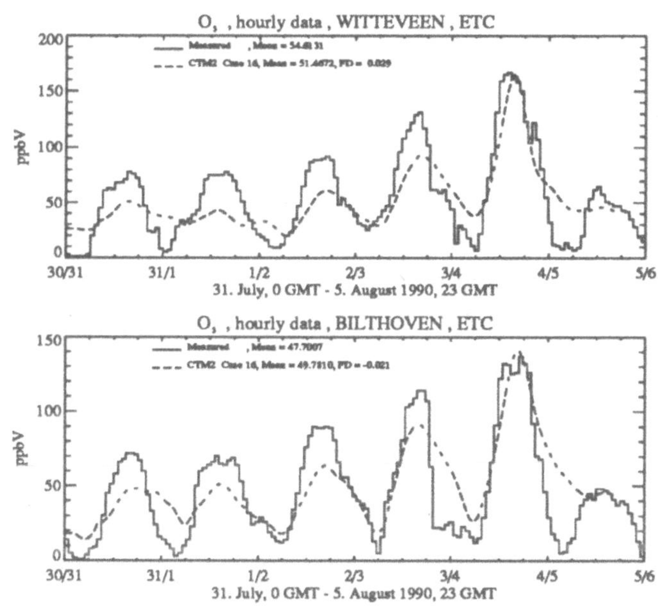

Figure 6: Comparison of measured (broken lines) and calculated ozone concentrations (continuous lines, in ppbv) at two Dutch stations (Witteveen and Bilthoven) during the episode from July 31, 00 GMT till August 6, 00 GMT, 1990.

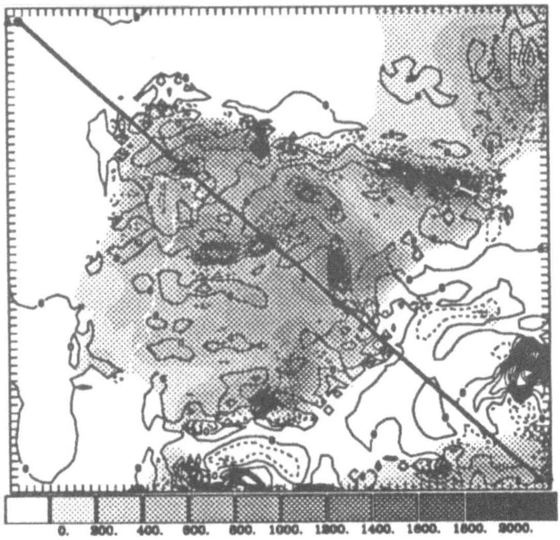

Figure 7: Simulated distribution of vertical wind over the Iberian Peninsula. Vertically integrated values are shown. Height range of integration: surface – 100 hPa. Upward and downward motion is represented by continuous and broken contour lines, respectively. The contour interval is 50 mm/s. The topography is shown by shading (contour interval 200 m). The position of the cross section exhibited in Fig. 8 is marked by the straight line. Nested calculation (20 km x 20 km).

3.2 Iberian Peninsula: subregional simulations

The nesting option of the EURAD system allows detailed studies of smaller regions, e.g. the former German Democratic Republic with high anthropogenic emissions (Jakobs et al., 1995) or the North Sea with nitrogen deposition problems (Memmesheimer and Bock, 1995). An interesting part of Europe regarding also specific Mediterranean features of air pollution is the Iberian Peninsula. Milan and Artinano (1992) speculated that the panshaped landmass may cause strong updrafts under frequent anticyclonic weather conditions and thus be a major source of pollutants in the free troposphere due to export of anthropogenic tracers from the atmospheric boundary layer. Stimulated by this hypothesis we have started an analysis of chemistry and transport over Spain and Portugal carrying out nested calculations for the late summer episode 1990. First results are discussed here.

The meteorological model run with a resolution of 20 x 20 km^2 reveals weak downward motion over most of the analysed domain as shown in Fig. 7 employing the vertically integrated vertical wind for a qualitative evaluation of this quantity. The large–scale slow downward motion, i.e. subsidence caused by anticyclonic flow, is interrupted by spots of intense updraft preferrably over land. The complex structure of the vertical motion field my be judged from a cross section (Fig. 8) cutting through the domain from northwest to southeast.

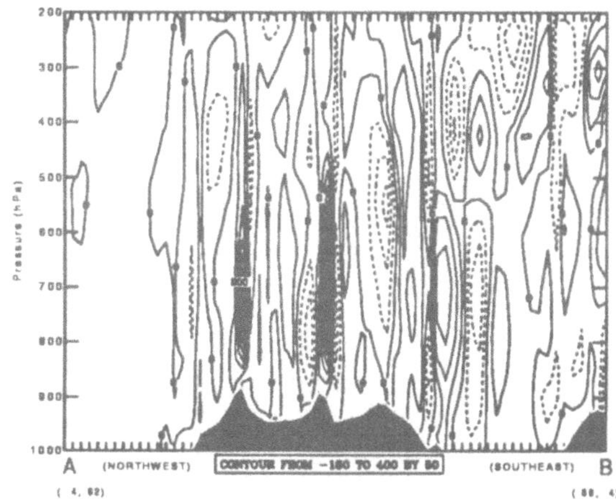

Figure 8: Simulated vertical wind between the surface and 200 hPa along the line indicated in Fig. 7. Up– and downdraft is represented by continous and broken lines, respectively. The contour interval is 25 mm/s.

As a first consequence of pure meteorological simulations we can conclude that complex flow with up– und downward transport can be expected over the Iberian Peninsula and that, therefore, not only tracer export from but also tracer import into lower tropospheric levels may occur.

Presently, chemistry and transport calculations are only available with coarse resolution (60 km). Nevertheless, they nicely illuminate the complexity of the problem of transport over Spain und Portugal and are therefore used for this analysis. Vertical cross sections of ozone, NO_x and PAN concentrations are compiled in Fig. 9. The cross sections cut from northwest to southeast as in Fig. 8. Only for such tracers which are mainly controlled by sources (emssions for NO_x or chemical production for PAN) at lower levels it is immediately evident that upward transport is effective in the expected sense, namely to carry tracers to higher tropospheric levels.

Yet, even when a clear plume as for PAN and NO_x is generated over the center of the peninsula during the day the distribution along the complete cross section indicates a rather complex behaviour of transport. One has to keep in mind that this is a three–dimensional process. Otherwise, it would not be possible to understand the growth and displacement of the PAN cloud appearing on the right side of the cross sections of 15 and 21 GMT. The three–dimensionality of transport also makes it difficult to identify the impact of the land–sea breeze on the tracer distribution though it is present in the calculated tracer fields. The peculiar features of PAN and ozone distributions found on the southeastern side at lower levels around 21 GMT seem to be generated by outflow of air from the land to a certain degree. In the case of ozone one has to take into account that the distribution is not only controlled by processes in the lowest tropospheric layers but also by its background in the free troposphere as indicated by the highly

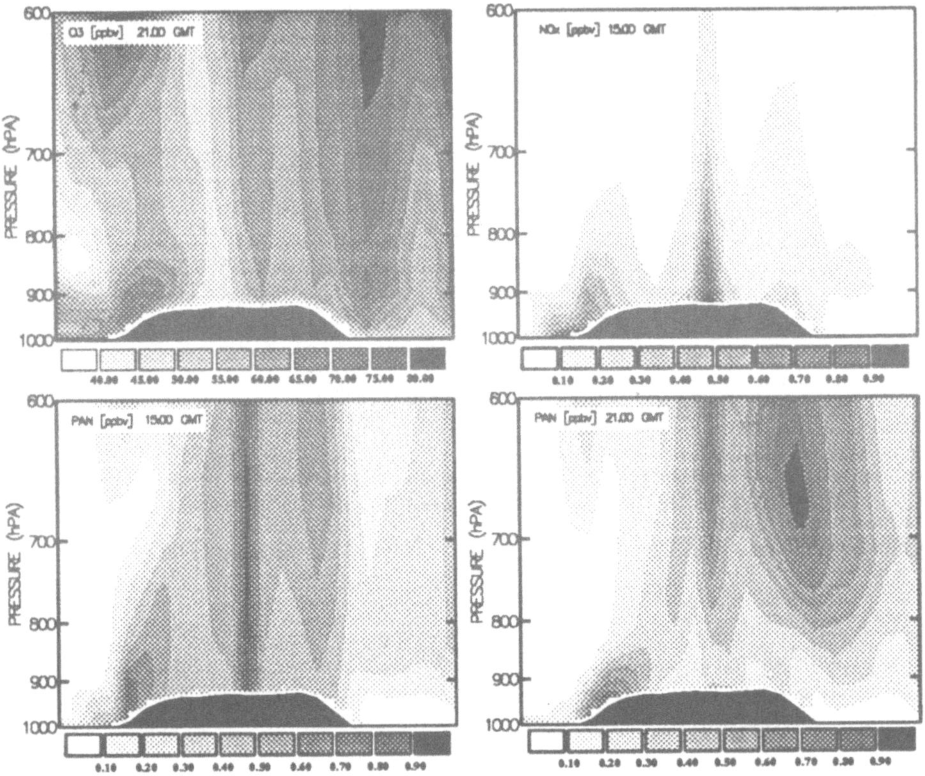

Figure 9: Northwest–southeast cross section of the vertical distribution of ozone (upper panel, left), NO_x (upper panel, right) and PAN (lower panels) over the Iberian Peninsula. The cross sections represent results obtained for August 4, 1990, at 15 GMT (NO_x and PAN, left panel) and 21 GMT (ozone and PAN, right panel). The concentrations are given in ppbv (see scaling below the panels). Horizontal resolution is 60 km.

variable values in the upper layers shown in the cross section of Fig. 9. This actually leads to a net downward flux of ozone into the lowest troposphere over the Iberian Peninsula. The contribution of this process to the total ozone budget will be discussed in the following section together with other tracer budgets in various subdomains of the total model area.

4 Budgets of pollutants

Using the simulations of the late summer episode of 1990 we have carried out calculations of contributions of the various processes to the budgets of selected pollutants over five subdomains of the model area. The aim has been to analyse the relative role of gas phase chemistry, cloud effects, horizontal and vertical

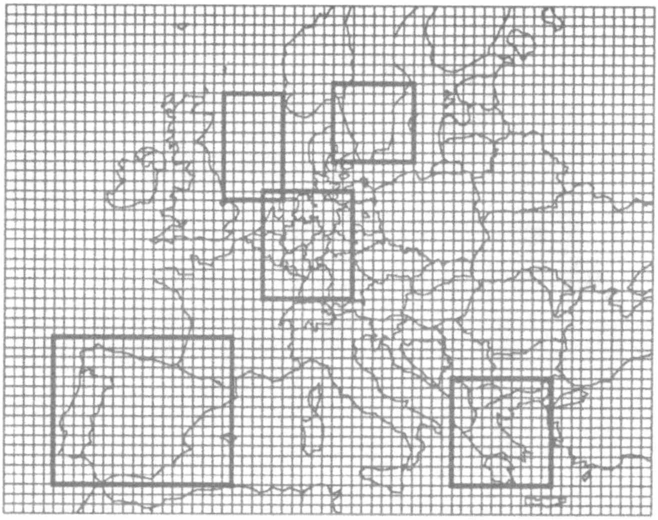

Figure 10: Position of subdomains for which budgets of O_3, NO_x and SO_2 were calculated.

advection, turbulent diffusion, deposition and emission under changing conditions (meteorology, source strength and distribution, land type etc.). Five areas have been studied (Fig. 10): Spain/Portugal and Greece as southern European regions, central Europe suffering from photo–smog, the North Sea area with very low emissions and southern Sweden under the influence of the heavily polluted central Europe during the episode.

Budget calculations were carried out integrating the loss and production due to the mentioned processes in a box covering the areas sketched in Fig. 10 and extending to a height of about 3600 m. This altitude is only seldom exceeded by the mixing layer height so that the planetary boundary layer is usually contained in the chosen height range. The period of integration was determined by the build–up phase of the photo–smog event in central Europe, i.e. the time between July 31, 00 GMT, and August 5, 00 GMT, 1990. Results obtained for ozone, NO_x and sulfur dioxide are shown. The contributions to the ozone budget by individual processes are plotted in Fig. 11 for the Iberian, Greek and Swedish subdomains. All estimated budget data are compiled in Table 1 containing changes of the three species obtained in the total subdomain over the 5–day episode (i.e. tendencies in kton per subdomain per episode). Table 2 exhibits the same tendencies but normalized to a volume with a horizontal surface of 360 000 km^2 (100 gridboxes using coarse resolution of 60 x 60 km^2).

As regards ozone it is clear that chemistry is a dominant process for the budget of any region. Yet, what is especially surprising is the strong control of the ozone budget by vertical advection. This fact was already mentioned when Iberian conditions were discussed in the previous section. Fig. 11 and Table 1 and 2 contain the quantification of this finding. It should be noted that the tendency

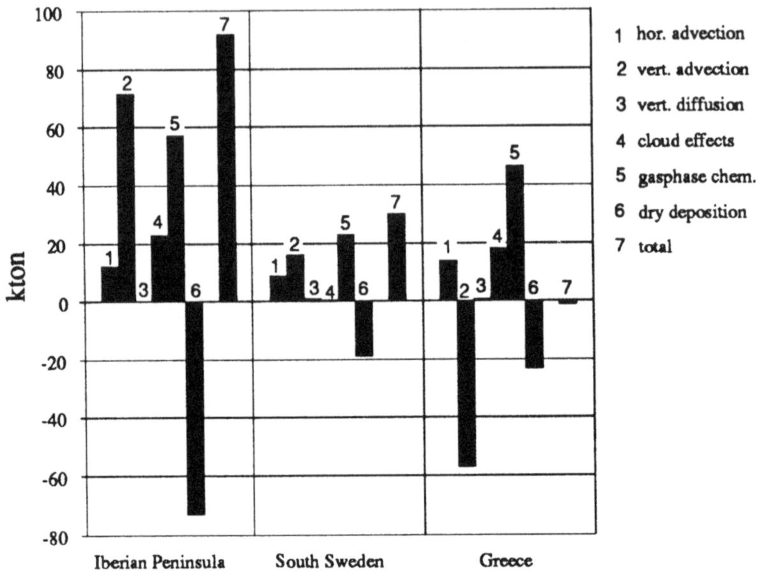

Figure 11: Contribution to the ozone budget by horizontal and vertical advection, vertical diffusion, cloud effects, gas phase chemistry and dry deposition. Results for the Iberian Peninsula, southern Sweden and Greece. Period between July 31, 00 GMT and August 5, 00 GMT, 1990. The vertical axis gives values in kton per subdomain per episode.

of ozone accumulation due to downward motion (subsidence) may nearly equal or exceed chemical production as found for the central European and Iberian domain. It is particularly noticeable that this happened during an episode of rather strong chemical ozone production in the former region.

The main process of ozone loss was dry deposition in Spain and Portugal and horizontal transport in central Europe though the simulations resulted in dry deposition fluxes for this subdomain which were stronger than in the Iberian region. There, horizontal advection slightly contributed to the ozone accumulation found during the episode. This different behaviour can be explained by horizontally divergent air mass flow over central Europe with strong subsidence (downward Ekman pumping, Wyngaard, 1988) whereas convergent ozone flow still was possible in the presence of relatively weak downward ozone flux over parts of the Iberian Peninsula (about 10% of the amount found in central Europe). The Greek subdomain exhibits accumulation through horizontal advection and loss through vertical flow thus pointing towards horizontally convergent winds in this region. Obviously, transport by eddy diffusion through the upper level of the five subdomains used for budget calculations is insignificant. Cloud effects used to be relevant in the southern part of Europe but not in the north as far as ozone is concerned.

Budget calculations clearly demonstrate the crucial role of free tropospheric ozone during accumulation phases when increased chemical production occurs. For

Subdomain	Iberian Peninsula (972)			Greece (396)			Central Europe (356)			North Sea (238)			South Sweden (230)		
Species	O_3	NO_x	SO_2	O_3	NO_x	SO_2	O_3	NO_x	SO_2	O_3	NO_x	SO_2	O_3	NO_x	SO_2
Horiz. adv.	13	—	0.1	16	0.3	7.5	−201	−0.9	3.4	−19	2.2	1.6	10	—	3.6
Vert. adv.	71	−0.2	−2.2	−58	−0.2	−3.3	149	—	−0.7	−7	—	1.7	16	—	−0.1
Vert. diff.	—	0.1	—	2	—	—	—	—	—	—	—	—	1	—	—
Gas ph. chem.	57	−4.3	−3.6	43	−3.0	−3.2	168	−11.5	−5.7	28	−3.4	−1.8	22	−1.1	−0.5
Clouds	23	−0.1	−3.6	18	−0.1	−2.8	—	—	−0.9	1	—	−1.2	—	−0.1	−0.8
Dry Dep.	−74	−0.2	−2.2	−23	−0.1	−1.2	−58	−0.9	−4.3	−4	−0.1	−0.8	−19	−0.1	−0.5
Emission	—	4.7	12.8	—	3.1	3.9	—	13.3	11.3	—	1.2	2.0	—	1.3	1.0
Total	90	0.0	1.3	−2	0.0	0.9	58	0.0	3.1	−1	−0.1	1.5	30	0.0	2.7

Table 1: Budgets of ozone, nitrogen oxides (NO_x) and sulfur dioxide for five subdomains shown in Fig. 10. Numbers in brackets indicate the covered area in units of 1000 km^2. Height of boxes is 3600 m. Episode July 31, 00 GMT - Aug. 5, 24 GMT, 1990. Contributions to the total budget by horizontal and vertical advection, vertical diffusion, gas phase chemistry, cloud processes, dry deposition and emission. Values represent kton per subdomain per episode (calculated as NO_2 in the case of NO_x).

Subdomain	Iberian Peninsula (972)			Greece (396)			Central Europe (356)			North Sea (238)			South Sweden (230)		
Species	O_3	NO_x	SO_2	O_3	NO_x	SO_2	O_3	NO_x	SO_2	O_3	NO_x	SO_2	O_3	NO_x	SO_2
Horiz. adv.	5	—	—	15	0.3	6.8	−203	−0.9	3.4	−29	3.3	2.4	16	—	5.6
Vert. adv.	26	−0.1	−0.8	−53	−0.2	−3.0	151	—	−0.7	−11	—	2.6	25	—	−0.2
Vert. diff.	—	—	—	2	—	—	—	—	—	—	—	—	2	—	—
Gas ph. chem.	21	−1.5	−1.3	39	−2.7	−2.9	170	−11.6	−5.8	42	−5.1	−2.7	34	−1.6	−0.8
Clouds	8	—	−1.3	16	−0.1	−2.5	—	—	−0.9	2	—	−1.8	—	−0.2	−1.2
Dry dep.	−27	−0.1	−0.8	−21	−0.1	−1.1	−59	−0.9	−4.3	−6	−0.2	−1.2	−30	−0.2	−0.8
Emission	—	1.7	4.7	—	2.8	3.5	—	13.4	11.4	—	1.8	3.0	—	2.0	1.6
Total	33	0.0	0.5	−2	0.0	0.8	59	0.0	3.1	−2	−0.2	2.3	47	0.0	4.2

Table 2: As Table 1, but all budget values normalized to an area of $36 \cdot 10^4 km^2$.

these are usually periods with anticyclonic conditions characterized by horizontally divergent air mass flow and subsidence. This effect has gained little attention in the past and needs careful research in the future, particularly regarding feedback processes with atmospheric chemistry during photo–smog events.

As regards nitrogen oxides (NO_x) the budget is dominated by emission and chemical transformation in the gas phase in all subdomains during the photo–smog episode. Only the North Sea exhibits a significant gain through horizontal advection.

To understand the role of chemical characteristics for subdomain budgets in the presence of meteorological control it is worthwile to briefly consider the behaviour of sulfur dioxide. Since only little SO_2 is found at higher levels it is expected that net vertical transport (by advection and diffusion) always results in sulfur export even when a strong downward flux is found for ozone, e.g. over Spain and Portugal. Furthermore, cloud effects which are negligible in the case of ozone during the episode except for the Greek subdomain are significant for SO_2 losses (chemical transformation and wet deposition). Whereas all processed SO_2 over the Iberian Peninsula is supplied by emissions from the subdomain, southern Sweden and Greece considerably suffer from sulfur import leading to significant accumulation in the Swedish box. Large accumulation is also found in central Europe due to high emission rates but also as a result of considerable advection from outside during the episode. Evidently, the budget estimates enable an interesting characterization of regional differences regarding processing of chemical species in the polluted lower troposphere. Yet, information about internal processes may be lost through time – space integration. For instance, turbulent diffusion internally plays an important role for the exchange of matter between different layers. It essentially contributes to the deposition flux at the surface which is treated as a separate process in this study but should be regarded in the connection with vertical diffusion. Furthermore, up– or downward transport through vertical wind is to a good deal compensated by horizontal transport at higher levels.

Nevertheless, vertical fluxes of ozone due to vertical wind are a prominent feature in the boundary layer of the model and, apparently, also in reality. This has been investigated in more detail by Memmesheimer et al. (1996) for central and total Europe dividing the budget box into several layers. Based on their ideas the dependence of the ozone budget terms on height is shown for the Iberian Peninsula and central Europe in Table 3 which confirms the important role of turbulent diffusion for vertical redistribution of ozone in the boundary layer. As expected, the major impact of advection is found in the transition region between the free troposphere and the planetary boundary layer.

height above surface (m)	Iberian Peninsula			Central Europe		
	0–75	75–100	1800–3600	0–75	75–1800	1800–3600
hor. adv.	0.1	-8.7	20.9	-5.4	-147.9	-47.8
vert. adv.	7.4	32.7	31.4	+3.8	76.1	68.6
vert. diff.	58.3	-64.3	6.1	60.5	-61.8	1.7
gas phase ch.	8.6	74.8	-26.3	-0.4	167.8	0.2
clouds	0.8	10.9	11.3	+0.2	-0.4	–
dry dep.	-73.8	–	–	-57.6	–	–
total	1.4	45.4	43.4	1.1	33.8	22.7

Table 3: Ozone budget terms for three height intervals as derived with the EURAD model for the Iberian Peninsula and central Europe. Values for the subdomain in kton per episode (July 31, 00 GMT – August 5, 00 GMT, 1990).

Another approach to budget studies was chosen by Feldmann et al. (1996) who were interested in temporal variations of the different terms as a function of height and, therefore, only carried out a horizontal integration of equ. (1). In this way it can be nicely shown that horizontal and vertical advection of tracers is clearly anticorrelated under anticyclonic conditions in the lower troposphere.

5 Final remarks

The examples of episode simulation and budget studies show the broad range of problems to which regional models can and have to be applied. They also provoke the question as to how reliable such simulations and estimates are. Models like EURAD have to be subjected to evaluation exercises persistently since their complexity always contains the risk of erroneous results. Therefore, development of the next model generation and improvement and exploitation of the existing ones also means model testing and evaluation to a very high degree.

On the other hand studies requiring high information density in space and time like budget assessment would not be possible without employing air quality models like the EURAD system. The sensitivity of simulation results to input data, in particular to emission estimates, is obvious. Therefore, evaluation studies cannot be confined to the numerical model alone. They have to be complemented by verification of emission and other input data and critical reviewing of measurements used in the process of evaluation.

A particular parameter to be mentioned when considering future application of air quality simulations to impact studies is deposition velocity. Calculation of surface uptake of minor constituents (or their exhalation) becomes the more demanding the higher the horizontal and vertical resolution is. This and other aspects put serious constraints to the application of the method of internal nesting or zooming of regional models though is proves to be a valuable advanced technique as demonstrated here and elsewhere with the EURAD model. The study of regional budgets of pollutants has revealed that vertical fluxes, i.e. the exchange between the heavily polluted planetary boundary layer and the less but significantly contaminated free troposphere, can be critical and that better understanding of this process is needed.

Acknowledgement: The study was financially supported by the Ministry for Education and Research of the FRG (BMBF) and the Land Nordrhein–Westfalen (MWF) in the framework of EUROTRAC I (subproject EUMAC). Data provided by the ECMWF (via the German Weather Service), EMEP, the EUROTRAC subproject GENEMIS and TOR (via RIVM) were used. In this context, we owe P. Speth, J. Schaug, H. Styve, R. Friedrich, J. Beck and their colleagues special thanks for valuable support and advice. The computations were carried out at the Research Center Jülich in collaboration with the Institute for Chemistry and Geodynamics (ICG2, ICG3). As regards the layout of the manuscript we gratefully acknowledge the help of J. Reusch, G. Piekorz and H.J. Bock.

References

- Ackermann, I.J., H. Hass, M. Memmesheimer, C. Ziegenbein, and A. Ebel, The parameterization of the sulfate–nitrate–ammonia aerosol system in the long–range transport model EURAD, Meteor. Atmos. Phys., 101–114, 1995.
- Arakawa, A., and V.R. Lamb, Compatutional design of the basic dynamical processes of the UCLA general circulation model, Math. Comput. Phys., 17, Academic Press, 174–264, 1977.
- Binkowski, F.S., and U. Shankar, The regional particulate model: part I, model description and preliminary results, J. Geophys. Res., 100, D12, 26191–26209, 1995.
- Bott, A., A positive definite advection scheme obtained by nonlinear renormalization of the advective fluxes, Mon. Wea. Rev., 117, 1006–1015, 1989a.
- Bott, A., Reply, Mon. Wea. Rev., 117, 2633–2636, 1989b.
- Builtjes, P.J.H., The LOTOS – LOng Term Ozone Simulation project; summary report, TNO report R92/240, Delft, The Netherlands, 1992.
- Chang, J.S., R.A. Brost, I.S.A. Isaksen, S. Madronich, P. Middleton, W.R. Stockwell, and C.J. Walcek, A three–dimensional Eulerian acid deposition model: physical concepts and formation, J. Geophys. Res., 92, 14681–14700, 1987.
- Ebel, A., H. Elbern, H. Hass, H.J. Jakobs, M. Memmesheimer, and H.J. Bock, Meteorological effects on air pollutant variability on regional scales,

In: Air Pollution III, Vol. 4, eds. A. Ebel, N. Moussiopoulos, Computational Mechanics Publication, p. 103–110, 1995a.

- Ebel, A., H. Feldmann, F. Fiedler, H. Hass, H.J. Jakobs, O. Klemm, K. Nester, E. Schaller, A. Schwartz, and J. Werhahn, Contributions to the evaluation of chemical transport models within the SANA Project, In: Air Pollution III Vol. 4, eds. A. Ebel, N. Moussiopoulos, Computational Mechanics Publications, p. 103–110, 1995b.

- Feldmann, H., H. Hass, H.J. Jakobs, and A. Ebel, Fluxes of air pollutants from local to continental scale simulated with the EURAD model, In: Air Pollution II, Vol. 2, ed. C.A. Brebbia, Computational Mechanics Publications, p. 3–11, 1994.

- Feldmann, H., H. Hass, M. Memmesheimer, and H.J. Jakobs, Budgets of atmospheric sulfur for East Germany based on meso–α–scale simulations, Meteorol. Zeitschrift, in press, 1996.

- Friedrich, R., Generation of time–dependent emission data, in: Photo–Oxidants–Precursors and Products, Proc. EUROTRAC Symp. '92, eds. P.M. Borrell, P. Borrell, T. Cvitas and W. Seiler, SPB Acad. Publ., bv, p. 255–268, 1993.

- Grell, G.A., J. Dudhia, D.R. Stauffer, A description of the fifth generation Penn State/NCAR Mesoscale Model (MM5), NCAR/TN–398+IA, National Center for Atmospheric Research, Boulder, CO, 117pp., 1994.

- Hansen, D.A., A. Ebel, S. Hanna, J. Kage, and R. Thuillier, CAMRAQ: the quest for modern solutions to modern air quality problems, Environ. Sci. Technol., Vol 28, No. 2, 71–77, 1994.

- Hass, H., Description of the EURAD Chemistry–Transport–Model Version2 (CTM2), Mitteilungen aus dem Institut für Geophysik und Meteorologie der Universität zu Köln, No. 83, 1991.

- Hass, H., A. Ebel, H. Feldmann, H.J. Jakobs, and M. Memmesheimer, Evaluation studies with a regional chemical transport model (EURAD) using air quality data from the EMEP monitoring network, Atmos. Environ., 27A, 867–887, 1993.

- Hass, H., E. Berge, P. Builtjes, A. Ebel, H.J. Jakobs, M. Memmesheimer, D. Simpson, and R. Stern, A comparison of long–range transport models applied for a European summer episode, in: EUROTRAC Symp. '94, eds. P. Borrell et al., SPB Academic Publishing, p. 857, 1994.

- Hass, H., P.J.H. Builtjes, D. Simpson, R. Stern, H.J. Jakobs, M. Memmesheimer, G. Piekorz, M. Roemer, P. Esser, and E. Reimer, A diagnostic comparison of air quality model results for an episode in July/August 1990, EUROTRAC Special Publications, International Scient. Secr., Garmisch–Partenkirchen, FRG, 1996 (in press).

- Jakobs, H.J., H. Feldmann, H. Hass, and M. Memmesheimer, The use of nested models for air pollution studies: an application of the EURAD model to a SANA episode, J. Appl. Met., 34, No. 6, 1301–1319, 1995.
- Lübkert, A., and M. Schöpp, A model to calculate natural VOC emissions from forests in Europe, IIASA, WP–89–082, 1989.
- Memmesheimer, M., J. Tippke, A. Ebel, H. Hass, H.J. Jakobs, and M. Laube, On the use of EMEP emission inventories for European scale air pollution modeling with the EURAD model, In: Proceedings of the EMEP workshop on photooxidant modelling for long–range transport in relation to abatement strategies, Berlin, April 16–19, p. 307–324, 1991.
- Memmesheimer, M., and H.J. Bock, Deposition of nitrogen into the North Sea for air pollution episodes calculated with the EURAD model, In: Air Pollution Studies with the EURAD Model System (2), Mitteilungen aus dem Institut für Geophysik und Meteorologie, University of Cologne, No. 105, p. 57–70, 1995.
- Memmesheimer, M., A. Ebel, and M. Roemer, Budget calculations for ozone and its precursors: seasonal and episodic features based on model simulations, submitted to J. Atmos. Chem., 1996.
- Millan, M.M., and B. Artinano, Mesometeorological cycles of air pollution in the Iberian Peninsula, Air Poll. Res. Report 44, Commission of the European Communities, CEC–DG XII, Brussels, 219 pp., 1992.
- Mölders, N., H. Hass, H.J. Jakobs, M. Laube, and A. Ebel, Some effects of different cloud parameterization in a mesoscale model and a chemistry transport model, J. Appl. Met., 33, 527–545, 1994.
- Moussiopoulos, N., P. Sahm, Ch. Kessler, and R. Kunz, Application of the EUMAC zooming model to the greater Athens area in the frame of APSIS, In: Air Pollution II, Vol. 1, Computer Simulation, eds. J.M. Baldasano, C.A. Brebbia, H. Power and P. Zanetti, Comp. Mechanics. Publ., p. 311–318, 1994.
- Peters, K.L., C.M. Berkowitz, G.R. Carmichael, R.C. Easter, G. Faiweather, S.J. Ghan, J.M. Hales, L.R. Leung, W.R. Pennell, F.A. Potra, R.D. Saylor, and T.T. Tsang, The current state and future direction of Eulerian models in simulating the tropospheric chemistry and transport of tracer species: a review, Atmos. Environ., No. 2, 29, 189–222, 1995.
- Petry, H., H. Elbern, E. Lippert, and R. Meyer, Three–dimensional mesoscale simulations of airplane exhaust impact in a flight corridor, In: Impact of Emissions from Aircraft and Spacecraft upon the Atmosphere, Proceed. Inform. Sci. Coll., Cologne, April 1994, eds. H. Schumann and D. Wurzel, p. 329–335, 1994.
- Stauffer, D.R., N.L. Seaman, Use of four–dimensional data assimilation in a limited area mesoscale model. Part I: experiments with synoptic–scale data, Mon. Wea. Rev., 118, 1250–1277, 1990.

- Stockwell, W.R., P. Middleton, J. Chang, and X. Tang, The second generation regional acid deposition model chemical mechanism for regional air quality modeling, J. Geophys. Res., 95, 16343–16367, 1990.
- Walcek, C.J., G.R. Taylor, A theoretical method for computing vertical distributions of acidity and sulfate production within cumulus clouds, J. Atmos. Sci., 43, 339–355, 1986.
- Wyngard, J.C., Structure of the PBL, In: Lectures on Air Pollution Modelling, eds. A. Venkatram and J.C. Wyngard, Americ. Met. Soc., Boston, 9–61, 1988.
- Ziegenbein, C., I.J. Ackermann, and A. Ebel, The treatment of aerosols in the EURAD model: results from recent developments, In: Air Pollution II, Vol. 1, Computer Simulations, eds. J.M. Baldasano, C.A. Brebbia, H. Power, P. Zannetti, Computational Mechanics Publications, p. 229–237, 1994.

Weather-Pattern Dependent Total Column Ozone Trends in Central Europe

W. Enke, G. Vogel and D. Spänkuch

Deutscher Wetterdienst, Meteorologisches Observatorium Potsdam, Postfach 60 05 52, D-14405 Potsdam, Germany

Abstract. Trends of total column ozone, Ω, based on TOMS, version 6.0 data are studied for the summer season in dependence of characteristic weather patterns (CWP) in Central Europe. The CWPs are objectively defined by cluster analysis on the basis of the 1000 hPa geopotential and 500/1000 hPa thickness fields.

There is not only significant difference in total ozone amount between the CWP, but also quite different trends. Considerable decrease in total ozone, up to 30 to 40 Dobson units (DU) from 1980 to 1994, were found for the CWPs Westerly Type (WT), Anticyclone Ridge (AR) and Cyclone France (CF). All these CWPs are characterized by low tropopause height in Central Europe. However, positive trends were found, too, but for the least frequent CWPs and of less amount, about 10 DU within the last 15 years.

Keywords. Total ozone, Großwetterlage, circulation pattern, cluster analysis

1 Introduction

The overall decline of stratospheric and total column ozone as a man-made perturbation and disturbance of the delicate balance of ozone producing and ozone disturbing processes in the atmosphere is now world wide accepted and documented (e.g. Reinsel et al. 1994, WMO 1994, Bojkov 1995, Bojkov and Fioletov 1995, Bojkov, this issue).

NATO ASI Series, Vol. I 53
Atmospheric Ozone Dynamics
Observations in the Mediterranean Region
Edited by Costas Varotsos

This decline, although of considerable amount - 5% on global average during the last years and 60% in late winter over Antarctica compared to the pre-ozone-hole averages - is hidden in inter-diurnal changes which are mainly the result of dynamic processes within the atmosphere (e.g. Dobson 1930, Tung and Yang 1988, Rabbe and Larsen 1992).

Local and regional changes in total column ozone are the product of both causes. The aim of this contribution is the analysis of dynamically induced trends of total column ozone in Central Europe. With other words: Is the decrease in total ozone above Central Europe within the last decade (Vogel et al., this issue) mainly the result of a changing circulation regime? Is the trend in total column ozone weather-pattern dependent?

2 Objective Weather Pattern Classification

The objective weather pattern classification is based on the 1000 hPa geopotential and 500/1000 hPa thickness fields. This is in contrast to the well-known subjective classification of weather patterns (Großwetterlagen) by Hess and Brezowsky (1952, 1969, 1977), actualized by Gerstengarbe et al. (1993) on the basis of the surface geopotential field only. Starting with any synoptic realization you like the algorithm looks for the least similar synoptic situation of the whole ensemble (about 30 years) by calculating the SI-score

$$SI_{l} = \frac{\sum_{l=1}^{m} \left(|d_{p,l} - d_{l,l}| * \alpha_l \right)}{\sum_{l=1}^{m} \left(\max\left(|d_{p,l}|, |d_{l,l}| \right) \right)}$$

where

$d_{p,j}$ and $d_{j,i}$ - horizontal differences of the 1000 hPa geopotential and the 500/1000 hPa thickness fields for the class average and the individual case, respectively,

α_i - weights, giving the inner differences at the reference grid point more weight than the outer differences,

m - number of horizontal geopotential differences, 24 for each pressure level, 48 all at all for the gridnet used.

The procedure is repeated up to the prefixed number of weather patterns, about 10. The second step assigns all realizations (synoptic situations) to their most similar class. In a third step the class assignment is optimized by an iterative change of the cases up to a stable state. The classification algorithm is described in more detail in Enke and Spekat (1995, 1996). Cloud classification schemes of satellite pictures use the same algorithm in defining cloud clusters (Rosenow et al. 1992, 1993).

3 Characteristic Weather Patterns (CWP)

Fig. 1 characterizes the weather patterns in summer (June to August) by means of their mean 1000 hPa and 500 hPa geopotential fields, derived from daily 12.00 UTC gridmaps of the German Weather Service. In Enke and Spekat the 0.00 UTC and 12.00 UTC data were used for classification to take the weather process to some degree into account resulting in slightly different mean geopotential fields of the CWP. The CWP are similar to some of the subjective classes. Their mean column ozone amount at Potsdam, Germany, derived from the TOMS data 1980 to 1994, are compiled in Table 1 for all seasons. Lowest ozone amounts occur for Anticyclone Southeast Europe (ASE) in winter and autumn. Permanently low ozone values are found throughout the year for the CWP Anticyclone Cen-

Figure 1 1000 hPa (left) and 500 hPa height fields of the objectively defined Characteristic Weather Patterns of Central Europa

ACE	Anticyclone Central Europe
ABI	Anticyclone British Isles
AS	Anticyclone Scandinavia
AR	Anticyclonic Ridge
SWT	Southwesterly Type
CF	Cyclone France
CCE	Cyclone Central Europe
WT	Westerly Type
CG	Cyclone Genoa

ACE

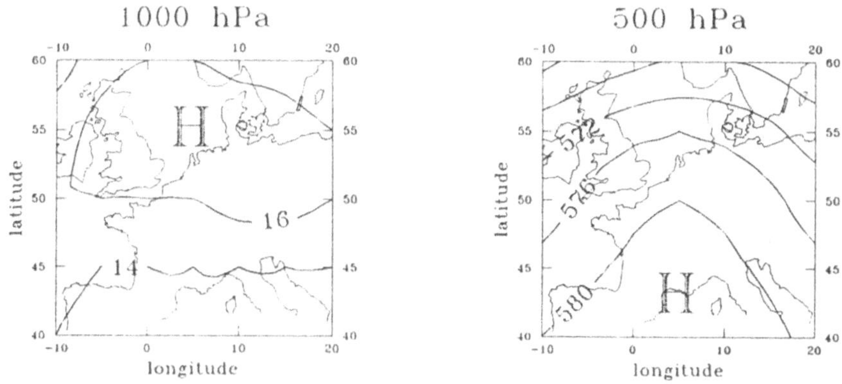
1000 hPa
H
16
14
latitude
longitude
500 hPa
572
576
580
H
latitude
longitude

ABI

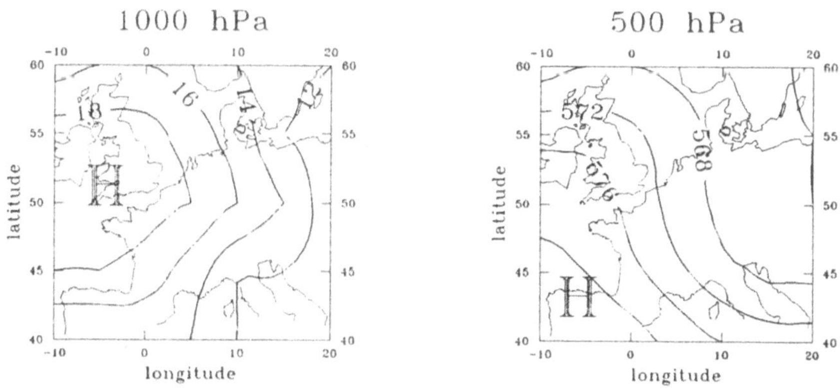
1000 hPa
18
16
14
H
latitude
longitude
500 hPa
572
576
568
H
latitude
longitude

AS

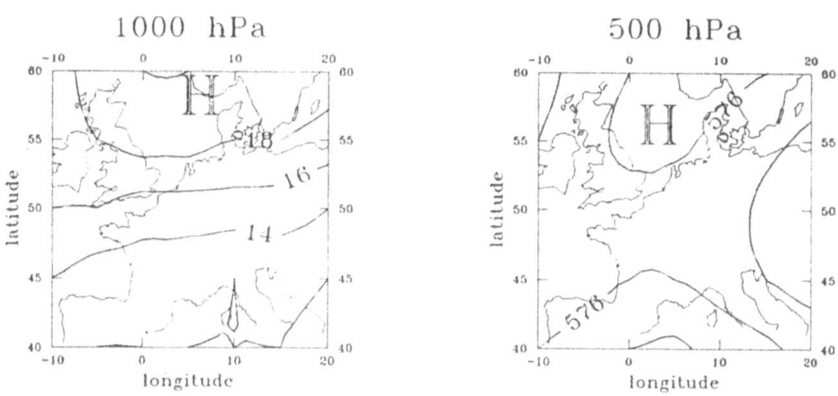
1000 hPa
H
18
16
14
latitude
longitude
500 hPa
H
576
576
latitude
longitude

AR

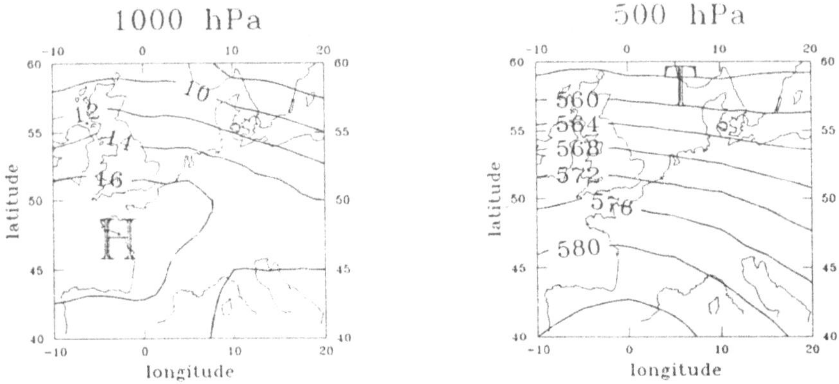
1000 hPa
500 hPa
10
12
14
16
H
T
560
564
568
572
576
580
latitude
longitude

SWT

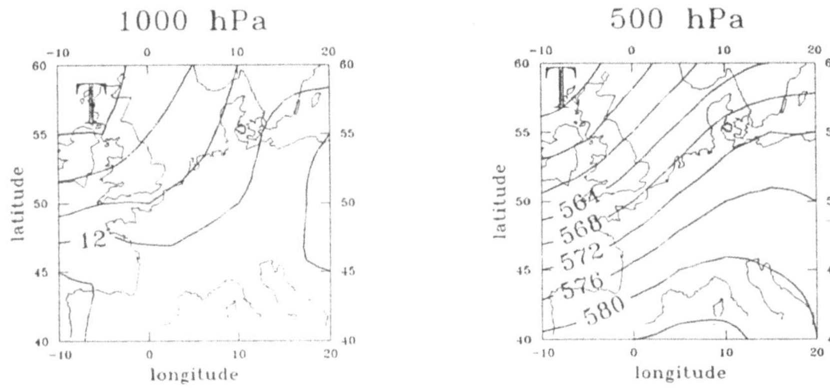
1000 hPa
500 hPa
T
12
564
568
572
576
580
latitude
longitude

CF

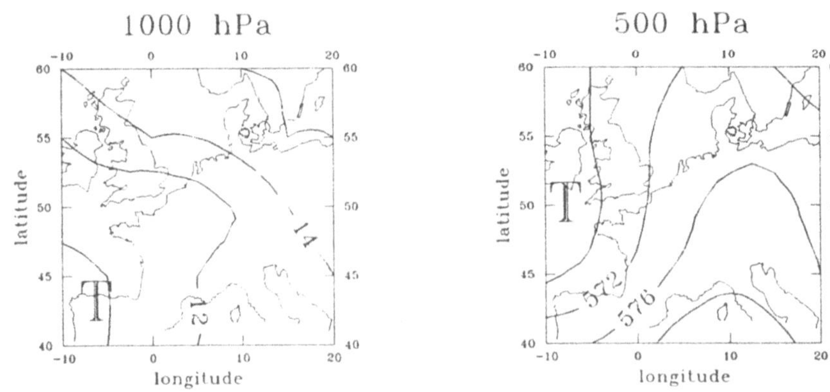
1000 hPa
500 hPa
T
14
12
572
576
latitude
longitude

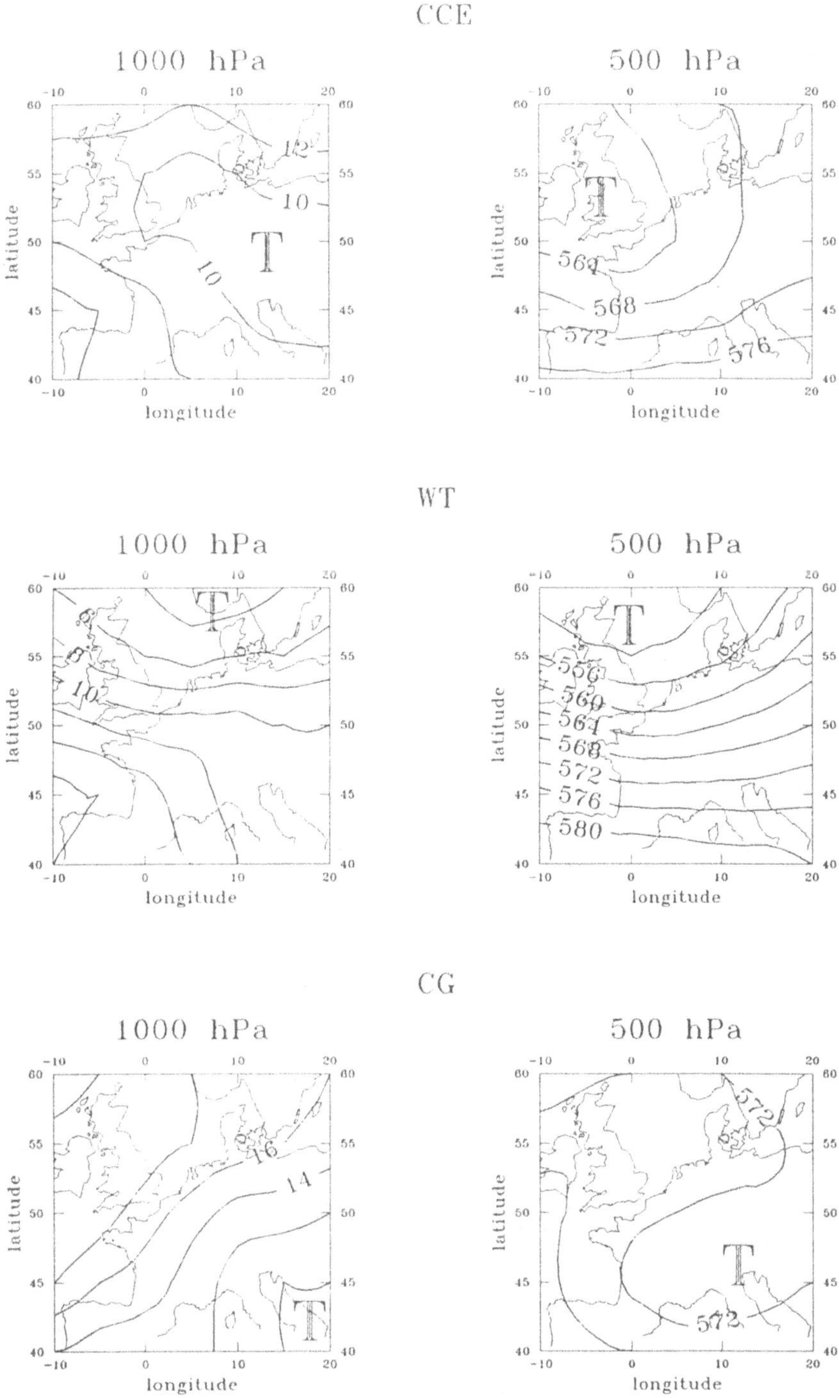
CCE
1000 hPa
500 hPa
latitude
longitude
12
10
10
T
T
560
568
572
576
WT
1000 hPa
500 hPa
latitude
longitude
6
8
10
T
T
556
560
564
568
572
576
580
CG
1000 hPa
500 hPa
latitude
longitude
16
14
T
572
T
572

Table 1 Mean amount of total ozone (DU) at Potsdam for objectively classified synoptic patterns

Synoptic pattern	Abbreviation	winter	spring	summer	autumn
Anticyclone British Isles	ABI	-	416	366	300
Westerly Type	WT	343	392	363	297
Cyclone Central Europe	CCE	-	401	359	314
Anticyclonic Ridge	AR	308	361	346	-
Anticyclone Scandinavia	AS	352	397	346	-
Cyclone Genoa	CG	-	407	341	320
Cyclone France	CF	373	378	338	289
Southwesterly Type	SWT	351	376	332	291
Anticyclone Central Europe	ACE	324	370	332	296
Southerly Type	ST	356	-	-	283
Anticyclone Southeast Europe	ASEE	291	-	-	279
Anticyclone Southwest Europa	ASWE	370	-	-	300

tral Europe (ACE) and for the CWP Anticyclonic Ridge (AR) in winter and spring, confirming the findings of Bekoryukov et al. (1995) of low ozone values in Western Europe when the center of the Azores High at the surface is north of 40°. The occurence of maximum ozone amounts in dependence of the CWPs are less uniform. Generally, high ozone values are observed in Central Europe when this region is under cyclonic influence (Cyclone France in winter, Anticyclone British Isles in spring and summer, Cyclone Genoa and Cyclone Central Europe in autumn), confirming the well known high anticorrelation between tropopause height and total column ozone.

More interesting than these climatological facts are, however, the differences in the trends of total column ozone for the CWPs as shown in Fig. 2 for the summer season. As a result of these

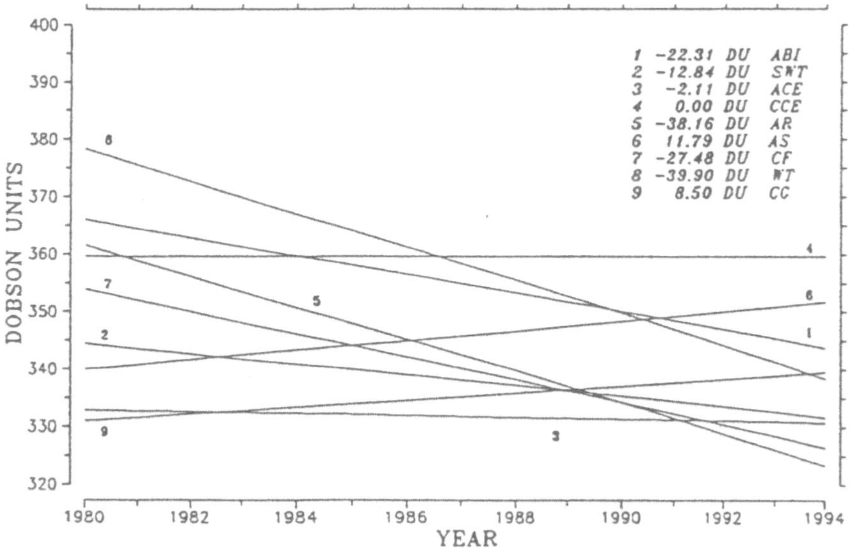

Figure 2 Trends of total column ozone in Central Europe for different Characteristic Weather Patterns. See legend of Fig. 1 for explanation of the abbreviations used.

differences, in 1980 highest ozone values were observed for the Westerly Type, in 1994 for the Cyclone Central Europe. There is a considerable decrease in total ozone in the period 1980 to 1994 for the Westerly Type (-39.9 DU), Anticyclonic Ridge (-38.2 DU), Cyclone France (-27.5 DU) and Anticyclone British Isles (-22.3 DU), that is to say for that CWPs except AR when the tropopause height is low in Central Europe. A positive tendency is found only for Anticyclone Scandinavia (+11.8 DU) and Cyclone Genoa (+8.5 DU). The latter two CWPs are by far the least frequent CWP in summer (see Table 2). Weather pattern dependent trends were also found for tropospheric and stratospheric temperatures (not shown).

The weather pattern dependent trends in Ω rise the question of trends in the frequency of the CWP. A first search does not show such a trend as shown in Fig. 3 for the CWPs with the extreme Ω values on average (ABI and ACI), but a considerable variation of the frequency of occurence throughout the years.

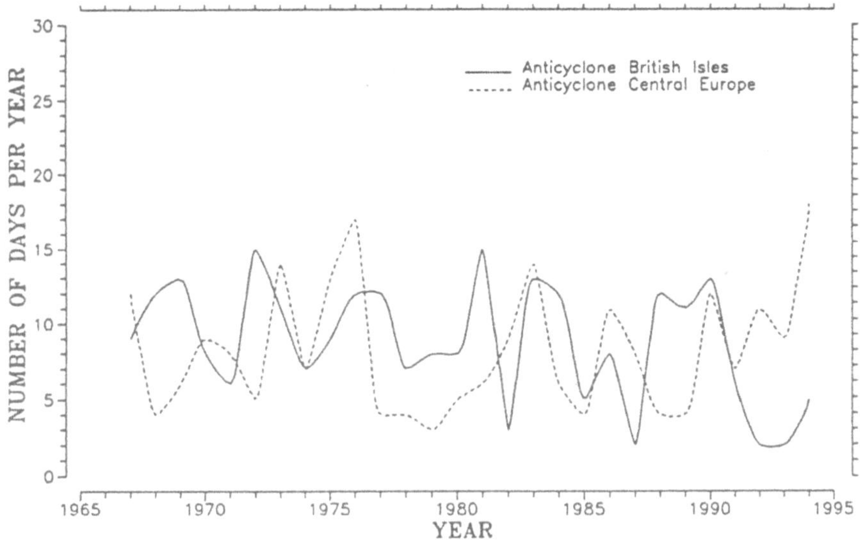

Figure 3 Frequency of the CWPs Anticyclone British Isles ans Anticyclone Central Europe in summer throughout the years

Table 2 Results of 24hr-Total Column-Ozone-Forecast for different weather patterns in summer

Weather pattern	ABI	SWT	ACE	CCE	AR	AS	CF	WT	CG
Frequ. since 1966 (days)	119	160	113	80	444	41	66	298	44
mean total column ozone (DU)	366	332	332	359	346	346	338	363	341
mean forecast error (DU)	1.0	-2.3	-3.4	1.1	-5.2	-2.1	1.9	-8.4	-2.6
RMSE original forecast (DU)	12.0	11.1	10.5	8.4	13.8	12.0	8.0	15.2	12.1

Table 2 gives the frequency for all CWPs in summer for the last 30 years (1966-1994). The most frequent CWP is the Anticyclonic Ridge with 10 times more days of occurence (444 days) within the last 30 years than for the CWPs Anticyclone Scandinavia and Cyclone Genoa. Table 2 contains, too, information relevant to the short-term Ω forecast, showing different quality of Ω prediction in dependence on CWP. Minimum mean Ω forecast errors are 1 DU for ABI and CCE, eight times less than the maximum mean Ω forecast error for WT.

4 Conclusions

Regional trends in total column ozone are not independent on weather. Within the last 15 years a significant decrease in total column ozone amount - up to 40 DU - was found for the CWPs Westerly Type and Anticyclonic Ridge in summer, the two most frequent CWPs within the last 30 years. Positive trends of less amount, however, (about 10 DU) were found for the least frequent CWPs (Anticyclone Scandinavia, Cyclone Genoa). There seems to be no substantial change in the long-term frequency distribution of CWPs despite a considerable variation of the frequency througout the years.

As a byproduct it was found, that the quality of short-term ozone forecast is dependent on the characteristic weather pattern. The lowest accuracy in forecast was reached for the CWP Westerly Type.

Acknowledgement: The study was in part supported by the Bundesministerium für Forschung und Technologie, Bonn, under grant 01 LO 9204/0.

References

Bekoryukov V. I., Bugaeva I. V., Zakharov G. R., Kiryushov B. M. and Tarasenko D. A. (1995) Azores High, ozone, and meteorological parameters of the troposphere and stratosphere. *Meteorol. i Gidrol. No.7*, 40-47.

Bojkov R. D. (1995) The changing ozone layer. *WMO-UNEP* 26 pp.

Bojkov R. D. and Fioletov V. E. (1995) Estimating the global ozone characteristics during the last 30 years. *J. Geophys. Res. 100*, D8, 16,537-16,551.

Bojkov R. D. (1996) The changing ozone layer: a review. *(this issue)*

Dobson G. M. B. (1930) Measurement of the amount of ozone in the earth's atmosphere and its relation to other geophysical conditions. *Part 4.Proc. Roy. Soc. London 129*, 411.

Enke W. and Spekat A. (1995) A two-step method to transform climate model outputs into regional and local weather elements, in: *Proc. 6th International Meeting on Statistical Climatology, 19-23 June 1995, Galway, Ireland*, S. 339-342.

Enke W. and Spekat A. (1996) A two-step downscaling method to transform climate model outputs into local and regional weather elements. Submitted to *Theor. Appl. Climatol.*

Gerstengarbe F.-W., Werner P. C., Busold W:, Rüge V. and Wegener K.-O. (1993) Katalog der Großwetterlagen Europas nach Paul Hess und Helmut Brezowsky 1881-1992. *Ber. Dtsch. Wetterdienst 113*, 249 pp.

Rabbe Å. and Larsen S. H. H. (1992) Ozone variations in the Northern Hemisphere due to dynamic processes in the atmosphere. *J. Atmos. Terr. Phys. 54*, 1107-1112.

Reinsel G. C., Tiao G. C., Wuebbles D. J., Kerr J. B., Miller A. J., Nagatari R. M., Bishop L. and Ying L. H. (1994) Seasonal trend analysis of published ground-based and TOMS total ozone data through 1991. *J. Geophys. Res. 99*, D3, 5449-5464.

Rosenow W., Güldner J., Spänkuch D. and Vogel G. (1992) Estimation of the meso-beta vertical cloud structure using Meteosat, conventional and model output data. 9th Meteosat Scientific Users' Meeting. Locarno, Switzerland, 15th-18th September 1992. *Proceedings EUMP 11*, EUMETSAT 1992, 433-422.

Rosenow W., Güldner J. Spänkuch D. and Vogel G. (1993) A 3d-cloud analysis for the validation of diabatic components in numerical models. In Keevallik S. and Kärner O. (Eds.): *IRS'92: Current Problems in Atmospheric Radiation. Proc. Intern. Radiation Symposium Tallinn, Estonia, 3-8 August 1992. A. Deepak Publ.*, 282-288.

Tung K. K. and Yang H. (1988) Dynamic variability of column ozone. *J. Geophys. Res. 93*, D9, 11,123-11,128.

Vogel G., Schubert U. and Spänkuch D. (1996) Regional distribution of total column ozone changes in Central Europe *(this issue)*

WMO (1994) Scientific assessment of ozone depletion: 1994. Clobal Ozone Research and Monitoring Project - *Report No. 37*

Ozone change in the polar atmosphere

H. Gernandt, P. von der Gathen, A. Herber

Alfred Wegener Institute for Polar and Marine Research
Research Unit Potsdam
Telegrafenberg A 43, D-14473 Potsdam, Germany

Abstract. Ozone is one of the most sensitive minor constituents in the atmosphere indicating regional and global changes in stratospheric and tropospheric chemistry and dynamics. Balloon-borne ozone observations from antarctic and arctic stations were used to assess the mean interannual as well as annual variations of ozone in the polar atmosphere at latitudes where regular observations by optical satellite or ground-based instruments are limited. Evidences for chemical ozone loss in the arctic stratosphere are presented which could be retrieved from balloon-borne observations. The impact of volcanic aerosols on stratospheric ozone as well as tropospheric ozone changes and the occurrence of anthropogenic aerosols are discussed. The current differences between both polar regions as well as altitude dependent trends partially declining strongly in the middle stratosphere are retrieved from balloon-borne data in the Antarctic and Arctic. The results demonstrate the complex impact of global dynamics and regional chemical processes on ozone in the polar stratosphere and troposphere.

Keywords. Polar ozone, vertical distribution, volcanic aerosols, anthropogenic aerosols, mean annual variation, long-term trends.

1 Introduction

The total ozone content in the atmosphere on a seasonal time scale is controlled by both the transport of ozone within the stratosphere and photochemical processes. Kaye (1993) summarized this global pattern. Net ozone is produced in the tropical upper stratosphere. This ozone is transported polewards and downwards, particularly in fall and winter. In the lower stratosphere, the lifetime of ozone is long during spring and winter. As ozone is transported downwards it begins to accumulate at high latitudes. The transition from winter to summer circulation in the stratosphere dramatically slows the accumulation of ozone and ozone decreases due to increased photochemical destruction in summer. Schoeberl (1993) has shown by TOMS records that the transport towards the North is greater and extends more polewards than in the Southern Hemisphere. In the stratosphere the planetary wave activity is responsible for most of the north-south transport of ozone and heat. Since the Southern Hemisphere lacks the significant orographic features found in the Northern Hemisphere planetary waves are weaker. As a consequence the winter antarctic stratosphere is on average more than 10 K colder than the arctic stratosphere, and the antarctic polar vortex lasts longer, and the winter accumulation of ozone is weaker. This principal global dynamical pattern and its hemispheric differences are the basic frame for the variation of the vertical ozone distribution caused by dynamical effects as well as by chemical and radiative processes.

The global pattern of the ozone layer is finally produced by photochemical and catalytic reactions as well as by dynamical transport (Brasseur and Solomon 1984, Røth

NATO ASI Series, Vol. I 53
Atmospheric Ozone Dynamics
Observations in the Mediterranean Region
Edited by Costas Varotsos

1994, Kaye 1993, WMO 1992). Ozone (O_3) is formed by the UV photolysis of molecular oxygen (O_2) followed by the recombination of atomic oxygen (O) with O_2.

$$O_2 + hv\,(UV) \dashrightarrow O + O$$
$$O + O_2 + M \dashrightarrow O_3 + M$$

Although mixing ratios of ozone rarely exceed 10 parts per million by volume (ppmv) in the stratosphere, its total amount comprises some 90% of the atmospheric ozone. The absorption of UV radiation by stratospheric ozone leads to heating of the stratosphere when the energy of radiation is converted to kinetic energy of air molecules (M*) by the recombination of O and O_2.

$$O_3 + hv\,(UV) \dashrightarrow O + O_2 \text{ (UV absorption)}$$
$$O + O_2 + M \dashrightarrow O_3 + M^* \text{ (heating of stratospheric air)}$$

This heating of stratospheric air causes increasing temperature with altitude. So the stratosphere becomes more stable with respect to vertical mixing, and large vertical gradients may occur in trace constituents. They can only be mixed vertically by the global dynamics associated with large-scale planetary waves, or by meso-scale wave disturbances of the stratosphere, e.g. rising vertical motion in the tropics balanced by descent at high latitudes.

While the stratospheric production of ozone is only based on the photolysis of oxygen, it is mainly destroyed by a large number of catalytic chemical reactions involving free radical species containing either hydrogen, nitrogen, chlorine, or bromine atoms (Brasseur and Solomon 1984, WMO 1992, WMO 1995). The general two step reaction chain is

$$O_3 + X \dashrightarrow XO + O_2$$
$$O + OX \dashrightarrow X + O_2$$

The most important radicals X include hydroxyl (OH), nitric oxide (NO), chlorine (Cl), and bromine (Br). Their tropospheric sources are water (H_2O) and methane (CH_4) for OH, nitrous oxide (N_2O) for NO, methyl chloride (CH_3Cl) and anthropogenic CFCs for Cl, as well as methyl bromide (CH_3Br) and halocarbons (Halons) for Br. But some ozone is also directly destroyed by a process involving only oxygen by the following reaction chain

$$O_3 + hv\,(UV, visible) \dashrightarrow O + O_2$$
$$O + O_3 \dashrightarrow O_2 + O_2\,.$$

The natural and anthropogenic sources for these ozone destroying radicals are found in the troposphere. Transported into the stratosphere, these chemically inert gases are photolysed by solar UV radiation and reactive atoms are released. Most ozone production and destruction occurs in the tropical upper stratosphere, where UV amounts are largest. Photodissociation of ozone, however, extends lower into the stratosphere and to higher latitudes than production does. So transport processes play an important role in controlling the distribution of ozone, whose production and destruction is not in local balance. The separation between production and destruction and the global transport are yielding to the net vertical profile of stratospheric ozone and its latitudinal variation. Any change in global transport would change the ozone distribution significantly.

The reason for separation between the production and destruction regions is that ozone production is driven by UV with wavelengths below 240 nm, while ozone loss typically involves production of atomic oxygen which occurs at longer wavelengths in the UV range (220-350 nm) and to a lesser extent in the visible range (400-800 nm). Since longer wavelength photons penetrate deeper into the atmosphere, the difference in altitude between production and destruction occurs.

The catalytic ozone loss processes are in equilibrium with a large number of reactions between these catalytically active constituents themselves. Reservoir gases are formed by these reactions. They are unreactive towards ozone. The main reservoir gases are HNO_3, HCl, and $ClONO_2$. They are formed by the following reactions:

$$OH + NO_2 + M \dashrightarrow HNO_3 + M^*$$
$$ClO + NO_2 + M \dashrightarrow ClONO_2 + M^*$$
$$OH + NO + M \dashrightarrow HONO + M^*$$

Because the reservoir gases are only slowly photolyzed by UV and have a lifetime of the order of days to weeks, they can build up in large concentrations. So the equilibrium between the catalytic radicals and the photolysis of reservoir gases specifies another control for the net ozone loss and subsequently the vertical ozone distribution. Additionally in the polar stratosphere at extremely low temperatures heterogeneous chemical reactions at the surfaces of aerosol particles and polar stratospheric cloud particles can change the chemical composition and yield unusual losses in ozone there.

Thus the large number of chemical processes as well as the transport of source gases from the troposphere into the stratosphere, the transport of stratospheric species back into the troposphere, and the transport and mixing of chemicals within the stratosphere are all critical for understanding the spatial distribution of ozone and its temporal changes for a wide range of time scales. Ozone in the polar atmosphere is basically controlled by all of these processes. So detailed studies of the ozone layer have to refer to observations with a sufficient vertical resolution.

2 Data for ozone and aerosols at high polar latitudes

At polar latitudes regular long-term observations of ozone or aerosols by satellite instruments in the stratosphere and troposphere are limited due to polar night conditions, due to orbit parameters as well as to the optical properties of the atmosphere (Schoeberl 1993, Stolarski et al. 1992, McCormick and Trepte 1987, McCormick et al. 1995). Furthermore, profiles of vertical ozone distributions can be retrieved from those data mainly for stratospheric but not for tropospheric regions. These limitations for regular observations can be partially compensated by measurements with balloon-borne ozone sondes at high polar latitudes. Those balloon soundings are useful to measure the vertical distribution of ozone in the troposphere and stratosphere up to altitudes of about 30 to 35 km. They are independent of solar irradiance conditions and radiative properties of the atmosphere such as clouds at high altitudes or large total aerosol contents.

In order to discuss the variability of ozone and aerosols in the polar stratosphere and troposphere records of balloon-borne ozone data and remote sensing data of the optical depth of aerosols in the atmosphere obtained from ground-based sun photometers and satellite instruments have been used (Tab. 1).

Plots by altitude versus time for the mean annual variations of the vertical distribution of ozone concentration were calculated for the periods from 1992 to 1995 for two stations and monthly means for four stations for periods as shown in Tab. 2.

Table 1. Observations of the vertical ozone distribution and spectral optical depth of aerosols. Data from Syowa and Mirny are not available for all years, c.f. Fig. 8 and Fig. 5.

station	coordinates	data/instrument/sensor	period
Georg Forster (GF)	(71°S; 12°E)	vertical ozone distribution	1985 - 1991
		balloon-borne ozone sensor (OSE)	
		spectral optical depth	1988 - 1992
		sun photometer	
Mirny (MI)	(66°S; 93°E)	spectral optical depth	1981 - 1995
		sun photometer and actinometer	
Neumayer (NM)	(71°S; 08°W)	vertical ozone distribution	1992 - 1995
		balloon-borne ozone sensor (ECC)	
		spectral optical depth	1991 - 1995
		sun photometer	
Ny-Ålesund (NA)	(79°N; 12°E)	vertical ozone distribution	1989 - 1995
		balloon-borne ozone sensor (ECC)	
		spectral optical depth	1991 - 1995
		sun photometer	
Syowa (SY)	(69°S; 40°E)	vertical ozone distribution	1968 - 1995
		balloon-borne ozone sensor (KC 79)	

In Fig.1 the mean annual variation of the vertical ozone distributions with color coded ozone concentrations is shown for Neumayer station (upper panel) and for Ny-Ålesund station (lower panel). Ozone profiles contributing to the mean pattern are indicated above each panel. It should be mentioned that these mean annual variations of ozone concentration were calculated from profile records which are not regularly distributed over the considered years.

Table 2. Calculation of the mean annual variation, balance and monthly means of vertical distribution of ozone concentration ozone for arctic and antarctic stations. No data are available for Syowa station in 1972, 1973, and from 1975 till 1981.

station	period	years	data presentation	
Neumayer	1992 - 4/1995	3.5	mean annual variation	(Figs. 1 and 6a)
	1992 - 1995	4	October means	(Figs. 5 and 8)
Ny-Ålesund	1992 - 4/1995	3.5	mean annual variation	(Figs. 1 and 6a)
	1989 - 1995	7	March means	(Figs. 5 and 8)
Syowa	1969 - 1995	26	October means	(Figs. 5 and 8)
Georg Forster	1985 - 1991	7	October means	(Figs. 5 and 8)

The inter-polar balance of ozone concentration in percent between both stations located in the Antarctic and Arctic was obtained by the calculation of the ratio between the two mean annual variations of ozone concentration, i.e. Neumayer versus Ny-Ålesund. The arctic data has been shifted by six months to match corresponding seasons. Thus, the recent difference between both polar regions can be shown for tropospheric and stratospheric ozone in Fig. 1 (central panel), where 100 % (yellow color) means no difference, values greater than 100 % (red colors) indicate more ozone at the corresponding altitude in the Antarctic than in the Arctic and values less than 100 % (blue colors) indicate vice versa less ozone in the Antarctic.

The aerosol optical depths were obtained by using the same type of sun photometers operated in a long-term network. Subsequently, the aerosol optical depths obtained from different locations are comparable with each other (Herber et al., 1996). The spectral aerosol optical depths are measured by interference filters for 8 channels in the wavelength range between 390 and 1045 nm. Satellite data were obtained from the Stratospheric Aerosol Monitoring experiment (SAM II) and the Stratospheric Aerosol and Gas Experiment (SAGE II) (McCormick et al., 1995). The satellite instruments can measure the extinction coefficients of aerosols with high vertical resolution of about 1 km for altitudes from the middle troposphere up to the stratosphere. At high latitudes the calculation of a zonal mean with a latitudinal resolution of 2.5 degrees is useful for the retrieval of stratospheric aerosol extinction coefficients (Thomason and Poole, 1993).

Here we use sun photometer data to calculate the total aerosol optical depth of the atmosphere at 1000 nm as weekly means and as monthly means for October in the Antarctic and March in the Arctic. The results are shown at Figs. 5 and 6. Satellite data were used to calculate the stratospheric aerosol optical depth for the same wavelength as monthly zonal means between 67.5° S and 72.5° S for October and between 70.0° N and 75.0° N for March. The results are shown at Fig. 5.

By calculating the stratospheric aerosol optical depth from the SAGE II satellite data and removing this value from the total aerosol optical depth, recorded with the sun photometers at the ground, the obtained residual value is the tropospheric part of the measured total aerosol optical depth (Herber *et al.*, 1995). Weekly means are calculated for Ny-Ålesund from sun photometer and satellite observations from 1991 till 1995 and for the antarctic stations Georg Forster (1988 - 1992) and Neumayer (1991-1995). The mean seasonal pattern of the tropospheric optical depth of aerosols for spring, summer, and early fall is shown at Fig. 6a (lower panel) for the Antarctic and at Fig. 6b (lower panel).

3 Stratospheric ozone in the polar regions

3.1 The mean pattern of polar ozone

As outlined before the mean annual vertical ozone distributions for the Antarctic and Arctic are the result of ozone production and loss by the different chemical reactions and the transport of ozone towards the polar regions. At polar latitudes the general profile features a maximum of ozone concentration at altitudes around 20 km. In winter ozone concentration is high in both polar regions (Fig. 1).

In the lower stratosphere a well pronounced minimum of ozone concentration was recorded in the Antarctic (Fig. 1, upper panel). The ozone decline begins at altitudes between 20 and 22 km in late August, and it lasts until December. At altitudes between 14 and 18 km the ozone loss amounts to about -80%, and it is almost total by -100% at altitudes around 18 km. Due to the downward transport of air inside the vortex in late winter and the advection of ozone during the reversal of stratospheric circulation, the ozone content simultaneously recovers from higher altitudes to lower altitudes from October until the middle of December. Ozone is advected into the polar stratosphere at altitudes above 20 km during the transition from the winter to summer circulation in the Arctic. From mid-summer until the new formation of the stratospheric vortex in the fall, the ozone concentration again decreases by two effects. Firstly the poleward transport of ozone rich air masses decreases, and secondly photochemical reactions deplete ozone.

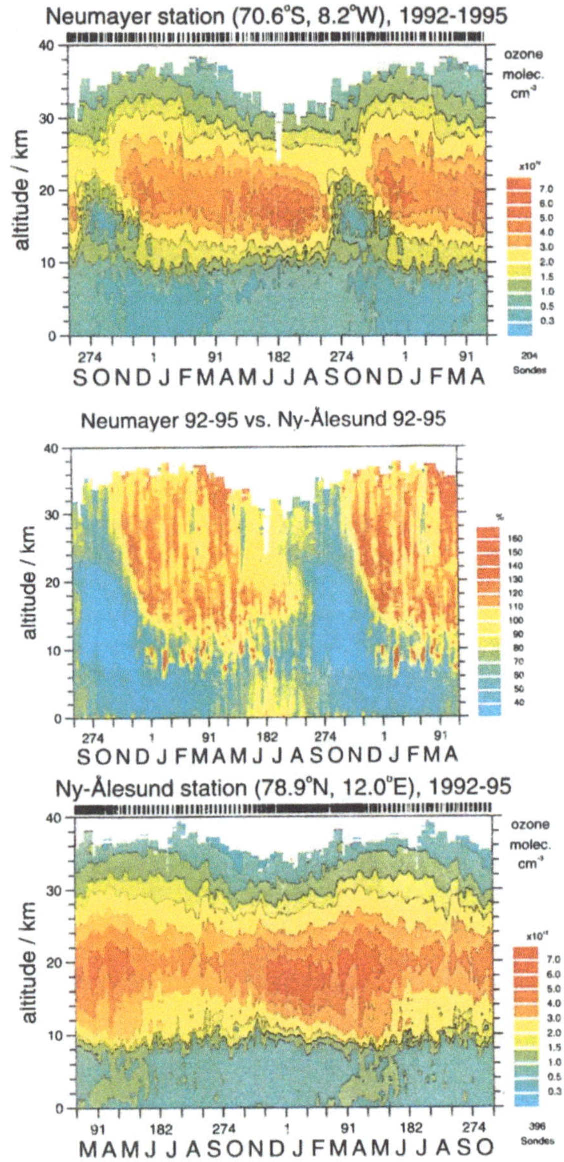

Fig. 1. Height-time cross sections of ozone concentration (color coded x10^{12} molec/cm^{-3}); mean annual variation; Neumayer station by 204 soundings (upper panel); Ny-Ålesund station by 396 soundings (lower panel); altitude range (0-40 km); time scale 20 months (day, month); contributing soundings are indicated above each panel.
Central panel: Height-time cross sections for percentage change (%) of the ozone balance for corresponding seasons (Neumayer versus Ny-Ålesund) Julian day and month are indicated for Southern Hemisphere seasons; values of 100% (yellow color) means no change, values greater than 100% (red color) indicate positive and values less than 100% (blue color) indicate a negative ozone balance in the Antarctic with respect to the Arctic.

Thus, the lowest ozone content occurs in the arctic stratosphere in late summer (Fig. 1, lower panel).

Compared to the antarctic mean (Fig. 1, upper panel), the arctic summer minimum extends to greater altitudes up to about 16 km. The arctic ozone shows maximum ozone content during the transition from winter to summer conditions (Fig. 1).

The polar balance (Fig. 1, central panel) shows less ozone at altitudes between 10 and 14 km in the Antarctic (blue color code). The opposite balance occurs above 15km altitude up to the balloon burst heights from December until April, i.e. in summer and fall, where the ozone concentration is higher in the Antarctic (red color code). The balance is equal (yellow color code) in the upper and lower stratosphere in winter (June, July) and in the lower stratosphere during late winter (August). That might be explained by corresponding variations in both polar regions at these altitudes and months. It means that during winter season the ozone content of the upper stratosphere is the same at altitudes above 20 km in the Arctic and Antarctic, and it remains equal between 16 and 22 km until August. At altitudes above 20 km the antarctic ozone balance becomes negative due to opposite development in both polar regions, i.e. the continued ozone loss inside the isolated stratospheric vortex in the Antarctic and increasing ozone concentrations in the arctic stratosphere, probably by ozone transport towards high latitudes. On the other hand the polar balance is also negative in the lower stratosphere between the tropopause and 14 km altitude, i.e. below the base of the stratospheric vortex. The spring season is dominated by the strong ozone loss in the antarctic lower stratosphere shifting the balance to significant negative values. It should also be mentioned that the ozone concentration is smaller throughout the troposphere and stratosphere up to altitudes greater than 30 km for September and October in the Antarctic.

A striking difference between both polar regions is also seen for the seasonal variation of tropospheric ozone. In spring, when the antarctic ozone decreases, a maximum ozone concentration is seen in the Arctic. A more detailed discussion on tropospheric seasonal variations is made in section 5.

3.2 Chemical ozone loss in the antarctic stratosphere

In the southern hemisphere the spring ozone depletion is one of the strongest signals for a changing atmosphere. This phenomenon was first noticed by Chubachi et al. (1984) in few balloon-borne data recorded at Syowa station showing an unexpected ozone depletion between 14 and 18 km altitude in 1982. Further striking evidences of this phenomenon were retrieved from the ground-based total ozone measurements taken at Halley station by Farman et al. (1985). The development of the antarctic spring ozone depletion has become evident in balloon-borne observations at Syowa station since 1979 (Gernandt et al. 1994). Since then the altitude range for a 60% loss in ozone concentration multiplied almost fourfold, i.e. it extended from 14 - 16 km to 12 - 22 km height until 1994. The largest vertical extent of depletion occurs in October each year. Later, by the end of October and in November, ozone concentrations recover continuously from upper to lower layers. Air masses enclosed in the polar vortex subside due to cooling, transporting non-depleted ozone from the upper stratosphere into the lower stratosphere until the final breakdown of the vortex in late spring. In the years 1992 till 1995 the spring depletion has been strongest so far.

The key to the ozone loss in late winter and spring is the meteorology of the southern polar stratosphere. The strong vortex circulation of air as a consequence of the relative lack of planetary wave activity yields strong radiative cooling of stratospheric air inside the vortex region during winter and spring. Temperatures may drop down to -90°C at altitudes between 18 and 22 km altitude. At these thermodynamic conditions the reservoir gas HNO_3 and water condense into sulfur acid droplets. This process leads to the formation of polar stratospheric cloud particles (PSC's). For example at 20 km altitude the formation of PSC I particles consisting of sulfur acid, nitric acid and water takes place when temperatures are colder than -78°C. The thermodynamic conditions of their existence as a frozen or liquid particle, where the liquid phase is understood as a ternary composition of sulfur, and nitric acid with water are recently still under discussion (Tabazadeh et al. 1994, Meilinger et al. 1995). At the lowest temperatures below thresholds between -83°C to -88°C water ice forms another type of polar stratospheric particles (PSC II) (Panegrossi et al., 1996).

At the surface of these PSC particle gaseous hydrochloric acid (HOCl) and chlorine gas (Cl_2) are formed from the reservoir gases $ClONO_2$ and HCl by heterogeneous reactions.

PSC I, II:	$ClONO_{2\ gas} + HCl\ gas$	---------> $Cl_{2\ gas} + HNO_{3\ liquid}$
PSC II:	$ClONO_{2\ gas} + H_2O$	---------> $HOCl + HNO_{3\ liquid}$
PSC I, II:	$HOCl + HCl$	---------> $Cl_2 + H_2O$

Among other controlling conditions the formation of HOCl and Cl_2 depends on the surface density of PSC particles and the duration of their existence. Both $ClONO_2$ and HCl are also relatively long lived by-products of the photolysis of the anthropogenic CFC source gases.

When sunlight returns to the Antarctic stratosphere, the molecular chlorine gas produced by PSC reactions photolyses to chlorine atoms which distract ozone. The photolysis takes place by solar light in the wavelength range 300-400 nm, which can deeply penetrate into the lower stratosphere.

$$Cl_2 + hn \quad (300\text{-}400\ nm) \text{ ----> } 2\ Cl$$

$$HOCl + hn \quad (300\text{-}400\ nm) \text{ ----> } OH + Cl$$

The atomic chlorine reacts with ozone and forms chlorine monoxide and molecular oxygen.

$$Cl + O_3 \text{ ----> } ClO + O_2$$

Due to the high concentrations of Cl the reaction with ozone yields to hundred fold elevated concentration of ClO compared to the global mean. If sufficient chlorine monoxide is present it can react with itself to form the dimmer Cl_2O_2. This dimmer photolysis into chlorine and molecular oxygen. So the chlorine is again free to react with ozone. This cycle is catalytic since it has the net effect of converting ozone to molecular oxygen without loss of chlorine.

$ClO + ClO + M \text{ ----> } ClOOCl + M^*$

$ClOOCl + hn\ (UV) \text{ ----> } ClOO + Cl$ net: $ClO + ClO + M \text{ ----> } 2\ Cl + O_2 + M^*$

$ClOO + M \text{ ----> } Cl + O_2 + M^*$

The ozone loss is roughly proportional to the square of the ClO concentration.

$$d[O_3]/dt \quad \propto \quad [ClO]^2$$

The increased formation of reactive chlorine species to sufficiently high concentrations became possible due to the continuous increase of man-made source gases which have yielded a positive trend of available chlorine in the stratosphere since the eighties (Engel et al. 1995). Normally this chlorine is mainly resided in chemically inactive reservoir gases as $ClONO_2$ and HCl. But the meteorological conditions in the polar stratosphere lead to the background for a strong activation of reactive chlorine and subsequent ozone depletion as summarized above. In general this pattern of chemical reactions qualitatively explains the strong ozone loss in the antarctic stratosphere after the return of sunlight in late winter and spring. However, as recently discussed, bromine compounds may also contribute to this ozone depletion by the formation of bromine monoxide and subsequent catalytic destruction of ozone (WMO 1995).

3.3 Chemical ozone losses in the arctic stratosphere

In the arctic stratosphere, the dynamically caused variations in ozone concentration at any given location can be large or greater than possible ozone losses due to the chemical reduction of ozone. Thus, until now the retrieval of long-term ozone trends from time records at one location is fraught with difficulties.

Approaches to identify chemical ozone loss in the arctic stratosphere have been indirect, typically relying on relationships between ozone and long-lived tracers. Manney et al. (1994) reported such an approach, based on satellite measurements. They concluded that the observed ozone decrease of about 20 % in the lower stratosphere in February and March 1993 was caused by chemical, rather than dynamic, processes. Von der Gathen et al. (1995a) reported on results obtained by calculating the chemical ozone destruction rates. Measured ozone concentrations in specific air parcels at different times have been used. During the winter 1991/92, about 1.200 electrochemical ozone sondes were launched as part of the European Arctic Stratospheric Ozone Experiment (EASOE) and from Canadian stations at middle and high northern latitudes. 10-day isentropic air parcel trajectories calculated from the wind and temperature analysis of the European Center for Medium Range Weather Forecasts (ECMWF) were used to identify cases where ozone measurements were made in the same air mass at two different times. The terminal points for these trajectories included 26 launching sites. An example for the observation of ozone change along an isentropic trajectory is shown in the upper panel of Fig. 2. That part of the trajectory, when the air mass was exposed by sunlight, is indicated by the dashed line of the trajectory. For the Arctic vortex of the 1991/92 winter von der Gathen et al. (1995a) and Rex et al. (1995) found that the loss is proportional to the exposure to sunlight (Fig. 2, central panel) and that, at 18 km altitude (between the isentropic levels 465 K and 485 K), chemical ozone loss occurred only between early January and mid February (Fig. 2, lower panel).

The correlation between ozone loss and sunlit exposure is the crucial evidence that chemical ozone loss has been measured. Fig. 2 (central panel) shows the percentage change in ozone mixing ratio with the number of sunlit hours on the isentropic surface 475 K inside the stratospheric vortex. The vortex edge is defined by the quantity of potential vorticity (PV), which is PV > 27 $Km^2kg^{-1}s^{-1}$ for 475 K. Between 18 to 27 trajectory matches inside the vortex contribute within ± 10 sunlit hours to the first four data points, and between three and five matches contribute to the last three data points. The exponential fit is shown by the solid line. The largest total ozone loss is found for matches which received the most sunlight.

Fig. 2 (lower panel) shows the time evolution of the ozone depletion rate (% per sunlit hour) during January and February 1992 inside the polar vortex in the 475 K potential temperature surface (black squares). In addition, the area of Northern Hemisphere where temperatures were cold enough to form PSC's (assuming NAT and constant volume mixing ratio profiles for H_2O = 5 ppm and HNO_3 = 5 ppb) is plotted (Rex et al. 1995).

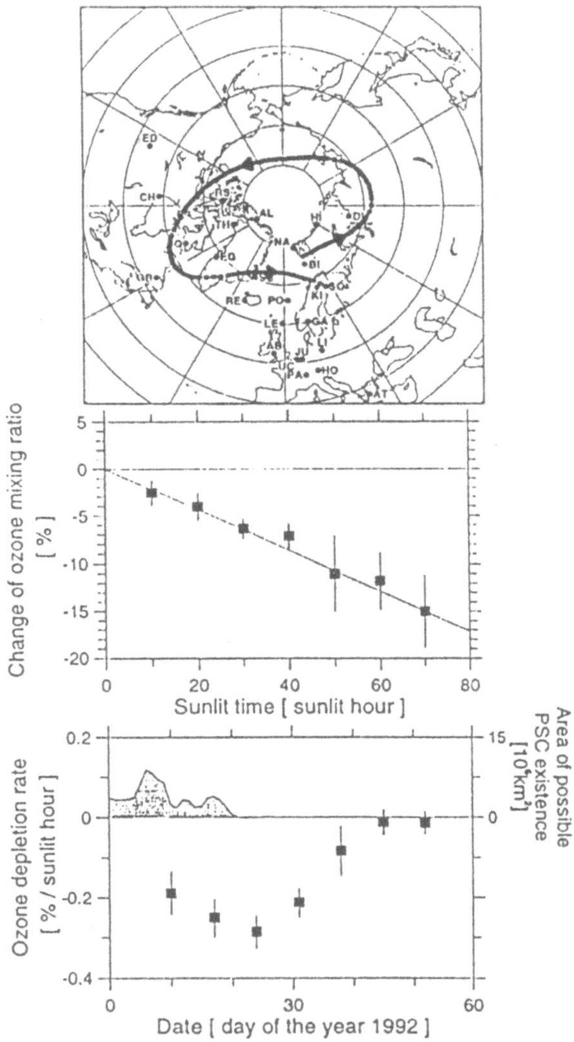

Fig. 2. Upper panel (reprinted with permission from Nature (von der Gathen et al., 1995a): Pattern of an isentropic trajectory inside the polar vortex at the 475 K level. Dashed trajectory parts indicate sunlit exposure.

Central panel (reprinted with permission from Nature (von der Gathen et al., 1995a): Running means over 20 sunlit hours (full squares) for percentage change in ozone mixing ratio (%) as a function of sunlit hours from January 9 to February 4, 1992.

Lower panel: Running mean (full squares) over 14 days for the ozone depletion rates (% per sunlit hour, left scale); area of possible PSC formation x10^6 km^2 (right scale) as function of time; time scale (Julian day for 1992).

There was found a mean ozone loss related to the chemical ozone depletion of about 0.23 ± 0.03 % per sunlit hour for the period January 4 to February 9, 1992. The timing and magnitude are broadly consistent with existing understanding of photochemical ozone depletion processes.

A more sophisticated approach was performed during the Second European Stratospheric and Mid-latitude Experiment in 1994/95 winter. The coordination of ozone soundings and trajectory calculations was performed in the MATCH campaign (Rex et al., 1995). About 35 ozone sonde launching stations have been participating in the MATCH campaign as the European ozone launching sites in mid and high latitudes as well as Canadian, Siberian and Japanese stations. Altogether about 1.000 ozone sondes were launched. These launches were on-line coordinated by using forward trajectories in order to achieve the matches actively. First preliminary results were reported by Rex et al. (1995). They show a similar mean ozone loss attributed to chemistry as it was obtained from the winter 1991/92. The mean ozone loss amounted 0.23 ± 0.045 % per sunlit hour for the period from January, 1 to February, 1.

In summary these results gave a more detailed picture of the chemical ozone loss in the arctic stratosphere. Chemical ozone loss was established in the Arctic polar vortex for short periods lasting about 10 to 20 days in January, February or March. It occurred only during sunlit periods along the trajectories, and it rised during periods when the probability for PSC existence was indicated by the related stratospheric temperatures. About 20% to 30% of total ozone change in the lower stratosphere can be attributed to the catalytic ozone destruction (von der Gathen et al., 1995b; Rex et al., 1995).

In the Antarctic at 70°S the catalytic ozone destruction appears when sunlight returns at this latitude after polar night in the end of July (Fig. 1). In the Arctic at 79°N air masses with chemically depleted ozone even occur in the middle of January although sunlight just returns later at this latitude in the end of February. This earlier onset of ozone depletion in the Arctic is connected with the stronger planetary wave activity moving air parcels towards lower latitudes where sunlight can force the chemical removal of ozone even earlier (Fig. 2, upper panel).

The winter 1994/95 has been an exceptional one for the arctic stratosphere. Lowest temperatures on record and a stable vortex circulation lasting until March have been the features (Naujokat 1995). The ozone losses add up over the winter, and by March ozone concentrations at about 18 km were as much as 50% below the values observed inside the vortex in previous years. During most of that time the Arctic station Ny-Ålesund was located inside the vortex. Thus, the ozone sounding data of this station show a similar vertical profile as it is known from antarctic observations with an altitude region in the lower stratosphere, where ozone is probably mainly chemically depleted. In comparison Fig. 3 shows observations of the mean ozone depletion for those months with the strongest ozone loss in spring 1995, i.e. for March in the Arctic (Ny-Ålesund station) and September in the Antarctic (Neumayer station).

The antarctic profiles observed on September 20, 23, and 27 in 1995 show a significant ozone loss related to the mean winter profile at 70°S for the period from May, 20 to July, 20 in 1995. The arctic profiles observed on March 20, 23, and 28 in 1995 show the ozone loss related to the mean winter profile for the period November 1, 1994 to January 31 in 1995.

This comparison for corresponding seasons, i.e. the spring months March and September, clearly shows the stronger ozone depletion in the South. Strong deviations of ozone from the winterly mean profiles can be seen between 380 K (14 km) and 500 K (20 km) in the Arctic, and between 350 K (13 km) and 620 K (24 km) in the Antarctic. The greater vertical extend of the ozone loss region by about 4 km in the Antarctic might be explained by the greater vertical extend of very low temperatures in the antarctic stratospheric vortex, by smaller exchange of air through the vortex margin, and, probably to some smaller part, also by the smaller vertical downward component of the residual circulation in the Antarctic compared to the Arctic (Geller et al., 1992).

Lowest ozone concentrations occur in the altitude range between 400 K and 450 K (16 km - 19 km) in both polar regions. Related to the ozone concentrations for polar night conditions at the 420K level the total ozone removal, i.e. by chemical loss and dynamic variation, amounts to about -87% at 71°S and -39% at 79°N, respectively. The ozone loss rate per sunlit hour for the corresponding latitude is as high as -0.11% at 79°N and -0.15% at 71°S. These values obtained by observations at fixed locations are smaller than those obtained for the actual chemical ozone loss in the arctic stratosphere along the trajectories of air masses (von der Gathen et al. 1995a; Rex et al. 1995).

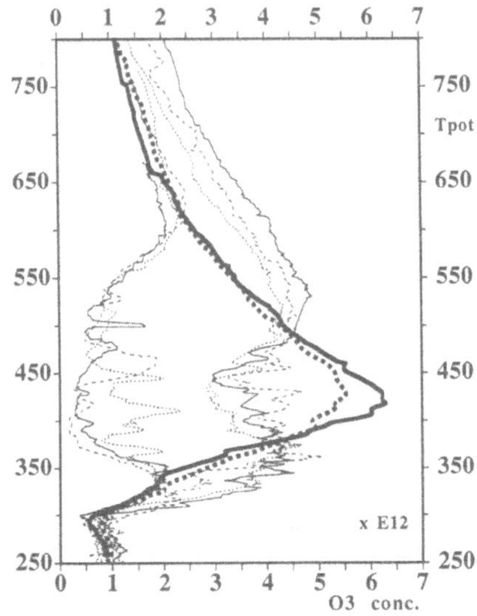

Fig. 3. Antarctic and arctic vertical distributions of ozone concentration (x10^{12} molec./cm^3). Individual profiles for corresponding days in spring: 20 September 1995 and 20 March 1995 (thin dotted lines), 23 September 1995 and 23 March 1995 (thin dashed lines), and 27 September 1995 and 28 March 1995 (thin full lines). Mean profiles for corresponding winter seasons: Ny-Ålesund (bold thick line), Neumayer (bold dashed line). Winter season for Ny-Ålesund station (1 November 1994 till 31 January 1995), and for Neumayer station (May 20 till July 20, 1995). Altitude coordinate is potential temperature T_{pot}(K).

Below the 500 K level intrusions of ozone within thin layers indicate rapid small scale variations within few days. This dynamic noise is caused by meso-scale exchange processes between polar and mid-latitudes. Those repeated poleward intrusions of mid-latitude ozone rich air lead to the filamentation of ozone in the arctic polar stratosphere (Orsolini et al. 1994, Orsolini 1995). Furthermore, this dynamic impact may cover chemical ozone depletion by the redistribution of chemical processed air masses throughout polar and mid-latitudes.

4 Volcanic aerosols and ozone in the polar stratosphere

4.1 Volcanic aerosol in the polar stratosphere

The ozone depletion events, especially during spring, are understood as results of human made emission of chlorine species in the form of CFC's into the atmosphere over decades (WMO, 1992; Engel et al., 1995). However natural events as volcanic eruptions may also modulate the ozone chemistry (McCormick et al., 1995). According to McCormick et al. (1995) a volcanic eruption can produce a significant perturbation to the Earth-atmosphere system, which may persist for several years. The injected material may include ash, which typically does not remain for more than a few months, and gaseous compounds including water vapor, sulfur dioxide and hydrochloric acid. Aerosols are produced when the sulfur dioxide (SO_2) is chemically transformed into sulfuric acid (H_2SO_4) which rapidly condenses into aerosols. This aerosol of volcanic origin increases the atmospheric albedo by reflecting solar radiation back into space, and can warm the stratosphere by absorbing upwelling infrared radiation.

The eruptions of Mt. Pinatubo in June 1991 and Cerro Hudson in August 1991 caused a strong perturbation of the stratosphere in a global scale (Fig. 4). The volcanic cloud has been globally monitored by the satellite-borne Stratospheric Aerosol Experiment (SAM II) as well as Stratopheric Aerosol and Gas Experiment (SAGE II) (Trepte et al. 1993; Trepte et al. 1994; McCormick et al. 1995).

The Mt. Pinatubo aerosol cloud immediately dispersed throughout the tropics between 30°N and 20°S (Fig. 4). Transport towards northern polar latitudes occurred about 4 months later from December 1991 to April 1992. It was forced by planetary waves in the zonal westerlies at mid-latitudes (Trepte et al. 1993; Herber et al. 1996). The appearance of aerosols in high arctic latitudes was recorded by lidar observations (Neuber et al. 1994, Beyerle and Neuber 1994, Beyerle et al. 1994). This aerosol transport was prevented after the transition to mid-latitude zonal easterlies in 1992. In the southern hemisphere the initial aerosol arriving the polar stratosphere was formed by the Cerro Hudson eruption (Herber et al. 1993; McCormick et al. 1995). Stratospheric aerosol from Cerro Hudson was concentrated at low altitudes below 16 km (Schoeberl and Doiron 1993). In 1991 during the southern hemispheric winter and spring transport of this aerosol through the prevailing zonal westerly wind was also forced by planetary waves in the lower stratosphere (Herber et al. 1996). So, it was able to penetrate deeply into high southern latitudes below the base of the stratospheric vortex soon after that eruption. During southern winter and spring in 1992 the Mt. Pinatubo aerosols also occurred in the lower stratosphere at high southern latitudes (McCormick et al. 1995).

At polar latitudes the appearance of volcanic aerosols has been monitored by ground-based sun photometers (Kaneto et al. 1994), lidars (Neuber et al. 1994; Beyerle et al. 1995), and balloon-borne aerosol sondes (Hofmann and Oltsmans 1993; Hofmann et al. 1994). Regular ground-based sun photometer measurements performed at three antarctic stations and one arctic station (Tab. 1) have completed these observations (Herber et al. 1996).

By using satellite and ground-based sun photometer data the interannual variation of the stratospheric aerosol optical depth and the total aerosol optical depth at 1000 nm were retrieved for October in the Antarctic and March in the Arctic (Fig. 5, upper panel). In the Antarctic sun photometers were operated at Georg Forster station from 1988 to 1992, and at Neumayer station since 1991. These records of the total aerosol optical depth of the atmosphere have been extended back to 1981 by including actinometric observations from Mirny station (Tab. 1). In the Arctic sun photometers are operated since 1991. Due to the contributions by tropospheric aerosols the values of the total aerosol optical depth of the atmosphere measured by ground-based sun photometers are approximately 10 times greater than the corresponding stratospheric values (Herber et al., 1993).

The interannual variation of the stratospheric aerosol optical depth as obtained from satellite data is shown for October since 1979 (Fig. 5, upper panel). As suggested by McCormick et al. (1995) the 1979 stratospheric aerosol optical depth values probably show the stratospheric background in both polar regions. Later on elevations of stratospheric aerosol by volcanic eruptions in 1980 to 1981, and the final eruption of El Chichon in April 1982, yielded the increase of stratospheric aerosol optical depth. The recovery almost lasted until 1988. In general the total atmospheric aerosol optical depths follows the pattern of the stratospheric values. The greater monthly mean values of total aerosol optical depth in the Arctic for March 1991 to 1995 are mainly caused by the regular maximum of the tropospheric aerosol concentrations during arctic spring (Herber et al. 1995). For October and March maximum stratospheric aerosol optical depths occurred after the eruptions of El Chichon in 1983 as well as after the eruptions of Cerro Hudson and Mt. Pinatubo in 1992. The latter one has been the greatest aerosol loading since the beginning of satellite record (McCormick et al. 1995). The early arrival of the Cerro Hudson aerosols in the antarctic lower stratosphere can be noticed by the elevated aerosol optical depth for October 1991. While the stratospheric aerosol optical depth is almost equal in both polar regions after the volcanic eruptions in 1991 a significant asymmetry with a greater aerosol optical depth for the artic stratosphere can be noticed in spring after the eruption in April 1982.

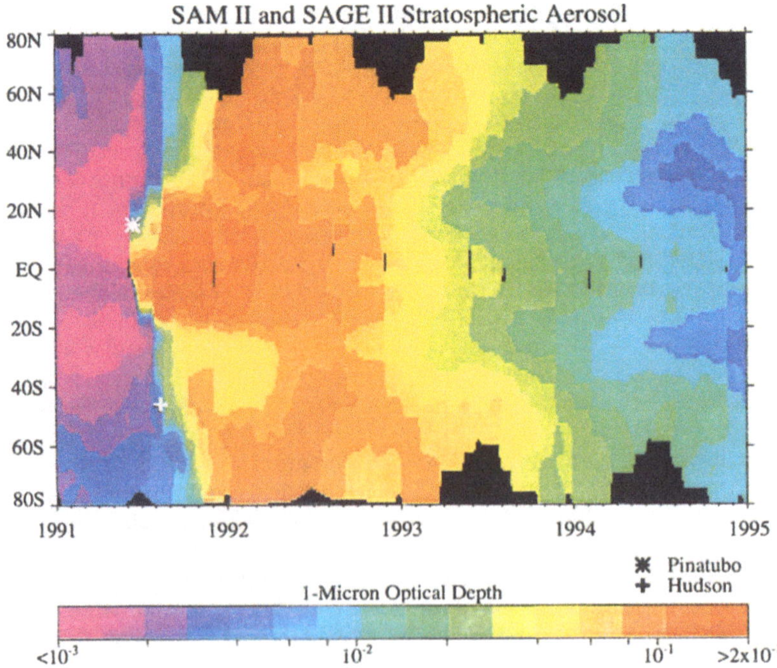

Fig. 4. The global dispersion of the Mt. Pinatubo and Cerro Hudson clouds as retrieved from stratospheric optical depths at 1000 nm monitored by the satellite-borne instruments SAM II and SAGE II.

4.2 Volcanic aerosols and ozone loss in polar regions

In polar latitudes, an early indicator during the post-Pinatubo period for a connection between aerosols and ozone loss was found in association with aerosol formed by the eruption of Cerro Hudson in August 1991 (Deshler et al. 1992). This aerosol appeared at southern polar latitudes below the base of the stratospheric vortex in late September 1991. In the densest portions of this Cerro Hudson aerosol layer between 11 and 13 km, ozone decreased by about 50% for a period of 30 days following the arrival of this aerosol above South Pole station in Antarctica (Hofmann et al. 1994). The aerosol surface-area density in this layer was as high as 100 mm^2/cm^3 or 20 -30 times higher than the background observed in 1990. Aerosol levels also remained unusually high at these altitudes for the following two years.

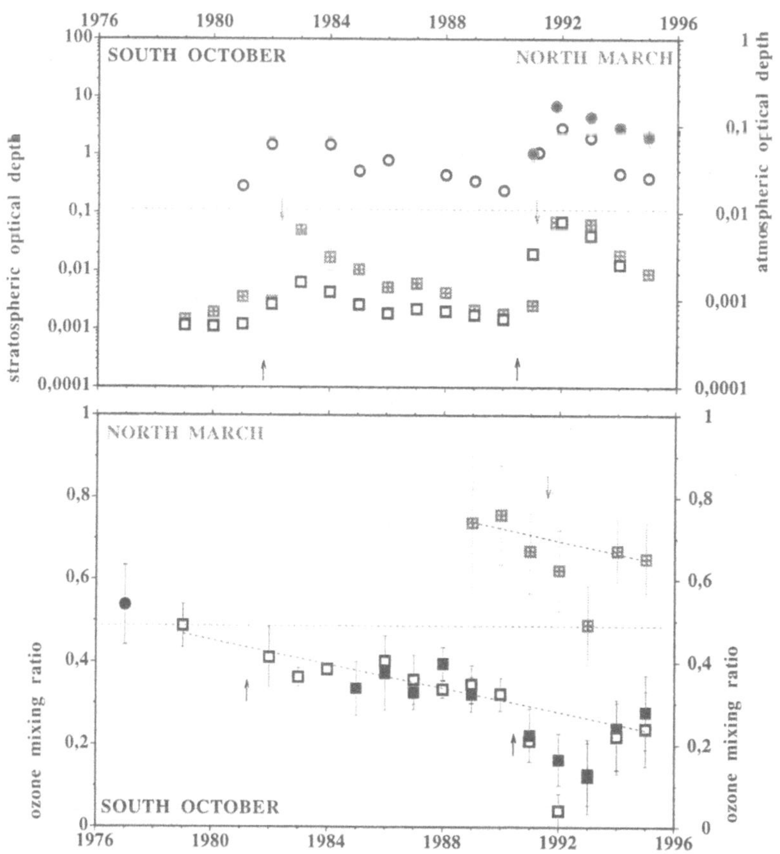

Fig. 5. Upper panel: monthly means of total aerosol optical depths by ground-based actinometers and sun photometers (right scale): Antarctic for October (open circles), Arctic for March (full circles); zonal monthly means of stratospheric aerosol optical depths by SAM II and SAGE II (left scale): Antarctic for October (open squares) Arctic for March (crossed squares). Lower panel: monthly mean ozone mixing ratios (ppmm) at 350 K (12-13 km); Antarctic for October: Syowa station (open squares), Georg Forster station from 1985 to 1992 and Neumayer station from 1992 to 1995 (full squares); Arctic for March: Ny-Ålesund station (crossed squares).

These antarctic observations can be confirmed for both polar regions by ozone sounding data from the stations Syowa, Georg Forster, Neumayer, and Ny-Ålesund (Fig. 5, lower panel). Monthly means of the ozone mixing ratios for the 350 K isentropic level, i.e.: about 12 to 13 km altitude, have been calculated from antarctic data for October each year since 1968. These values are compared with monthly means for March obtained from Ny-Ålesund ozone sounding data since 1989. Except for the years 1991, 1992, and 1993 when the volcanic aerosol was seen in the southern polar stratosphere as well as for the years 1992 and 1993 when volcanic aerosol was detected in the arctic stratosphere (Fig. 5, upper panel) the linear regressions were calculated for Syowa data for the period 1979 to 1995 and Ny-Ålesund data for the period 1989 to 1995 in order to estimate the long-term negative trends at these altitudes (Tab. 4). With respect to these linear regressions the monthly means of ozone mixing ratio decline within one standard deviation for October 1991, and within more than one standard deviation for October 1992 and 1993 in the Antarctic. In the Arctic the monthly mean only declines with more than one standard deviation for March 1993. So in the lower stratosphere at about 12 to 13 km altitude a significant negative correlation between the stratospheric aerosol optical depth and the monthly mean ozone mixing ratios can be noticed for both polar regions at the 350 K isentropic level during spring, i.e. in October and March, respectively. In the Antarctic the occurrence of the Cerro Hudson aerosol in late September 1991 leads to a first decrease of the mean ozone mixing ratio in October 1991.

In 1992 the antarctic the ozone decline was significantly strong, but different in magnitude for the stations Syowa and Neumayer. In 1993 a strong ozone decline is evident for both polar regions during spring. In 1994 mean ozone mixing ratios recovered with respect to the long-term negative trends for the same season, although the stratospheric aerosol optical depth values are still elevated, but already below the level for spring in 1991 (Fig. 5, upper panel).

A detailed two dimensional simulation of the Pinatubo aerosol distribution and its effect on stratospheric ozone (Tie et al. 1994) shows a significant change in ozone concentration at 60°N related to the appearance of volcanic aerosols. While ozone concentration increases at altitudes higher than approximately 25 km, below this level down to tropopause altitudes (9 -10 km) a decrease is obtained for late winter and spring in both years 1992 and 1993, respectively. The maximum of this depletion occurs at about 15 km altitude in March, i.e. in spring. This model result is consistent with the observations of depleted ozone in the Arctic. In the model this reduction is caused by an overall increase of OH and ClO concentrations. Fahey et al. (1993) suggested that elevated concentrations of reactive chlorine might be formed by the heterogeneous reactions

$$N_2O_5 + H_2O \xrightarrow{\text{aerosol}} 2\,HNO_3$$

$$ClONO_2 + H_2O \xrightarrow{\text{aerosol}} HOCl + HNO_3$$

on the surface of volcanic aerosols. Laboratory experiments (Fahey et al. 1993) have shown that nitrogen oxides, N_2O_5 and $ClONO_2$ react on the surface of sulfuric acid solutions the composition of which is similar to stratospheric aerosol particles. Both reactions might be important for reasonable water vapor partial pressures and temperatures in the lower stratosphere when the photolysis of HNO_3 is suppressed due to low solar elevations at high latitudes (Fahey et al. 1993). In that case the low photolysis rate of HNO_3 back to NO_2 decelerates the formation of $ClONO_2$. Thus, concentrations of chlorine species remain high and can additionally remove ozone at these altitudes even when it is too warm for PSC's to occur (McCormick et al. 1995).

The strong ozone reduction in spring 1993 at 350 K might also be associated with the formation of type Ib polar stratospheric cloud particles as well as volcanic aerosols which are composed of a ternary liquid solution of sulfuric and nitric acid as discussed by Beyerle et al. (1996). During the winter 1992/93 lidar observations at Ny-Ålesund showed that at all altitude levels between 350 K and 550 K volume densities consistently increased by a factor of 2-30 if temperatures dropped down to the frost point. The model results indicate more than 90% HNO_3 gas phase depletion below this temperature. This removal of gas phase HNO_3 might be more efficiently for maintaining high concentrations of reactive chlorine than its suppressed photolysis at low solar elevations.

The delay of strongest ozone loss by one year can not be completely explained. On the one hand there is a very sensitive dependence on temperatures and PSC particle formation at these altitudes. On the other hand this delay might be associated with the vertical redistribution of volcanic aerosols by gravitational sedimentation leading to higher aerosol concentrations at the 350 K level in 1993 than in 1992. That assumption is supported by lidar and sun photometer measurements from Ny-Ålesund in the Arctic where the height integrated backscatter coefficient and the total aerosol optical depth decreased exponentially from spring 1992 until spring 1993 with time constants of 0.94 ± 0.37 years and 0.89 ± 0.39 years, respectively (Beyerle et al. 1995). These time constants were obtained by a model which only included gravitational sedimentation as the removal process for aerosols from the stratosphere. This agreement between observed and calculated time constants shows that gravitational sedimentation might be a dominant process for the decay of Mt. Pinatubo aerosol in the arctic stratosphere.

As the ozone observations in both polar regions show, additional ozone losses in springs 1991, 1992, and 1993 can be attributed to the presence of volcanic aerosol particles particularly in the lower stratosphere. However, as briefly discussed the ozone loss probably depends on a critical level of particle surface area densities, age of the aerosols, and temperatures at these altitudes. Furthermore, a similar ozone loss could not be observed in the antarctic lower stratosphere in the years following the eruption of El Chichon in April 1982, although the aerosol loading has been sufficiently high. McCormick et al. (1995) argued that the amount of ozone losses probably depends not only on the parameters discussed above but also on the higher chlorine and bromine levels in the stratosphere about ten years later as a result of anthropogenic emissions of CFC's. Following that, there is an striking example of coupling between human and natural perturbations.

5 Ozone and aerosols in the polar troposphere

The recent mean annual variation of ozone concentration is significantly different between the Arctic and Antarctic (Fig. 1). Detailed plots for tropospheric ozone and tropospheric optical depth of aerosols are shown at Figs. 6a and b. Maximum ozone concentrations appear near the surface up to 2 km altitude in the Antarctic during polar night (Fig. 6a, upper panel). After the return of sunlight this ozone maximum disappears. Further decrease of tropospheric ozone in the altitude range up to 5 km coincides with the October minimum of stratospheric ozone. Lowest tropospheric ozone concentrations occur after summer solstice until the beginning of March. At altitudes between 6 and 7 km the seasonal variation is not significant. At tropopause altitudes a downward displacement of stratospheric ozone is recorded for late spring. This downward stretch extends down to about 8 km altitude. Below this altitude the control of tropospheric ozone seems not to be significant by downward transport of ozone from the stratosphere.

Except for polar night conditions the seasonal variation of weekly mean values of the tropospheric optical depth of aerosols is also shown at Fig. 6a (lower panel). The optical depths are very small throughout all seasons and no significant seasonal variation can be seen.

In contrast to the Antarctic, ozone concentrations are elevated in the arctic troposphere during spring. Lowest ozone concentrations occur in the whole troposphere in polar winter. Near the surface up to about 1 km altitude the seasonal variation is similar to that of the Antarctic with higher ozone concentration in winter and lowest ozone values after summer solstice. At altitudes between 4 to 6 km, i.e. in the middle troposphere, the seasonal variation is reversed to that near the surface. Maximum ozone concentrations occur at those altitudes in spring. At tropopause altitudes a downward displacement of stratospheric ozone is seen in late winter and spring lasting until May. This downward stretch is evident down to about 7 km altitude. Probably both, the downward transport of stratospheric ozone rich air as well as the elevated ozone from the lower parts of the troposphere control the seasonal variation in the middle troposphere at arctic latitudes.

That is consistent with the findings of Grudzev et al. (1992). Their analysis of balloon-borne ozone data from Amundsen-Scott Station (South Pole) and from Resolute Station (76°N; 95°W) show that ozone fluxes through the tropopause due to stratospheric-tropospheric exchange processes only affect the tropospheric ozone content down to the middle troposphere. Logan (1985) has shown that ozone concentrations in the arctic middle troposphere exhibit a maximum between March and May below the tropopause, earlier than in the middle troposphere. That also reflects, that the ozone content is only controlled by stratospheric-tropospheric exchange of air masses in the upper troposphere.

The asymmetry between the Arctic and Antarctic for the seasonal cycle of ozone in the lower troposphere was also reported by Logan (1985). Ozone at 700 hPa and 500 hPa is largest in winter and smallest in summer, as shown by Syowa data (69°S), while ozone is largest in late spring and early summer at Resolute Station. The arctic spring maximum is partially addressed to the transport of air mass from mid-latitudes assuming a tropospheric source to form additional ozone due to adequate concentrations of NO_X within those air masses. If the NO_X concentrations will be sufficiently high, i.e. greater as 200 ppt (Fishman et al. 1979) additional ozone can be formed by photochemical processes (Rφth 1994). In the opposite case, i.e. for low NO_X concentrations or non polluted clear air conditions as in the Antarctic, ozone will be reduced.

The optical depths of arctic tropospheric aerosols are smallest in summer, and equal in magnitude to the antarctic record (Fig. 6a). Despite of the large scatter of the weekly mean values and their standard deviations the tropospheric optical depth is significantly higher in spring than in late summer when smallest values and smallest standard deviations occur.

At Ny-Ålesund the ozone increase starts with the return of sunlight at the end of February. First ozone increases in the boundary layer. Later on it extends towards greater altitudes, while a reduction of ozone occurs in the boundary layer. High ozone concentrations are recorded by June, i.e. until the maximum of solar elevation. These elevated ozone concentrations coincide with the occurrence of the "Arctic Haze" phenomenon. During spring, with the return of solar light, photometer measurements show an considerably elevated tropospheric optical depth (Fig. 6b). It is caused by anthropogenic produced aerosols transported from mid-latitudes to the arctic region (Heintzenberg, 1989). The "Arctic Haze" phenomenon causes a pronounced seasonal variation of tropospheric optical depth. The artic aerosol consists of sea salt particles, sulfate aerosols and soot particles, where the latter both are mainly of anthropogenic origin. The arctic aerosol has a strong absorbing component due to its soot content.

Therefore the extinction coefficient of these tropospheric aerosols has got a scattering and an absorbing part. In the Antarctic the tropospheric aerosol content is, in general, very low and shows no significant annual variation (Fig. 6a). There is no anthropogenic contribution so far.

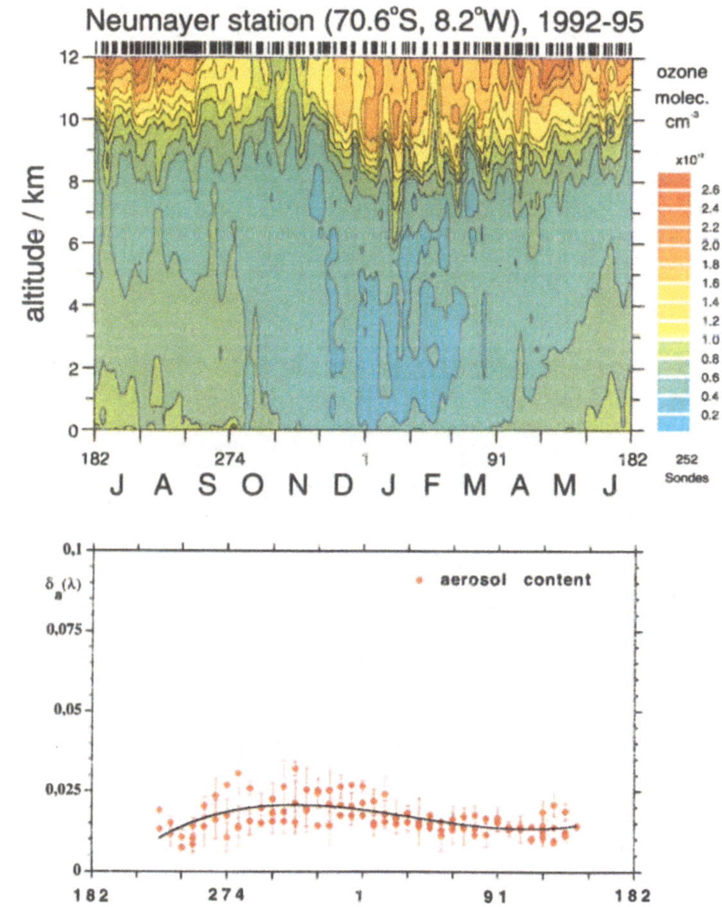

Fig. 6a. Mean annual variation of tropospheric ozone and tropospheric aerosol optical depth at Neumayer station.
Upper panel: height-time cross section of ozone concentration (color coded $x10^{12}$ molec./cm^3), mean annual variation for the period 1992 to 1995 by 252 soundings; altitude range (0 - 12 km); time scale one year (Julian day, month).
Lower panel: 5 year weekly means of the tropospheric aerosol optical depth for the period from 1991 to 1995; time scale (Julian day).

Both the greater ozone concentrations in the lower troposphere as well as the increased tropospheric optical depths coincide very well in their temporal behavior. That might be accidental, but it is striking. As discussed by Logan (1985) and Heintzenberg (1989) the formation of tropospheric ozone in the Arctic as well the appearance of elevated aerosol concentrations are connected with the transport of polluted air from lower latitudes towards the Arctic. Considering that, the simultaneous appearance of both events might be causal. The aerosol resides mainly in the boundary layer. Greater ozone concentrations appear in the lower troposphere above. So we guess that scattered short wave radiative flux might change photolysis rates and, subsequently, more ozone might be formed in the lower troposphere above this aerosol layer.

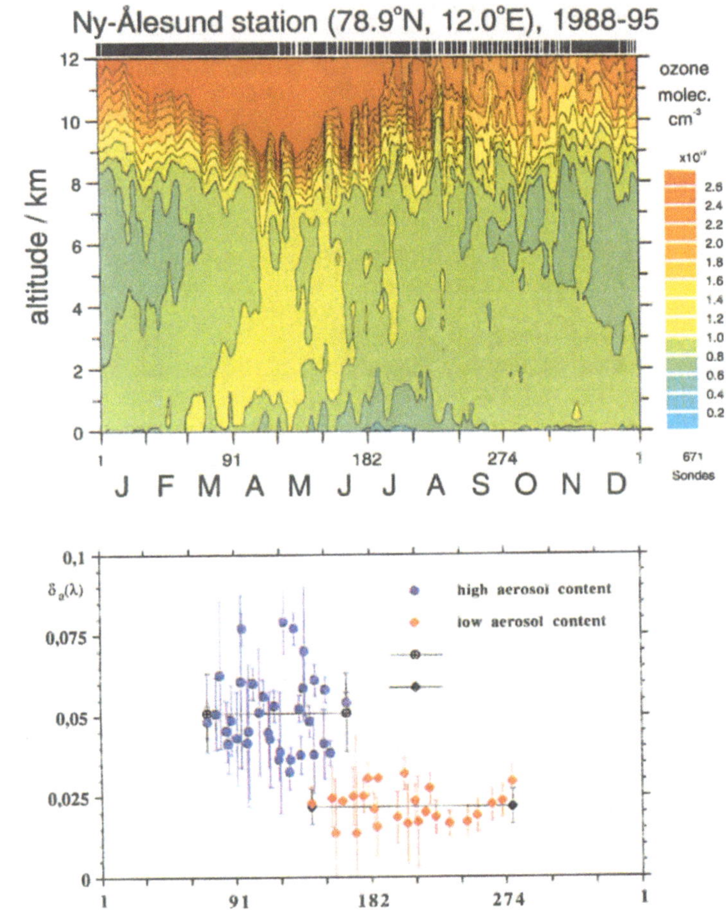

Fig. 6b. Mean annual variation of tropospheric ozone and tropospheric aerosol optical depth at Ny-Ålesund station.
Upper panel: height-time cross section of ozone concentration (color coded $x10^{12}$ molec./cm^3). mean annual variation for the period 1988 to 1995 by 671 soundings; altitude range (0 - 12 km); time scale one year (Julian day. month).
Lower panel: 5 year weekly means of the tropospheric aerosol optical depth for the period from 1991 to 1995; time scale (Julian day).

6 The development of polar atmospheric ozone towards the current state

As it could be shown by the observations at Neumayer and Ny-Ålesund (Figs. 6a and 6b), higher ozone concentrations in the arctic troposphere are recognized as the impact of air transported from mid-latitudes into the arctic region. There is no historical record of tropospheric ozone at high latitudes as Ny-Ålesund for estimating the development of tropospheric ozone over decades. Due to the tropospheric coupling one might conclude that the development of the arctic tropospheric ozone appears to be associated with the changes of tropospheric ozone in mid-latitudes. The study of Logan (1985) presents an assessment of the impact of human activities on the distribution of tropospheric ozone based on global observations by sondes and ground-based instruments. In the middle troposphere a positive trend was found for Resolute retrieved from data records between 1966 and 1979. This positive trend coincides with similar trends for the middle troposphere in mid-latitudes. This similarity of the seasonal cycle of ozone in the middle troposphere certainly supports the idea of a significant anthropogenic impact.

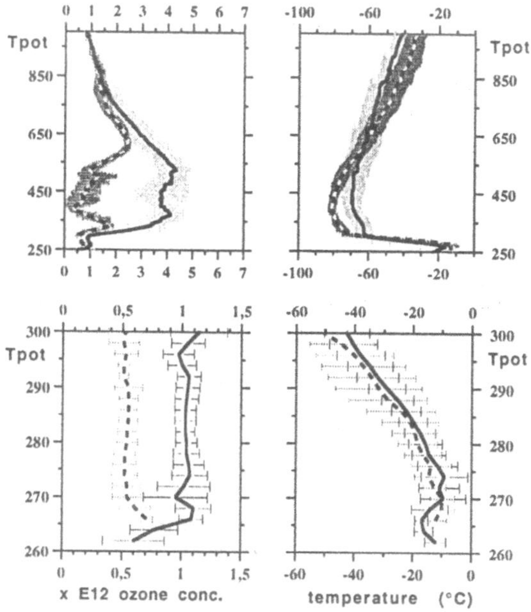

Fig. 7. The current mean vertical ozone distributions at high latitudes during spring
Upper panel: Mean ozone (left side) and temperature (right side) profiles with standard deviations from Neumayer for October (dashed lines) and Ny-Ålesund (full lines) for March in 1995. Lower panel: Mean tropospheric ozone (left side) and temperature (right side) profiles with standard deviations from Neumayer for November (dashed lines) and Ny-Ålesund for May (full lines) in 1994. Ozone concentration (x10^{12} molec./cm^3), temperature (°C), altitude coordinate is potential temperature (T_{pot}), two standard deviations are shadowed.

So the increase in middle tropospheric ozone over Resolute in spring and summer is also attributed to long-range transport of ozone or its precursors from those mid-latitudes. Having in mind that the Arctic Haze phenomenon is another evidence for long-range transport from mid-latitudes towards the Arctic the development of mid-tropospheric ozone in the Arctic can be understood as the response to changes in mid latitudes. In contrast the current tropospheric ozone distribution in the remote antarctic region shows the undisturbed state which was probably similar in the Arctic a couple of decades ago. So the intercomparison of current tropospheric vertical ozone distributions in the Arctic for May 1994 and in the Antarctic for November 1994, i.e. for corresponding months of seasons, might give an idea about the magnitude of changes in tropospheric ozone at high arctic latitudes (Fig. 7. lower panel). The tropospheric ozone in the remote unpolluted antarctic area is smaller by about 50% than in the arctic area. The potential temperature level between 275 and 280 K corresponds to altitudes between 5 and 6 km. Below this level a stratospheric impact on ozone is recognized to be very small at polar latitudes (Grudzev 1992).

As it could be shown by the observations at Neumayer and Ny-Ålesund chemical ozone loss occurs in the polar stratosphere in late winter and spring (Figs. 1 and 3). The current state of stratospheric ozone in the Antarctic for October 1995 and in the Arctic for March 1995, i.e. months with the strongest ozone loss signals in 1995, is shown at Fig. 7 (upper panel). Despite of different mean temperatures the ozone concentration is equal at altitudes above 800 K (28-30 km). A difference of ozone concentration by one standard deviation occurs between 600 K (24-25 km) and 800 K. At altitudes between 330 K (12 km) and 600 K the mean temperature in the Antarctic is significantly lower than in the arctic stratosphere. The altitude range with strong ozone depletion coincides with these lower temperatures in the Antarctic.

There is only a very limited long-term record of data to study the historical development in the vertical distribution of polar stratospheric ozone. In the Antarctic balloon-borne observations are available from Syowa station with some exceptions for October from 1968 to 1995. These data have been compiled together with the observations from Georg Forster (1985-1991)/Neumayer (1992-1995), and for the corresponding season, when chemical ozone losses may occur, for high arctic latitudes from Ny-Ålesund (1989-1995), in order to study the altitude dependent development of ozone for certain isentropic levels. These levels were selected with regard to the current state, c.f. Fig. 7, for 800 K (28-30 km) with no difference in the mean ozone mixing ratios, for 600 K (24-25 km) with less ozone in the Antarctic, for 500 K (20-21 km) and 350 K (12-13 km) as upper and lower altitudes between which chemical ozone losses are mainly detected in the Arctic and Antarctic. At Fig. 8 monthly mean mixing ratios are shown for October from Syowa (1968-1995) by open squares, and from Georg Forster/Neumayer (1985-1995) by crossed squares, as well as from Ny-Ålesund (1989-1995) by full dots.

The antarctic stations are all located near 70°S but at different longitudes (Tab. 1). Despite of this longitudinal difference the deviations of monthly mean mixing ratios are within one standard deviation up to 600 K altitude. At 800 K the interannual variations are more than one standard deviation. A significant longitudinal difference appears between SY and GF/NM after 1989. These differences in altitude might be explained with the temporal and spatial pattern of the stratospheric vortex. While at lower altitudes all stations are more or less located inside the vortex, the higher ozone mixing ratios at SY at greater altitudes indicate a tilted vortex towards western longitudes in October. The arctic mixing ratios for March are as high as in the Antarctic at 800 K. Due to the decreasing ozone in the southern lower stratosphere arctic ozone values are recently as high as they have been at those altitudes in the Antarctic in the late sixties. At 350 K the perturbation by volcanic aerosols, c.f. Fig. 5, caused a significant depletion of

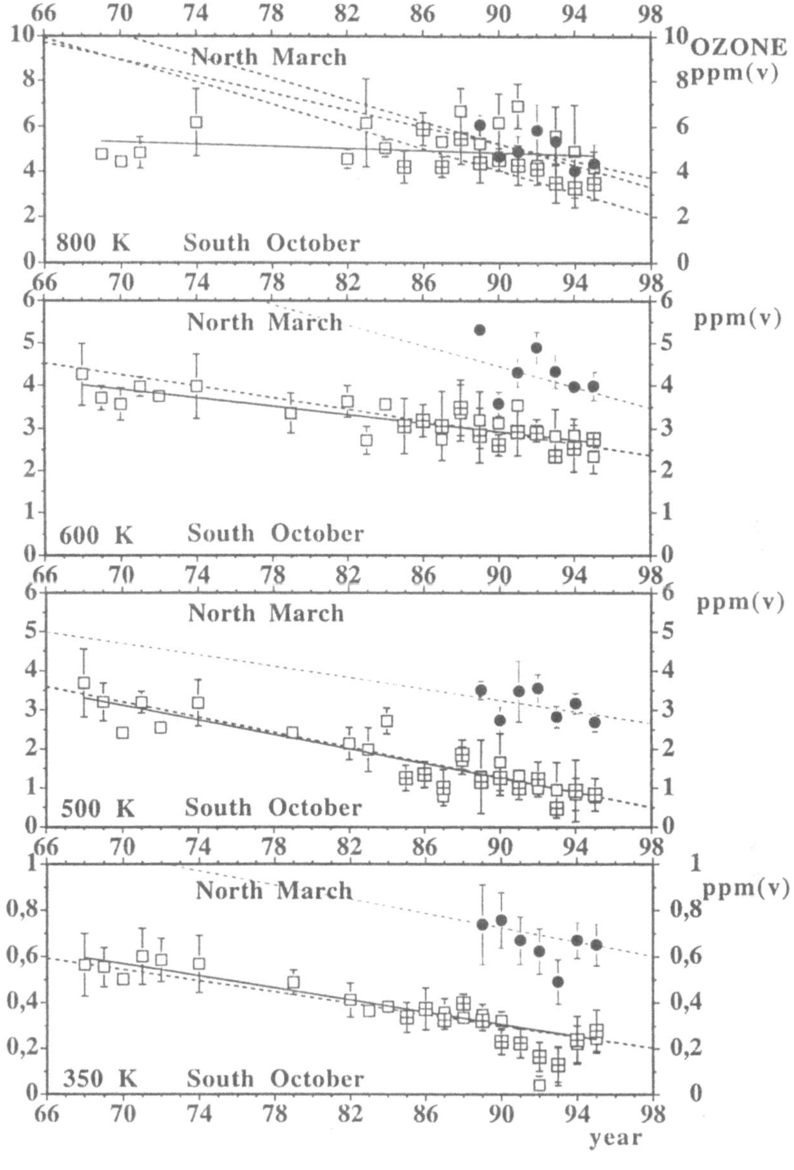

Fig. 8. Antarctic and arctic interannual ozone variations at isentropic levels 800 K (28-30 km), 600 K (24-25 km), 500 K (20-21 km), and 350 K (12-13 km). Monthly means of ozone mixing ratio (ppm(v)) with their standard deviations for October from Syowa (open squares), Georg Forster/Neumayer (crossed squares), and Ny-Ålesund (full dots). Linear regression for the period 1968-1995 (full lines) and for the period 1989-1995 (dashed lines).

ozone in the Arctic and Antarctic. At the 500 K level no significant response can be noticed. However relative maxima of ozone mixing ratios occur at 800 K over Syowa in 1991 and at 600 K and at 800 K above Ny-Ålesund in 1992. That is consistent with model calculations (Tie et al. 1994) where ozone is reduced below 20 km in spring while a slight increase occurs between 20 km and 30 km altitude for the same season.

Since the southern polar stratosphere was loaded with volcanic aerosols each October from 1991 to 1993 and also the northern polar stratosphere each March from 1992 to 1993 (Figs. 4 and 5), these years were not considered for the calculation of trends. All other data from SY and GF/NM were used to calculate the linear regressions for the period from 1968 to 1995. The retrieved ozone loss per year, the total ozone loss, and the ozone loss per decade are compiled in Tab. 3. The long-term trends are shown by full lines at Fig. 8. The strongest ozone loss occurred in the lower stratosphere. At 800 K, i.e. about 28 to 30 km altitude, the trend amounts about - 5% per decade, but it should be noticed that it mainly accounts for the low values of ozone mixing ratio after 1989. The main contribution for negative trends in total ozone at high latitudes is found in the lower stratosphere. (Stolarski et al. 1992. Hollandsworth et al. 1995).

The ozone loss per year as well as the total ozone loss for the period 1989 to 1995 are compiled in Tab. 4.

In order to compare both polar regions linear regressions from 1989 to 1995 were calculated for both from the Antarctic and the Arctic. The obtained regressions are shown by dashed lines at Fig. 8.

Table 3. Long-term ozone loss in the antarctic stratosphere 1968 to 1995 calculated from data obtained at Syowa and Georg Forster/Neumayer. The years of volcanic perturbation are omitted (1991, 1992, 1993).

level	O3 loss ppm(v)/year	O3 loss total (%)	O3 loss % per decade
350 K (12-13 km)	- 0.01	- 59	- 22
500 K (20-21 km)	- 0.09	- 76	- 28
600 K (24-25 km)	- 0.05	- 33	- 12
800 K 28-30 km)	- 0.02	- 12	- 5

Compared with the trends obtained for the period from 1968 to 1995 the short term trends for the period from 1989 to 1995 are similar in their ozone loss per year for the antarctic stations at the 350 K, 500 K, and 600 K level.. Furthermore the short term trends retrieved for the arctic station are also similar at the 350K and 500K level to those in the Antarctic . A strong change is seen at 800 K in the middle stratosphere. In the Antarctic the individual negative trends for both stations, Syowa and Georg Forster/Neumayer, changed significantly, and they are similar to the strong decline in the Arctic. This coincidence of decreasing ozone suggests a bipolar phenomenon perhaps caused by a large scale change of atmospheric dynamics, which might reduce the meridional transport towards polar latitudes.

Similar conclusions were made by von der Gathen et al. (1995b) for arctic ozone observations. Hollandsworth et al. (1995) retrieved long-term negative trends of ozone in the upper stratosphere for the period 1978 to 1994 by combining data from Nimbus 7 SBUV with data from the NOAA 11 SBUV/2 instrument. In the 2-4 hPa layer (about 40 km altitude) maximum negative trends of about - 16% per decade at 60°S and - 10% per decade at 60°N were retrieved during fall season. That might also partially explain the ozone decrease in the middle stratosphere. If less ozone is available when the stratospheric vortex forms then reduced ozone might be also measured in spring.

Table 4. Short-term ozone loss in the arctic (March) and antarctic (October) stratosphere 1989 to 1995, calculated from data obtained at Syowa, Georg Forster/Neumayer, and Ny-Ålesund. The years of volcanic perturbation are omitted: for the Antarctic 1991, 1992, and 1993; for the Arctic 1992 and 1993.

level	Antarctic (SY, GF/NM)		Arctic (NA)	
	O3 loss ppm(v)/year	O3 loss total (%)	O3 loss ppm(v)/year	O3 loss total (%)
350 K	- 0.01	- 22	- 0.02	- 11
500 K	- 0.10	- 43	- 0.07	- 13
600 K	- 0.07	- 13	- 0.12	- 16
800 K (for SY)	- 0.25	- 21	- 0.24	- 27
800 K (for GF/NM)	- 0.19	- 34	---	---

7 Conclusions

In late winter chemical ozone loss is now evident in the arctic lower stratosphere. Natural perturbations by extremely high aerosol loading after volcanic eruptions also cause additional ozone depletion in the lower stratosphere in both polar regions. That was not observed in the Antarctic after volcanic eruptions in the early eighties.

While the stratosphere in the polar regions shows a more similar behavior with respect to the ozone chemistry, seasonal variations of tropospheric ozone are significantly different. Due to the transport of polluted air masses from mid-latitudes to the arctic region, ozone is additionally produced by photochemistry, but it is reduced by photochemistry at clean air conditions in the Antarctic.

Having in mind the complex and global interactions between ozone production, destruction and transport into the polar stratosphere, the simultaneous decrease of ozone in the middle stratosphere, shown for spring in the Arctic and Antarctic, seems not to be related to the chemical ozone losses in the lower stratosphere. It might be associated with negative trends in the upper atmosphere or with a global change in meridional transport. The five year period showing this decline in ozone is still too short to get a reliable evidence on a changing trend. However these results might indicate that the long-term internal variability of the atmospheric system should be considered for getting a more detailed understanding on the global interactions controlling ozone in the polar stratosphere.

8 Acknowledgments

The performance of regular observations by the staffs of Neumayer and Ny-Ålesund stations is gratefully acknowledged. Thanks are to S. Kaneto from the Japan Meteorological Agency, Tokyo, for constructive cooperation and supplying of ozone sonde data of Syowa station. Special thanks are due to Larry Thomason from NASA Langley Research Center, Hampton for providing SAGE II data, and to V. Radionov from the Arctic and Antarctic Research Institute, St. Petersburg, for providing sun photometer data from Mirny station. Helpful discussions and assistance is sincerely appreciated to R. Neuber and M. Rex of the Alfred Wegener Institute for Polar and Marine Research, Research Unit Potsdam.

AWI contribution number: 1019

9 References

Beyerle, G., Neuber, R., and Schrems, O. (1994): Multiwavelength lidar measurements of stratospheric aerosols above Spitsbergen during winter 1992/93. Geophys. Res. Lett., 21, NO. 1, 57-60.

Beyerle, G., and Neuber, R. (1994): The stratospheric aerosol content above Spitsbergen during winter 1991/92. Geophys. Res. Lett. 21, NO. 13, 1291-1294.

Beyerle, G., Herber, A., Neuber, R., and Gernandt, H. (1995): Temporal development of Mt. Pinatubo aerosols as observed by Lidar and sun photometer at Ny-Ålesund, Spitsbergen. Geophys. Res. Lett., 22, NO. 18, 2497-2500.

Beyerle, G., Luo, B., Neuber, R., Peter, Th., and McDermid, I. S. (1996): Temperature dependence of ternary solution particle volumes as observed by lidar in the Arctic stratosphere during winter 1992/93. J. Geophys. Res., submitted.

Brasseur, G., and Solomon, S., (1984): Aeronomy of the Middle Atmosphere. D. Reidel Publishing Company, Dordrecht/Boston/Lancaster, ISBN 90-277-1767-2 (Netherlands).

Chubachi, S. (1984): Preliminary result of ozone observations at Syowa from February 1982 to January 1983. Mem. Natl Inst. Polar Res., Special Issue, 34, 13-19.

Deshler, T., Adriani, A., Gobbi, G.P., Hofmann, D.J., Di Donfrancesco, G. and Johnson, B.J. (1992): Volcanic aerosol and ozone depletion within the Antarctic polar vortex during the Austral spring of 1991. Geophys. Res. Lett., 19, 1819-1822.

Engel, A., Bauer, R., Buness, A., Klein, E., Kulessa, G., Maas, H., Schunk, E., and Schmidt, U. (1995): Chlorine trends in the stratosphere deduced from a decade of measurements of long lived tracers. Poster HA.8, International Conference on Ozone in the Lower Stratosphere, Halkidiki, Greece, May 1995.

Fahey, D.W., Kawa, S.R., Woodbridge, E.L., Tin, P., Wilson, J.C., Jonsson, H., Dyes, J.E., Baumgardner, D., Borrmann, S., Toohey, D.W., Avallone, M., Proffitt, M.H., Margitan, J., Loewenstein, M., Podolske, J.R., Salawitch, R.J., Wofsy, S.C., KO, M.K.W., Anderson, D.E., Schoeberl, M.R., Chan, K.R. (1993): In situ measurements constraining the role of sulphate aerosols in mid-latitude ozone depletion. Nature, Vol. 363, 509-514.

Farman, J.C., Gardiner, B.G. and Shanklin, J.D. (1985): Large losses of total ozone in Antarctica reveal seasonal ClOx/NOx interaction. Nature, 315, 207-210.

Fishmann, J., Solomon, S., and Crutzen P.J., (1979): Observational and theoretical evidence in support of a significant in-situ photochemical source of tropospheric ozone. Tellus, 31, 432-446.

Geller, M.A., Nash, E.R., WU, M.F., and Rosenfield, J.E. (1992): Residual Circulation Calculated from Satellite Data: Their Relations to Observed Temperature and Ozone Distributions. J Atmos. Sci., Vol. 49, No. 13, 1127-1137.

Gernandt. H.,, Dethloff. K., and Kanzawa, H. (1994): A qualitative assessment of height dependent interannual variability of polar stratospheric ozone part I: Long-term variability and stratospheric ozone depletion. Proceedings of the NIPR Symposium on Polar Meteorology and Glaciology, No. 8, 1-13.

Grudzev, A.N., and Sitnov, S.A. (1992): The annual Variation of Tropospheric Ozone and estimates of Tropospheric-Stratospheric Exchange in the Arctic and Antarctica Based on ozone Sounding Data, Atmospheric and Oceanic Physics, Vol 28, No. 9, pp 707-714.

Heintzenberg. J. (1989): Arctic Haze: Air Pollution in Polar Region. Ambio, Vol. 18. 50-55.

Herber, A., Thomason, L.W., Radionov, V.F., Leiterer, U. (1993): Comparison of Trends in the Tropospheric and Stratospheric Aerosol Optical Depths in the Antarctic. J Geophys. Res.. Vol. 98. NO. D10. 18,441-18,447.

Herber, A., Gernandt, H., Thomason, L.W., Schulz, K.H., Schwartz, U., and Graeser, J. (1995): Aerosol Optical Depth Measurements at Sptsbergen with Sun and Moon as Light Source. Poster A31C-03, AGU Spring Meeting, Baltimore, MD.

Herber, A.. Dethloff, K.. Thomason, L.. Viterbo. P., Radionov, V.F.. Leiterer. U.: (1996): Volcanic Perturbation of the Atmosphere in both Polar Regions: 1991-1994. JGR in press.

Hofmann, D.J., and Oltmanns, S.J. (1993): Anomalous Antarctic Ozone during 1992: Evidence for Pinatubo Volcanic Aerosol Effects. J. Geophys. Res.. 98, 18555-18561.

Hofmann, D.J., Oltmans, S.J., Lathop, J.A., Harris, J.M., Vomel, H. (1994): Record low ozone at the South pole in spring of 1993. Gephys. Res. Lett., Vol. 21, 421-424.

Hollandsworth, S.M., McPeters, R.D., Flynn, L.E., Planet, W., Miller, A.J., and Chandra. S. (1995): Ozone trencs deduced from combined Nimbus / SBUV and NOAA 11 SBUV/2 data. Geophys. Res. Lett., Vol. 22, 905-908.

Kaneto, S., Kishi, T., and Meteorological Staff of 33rd and 34th Japanese Antarctic Research Expedition (1994): Aerosol Optical Depth Observations by JARE-33 and JARE-34 in 1991 and 1992, Proceedings of the NIPR Symposium on Polar Meteorology and Glaciology, No. 8, pp 19-26.

Kaye, J.A. (1993): Stratospheric chemistry, temperatures, and dynamics. In Atlas of Satellite Observations Related to Global Change, Edited by R.J. Gurney, J.L.Foster, C.L.Parkinson, Cambridge. University Press, 41-57.

Logan, J. (1985): Tropospheric Ozone: Seasonal Behavior, Trends and Anthropogenic Influence. J. Geophys. Res., Vol. 90, NO. D6, 10,463-10,482.

Manney, G.L., Froidevaux, L., Waters, J.W., Zurek, R.W., Read, W.G., Elson, L.S., Kumer, J.B.. Mergenthaler, J.L.. Roche. A.E.. O'Neill. A., Harwood. R.S.. Mackenzie, I., and Swinbank, R. (1994): Chemical depletion of ozone in the Arctic lower stratosphere during winter 1992-93. Nature, 370, 429-434.

McCormick, P.M., and Trepte, C.R. (1987): Polar Stratospheric Optical Depth Observed Between 1978 and 1985. J. Goephys. Res.. 92 NO. D4. 4297-4306.

McCormick, P.M., Thomason, L.W.. Trepte, CH.R. (1995): Atmospheric effects of the Mt. Pinatubo eruption. Nature, Vol. 373, 399-404.

Meilinger, S.K., Koop, T., Huthwelker, T., Carslaw, K.S., Krieger, U., Crutzen, P.J. and Peter Th. (1995): Size-dependent stratospheric droplet composition in mesoscale temperature fluctuations and their potential role in PSC freezing. Geophys. Res. Lett., 22, NO. 22, 3031-3034.

Naujokat, B., Labitzke, K., Lenschow, R., Rajewski, B., Wiesner, M. and Wohlfahrt, R.-C.. (1995): The stratospheric winter 1994/95: A Cold Winter with a Strong Minor Warming. Beilage zur Berliner Wetterkarte, 81, SO 24/95, 1-24.

Naujokat, B. (1986): An updateof the observed Quasi-Biennial Oscillation of stratospheric winds over the tropics. J. Atmos. Sci., 43, 1873-1877.

Naujokat. B. (1995): Updated QBO data by private communication.

Neuber, R., Beyerle, G., Fiocco, G., Di Sarra, A., Fricke, K.H., David, CH., Godin, S., Knudsen, B.M., Stefanutti, L., Vaughan, G., Wolf, J.-P. (1994): Latitudinal distribution of stratospheric aerosols EASOE winter 1991/92. Geophys. Res. Lett., Vol. 21. 1283-1286.

Orsolini, Y.J., Cariolle, D., De_Que_, M. (1994): A GCM study of the late January 1992 "mini-hole" event observed during EASOE. Geophys. Res. Lett., Vol. 21, NO. 13, 1459-1462.

Orsolini. Y.J. (1995): On the formation of ozone laminae at the edge of the Arctic polar vortex. Quart. J. Royal Met. Soc., in press.

Panegrossi. G., Fua, D., Fiocco, G. (1996): A 1-D Model of the Formation and Evbolution of Polar Stratospheric Clouds. J. Atmos. Chem. 23, 5-35.

Rex, M., von der Gathen, P., Harris,N.R.P., Reimer, E., Beck,A., Alfier, R., Carstensen,R., Knudsen, M., Mikkelsen, I.S., Chipperfield, M., Lucic, D., Allaart, M., De Backer, H.. Braathen. G.O.. Reid, S.. Claude, H.. O'Connor. F.. Dier.H., Fast. H., Gamma, A., Gil,M., Godin, S., Guirlet, M., Kyr, E., Rummukainen, M., Litynska, Z., Kois,B., Murphy, G., Ravegnani, F., Varotsos, C., Wenger, J., Yushkov,V., Dorokhov,V., Zerefos, C., Balis , D. And Ziomas I. (1995): Chemical Ozone Loss in the Arctic Winters 1991/92 and 1994/95 (Match). Proceedings of the Third European Symposium on Polar Stratospheric Ozone Research, Schliersee.

Rφth, E.-P. (1994): Ozonloch - Ozonsmog: Grundlagen der Ozonchemie. Mannheim; Leipzig; Wien; Zórich: BI-Taschenbuchverlag (Meyers Forum; 26), ISBN 3-411-10421-X. 67-73, and 96-107.

Schoeberl, M.R. (1993): Stratospheric ozone depletion. in Atlas of Satellite Observations Related to Global Change, Edited by R.J. Gurney, J.L.Foster, C.L.Parkinson, Cambridge, University Press, 59-65.

Schoeberl, M.R.. And Doiron. S.D. (1993): A simulation of the Cerro Hudson SO_2 cloud. J. Geophys. Res., 98, 2949-2955.

Stolarski, R., Bojkov, R., Bishop, L., Zerefos, Ch., Staehelin, J., and Zawodny, J. (1992): Measured Trends in Starospheric Ozone. Science, Vol. 256, 342-349.

Tabazadeh. A.. Turco. R.P., and Jacobson. M.Z. (1994): A model for studiyng the composition and chemical effects of stratospheric aerosols. J. Geophys. Res., 99, 12897-12914.

Thomason, L.W., and Poole, L.R. (1993): Use the Stratospheric Aerosol Properties as Diagnostics of Antarctic Vortex Processes. J. Geophys. Res., 98. 23003-23012.

Tie, X., Brasseur, G.P., Briegleb, B., Granier, C. (1994): Two dimensional simulation of Pinatubo aerosol and its effect on stratospheric ozone. J. Goephys. Res., Vol. 99, NO. D10, 20,545-20,562.

Trepte, C.R., Veiga, R.E., McCormick, M.P.(1993): The polward Dispersal of Mount Pinatubo Volcanic Aerosol. J. Geophys. Res., Vol 98, 18563-18573.

Trepte, C.R., Thomason, L.W., and Kent, G.S. (1994): Banded Structure of Stratospheric Aerosol Distribution. Geophys. Res. Lett., 21, 2397-2400.

von der Gathen, P., Rex. M.. Harris. N.R.P.. Lucic. Knudsen, B.M.. Braathen. G.O.. De Backer,H., Fabian,R., Fast, H., Gil,M., Kyr,E., Mikkelsen, I.S., Rummukainen, M., Stachelin, J., and Varotsos,C. (1995a): Observational evicence for chemical ozone depletion over the Arctic in winter 1991-92, Nature, Vol. 375, 131-134.

von der Gathen, P., Gernandt. H.. Neuber, R., and Rex, M.. (1995b): Ozone deficits in the lower and middle stratosphere of the Arctic polar vortex, Submitted to the Proceedings of the 3rd Europen Symp. on Polar Stratosphereic Ozone Research, Schliersee.

World Meteorological Organization (1992): Scientific Assessment of Ozone Depletion: 1991, World Meteorological Organization, Global Ozone Research and Monitoring Project - Report No. 25.

World Meteorological Organization (1995): Scientific Assessment of Ozone Depletion: 1994. World Meteorological Organization. Global Ozone Research and Monitoring Project - Report No. 37.

Ozone and NO_2 monitoring in Southern Spain: The 1994/95 winter record low

Manuel Gil , Concepción Parrondo, Margarita Yela and Benito de la Morena

Departamento de Teledetección y Aeronomía
Instituto Nacional de Técnica Aeroespacial
C/ Ajalvir km 4, 28850 Torrejón de Ardóz, Madrid, Spain
Contact E-mail: gilm@inta.es

Abstract: During the winter 1994/95, three episodes occurred at El Arenosillo (37N, 6.7W) when the total ozone dropped below the 2σ level, in the months of November 94, January and February 95. These episodes are analysed in detail with the help of ozone sounding and ECMWF backtrajectories. The NO_2 column measurements have been used as additional tool to get more insight of the cause of the observed low ozone. Results show that the January episode is related to the elongation and unusual position of the Arctic vortex over Europe. Temperatures in the stratosphere were at their minimum annual values and the backtrajectories present a southward strong meridional component. In addition, measurements of NO_2 show values typical of winter high latitude. The episode in February, on the other hand, was of different origin. Although the total amounts were similar, the trajectory analysis shows typical air masses over the station, resulting in an absence of the secondary ozone maximum.

Keywords: Ozone, transport, stratosphere, NO_2.

1 Introduction

The ozone total column are routinely being measured over Spain at the El Arenosillo station (Huelva) from INTA by the Dobson spectrophotometer 120#. The instrument started its operation in 1982 but significant gaps in the data exist from that time. Since the intercomparison of June 1994, at Izaña observatory, the instrument is providing good quality continuous data. In may of 1993, a UV-Vis home made, automatic spectrometer operating in the short spectral range 430-450 nm was installed for daily NO_2 total column observations using the zenith sky technique at twilight.

Episodes with very low ozone have been reported in the winter 1994/95 at both low latitudes (Mims III et al., 1995) and high latitudes (Bojkov et al. 1995). Over Spain, the data have been one standard deviation below the normal values for the season. In particular, three episodes when the deviation from the mean values was significant, were subject of interest from our group. The absolute historical record over the station was reached in November when the column dropped to 230 DU (see figure 1), but during two others short periods in January and February the negative deviation from the mean was even higher. These two later

NATO ASI Series, Vol. I 53
Atmospheric Ozone Dynamics
Observations in the Mediterranean Region
Edited by Costas Varotsos

cases are studied in this paper. The average and standard deviation curves are obtained from TOMS version 6 with the data set 1978/91 to avoid the gaps of the Dobson spectrophotometer. It must be observed that the average contains the decreasing trend of last years and therefore it is not representative of pre-depletion situation.

Nevertheless the minima observed are at the -2σ level. These episodes occurred within the Second European Stratospheric Ozone Experiment (SESAME), and information on stratospheric temperature, potential vorticity and back-trajectories were available from ECMWF data for the station.

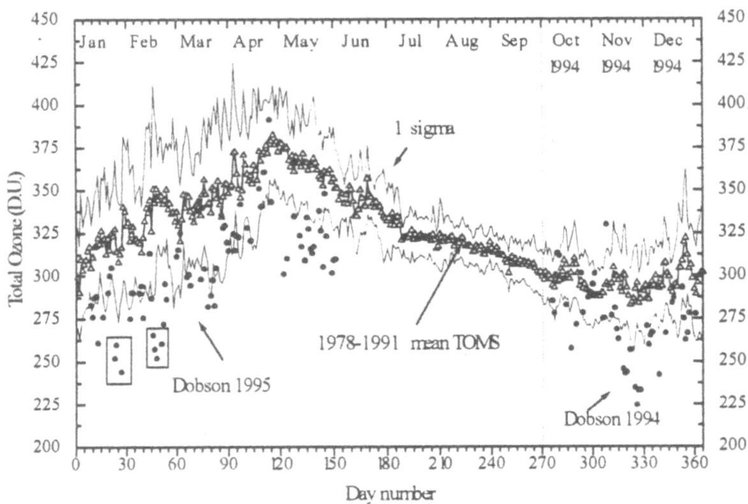

Fig. 1.- Ozone total column at El Arenosillo from Dobson spectrophotometer #120 during the winter 1994/95 (solid circles). The mean for the period 1979 to 1991 have been plotted for comparison (open triangles). Thin lines represent the deviation of the mean at the one sigma level. The two cases studied are shown inside a frame.

2 Results

The Ertel's potential vorticity (PV) is a good index to estimate the proximity of the polar vortex to the station. The ECMWF maps produced at the isentropic levels of 475K, 550K and 675K provides a good information on the location and shape of the vortex in the lower stratosphere. At the latitudes where the station is located, is not common to reach high values of the PV except when strong distortion of the polar vortex occurs. In figure 2 the PV at the three mentioned levels have been plotted. The shaded areas approximately represent the low ozone situation of study. The results show that for the second period, no change in PV occurs. If any, it would be in the decreasing direction. The three levels follow the same pattern. In the first period, however, the 550K and the 675K levels increase significantly as compared with the previous days. The lack of correspondence at

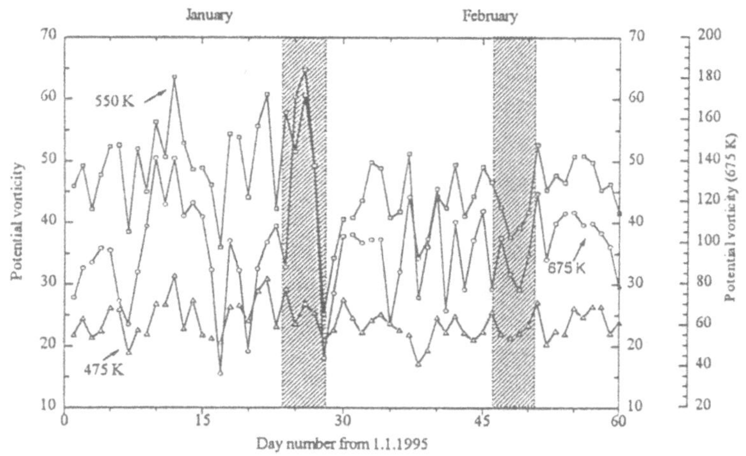

Fig 2.- Potential vorticity over El Arenosillo (Units x10^{-6} $Km^2kg^{-1}s^{-1}$) at the isentropic levels of 475K, 550K and 675K (approximately 60hPa, 40hPa and 20hPa). Note the different ordinate scale for the 675K data.

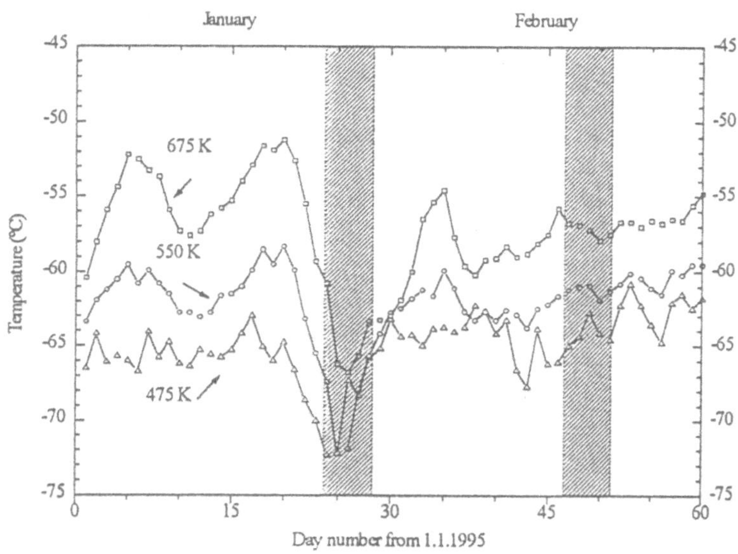

Fig 3.- Temperatures over El Arenosillo at the isentropic levels of 475K, 550K, and 675K.

the 475K level can be explained by the funnel-shape of the polar vortex. It is not uncommon to find in winter situations at the sub-polar belt, back-trajectories at different stratospheric levels from different air-masses coming from very different latitudes.

The temperature displays a similar behaviour, but in the January episode all three levels are affected by very low temperatures. It has been reported that the Arctic temperature during this winter has been the lowest in magnitude and one of the largest in extension since records are available (Naujokat et al. 1995). At El

Arenosillo site the situation is the same, the course of the temperature was well below the long term mean (-61.6° in January, -61° in February at 475 K) not only for the mentioned episodes, but during almost the whole winter. The departures from the mean have been maximum for the 475K level.

2.1 The case 20-25 January

We will consider separately the two episodes. It is helpful to look at the general picture of the shape and location of the polar vortex. In figure 4 PV maps at 475K for the January episode have been displayed. In January 12, the vortex was almost circumpolar. By the 20, a warming at higher levels produced a compression from the 90-270 meridian and a shifting from the pole to the European quadrant. Two days later a rotation has started, and in similar position remains during the next 3 days, when the lowest ozone was measured.

The temperature fields at the same isentropic level show that the areas of low temperatures match well with the vortex location. In January 25 the area of lowest temperatures was shifted to unusually meridional latitudes. Temperatures colder than -80°C at 475K were located over north of France through Finland and a much broader area embracing the -75°C covered even the north of Spain.

We tried at this point to search for the altitudes where the ozone deficiencies were observed. The closest available station with routine ozonesoundings is Madrid (40°N, 3°W). Unfortunately no sounding was available for the day when the minimum value of ozone was reached. In figure 6 we have shown the temperature and ozone profiles for days 18.1 and 23.1. In the later day the tropopause was 2 km higher and the temperature was colder during all the sounding. The ozone was lower at almost all altitudes, but more significantly in two layers from 17 to 20 km and from 24 km upward. Back-trajectories calculated from the ECMWF field are displayed in figure 7 for 475 K and 550 K (aprox. 20 and 23 km) showing that airmasses are of high latitudes origin.

An additional evidence of the polar origin of the airmasses is provided by the NO_2 total column record from the visible spectrometer located at the station (figure 8). While seasonally dominated by the photochemistry, short term variations in NO_2 are related to transport processes. The winter Arctic stratosphere is poor in NO_2 since photodissotation from the N_2O_5 reservoir is slow and is mostly converted to HNO_3 (Fried et al. 1994). The lowest NO_2 column values were measured in coincidence with the low ozone, showing that high latitude air was reaching the station.

In summary, the episode of low ozone in January is characterised by low temperatures and airmasses of polar origin with very low columns of NO_2 compatible with activated chlorine.

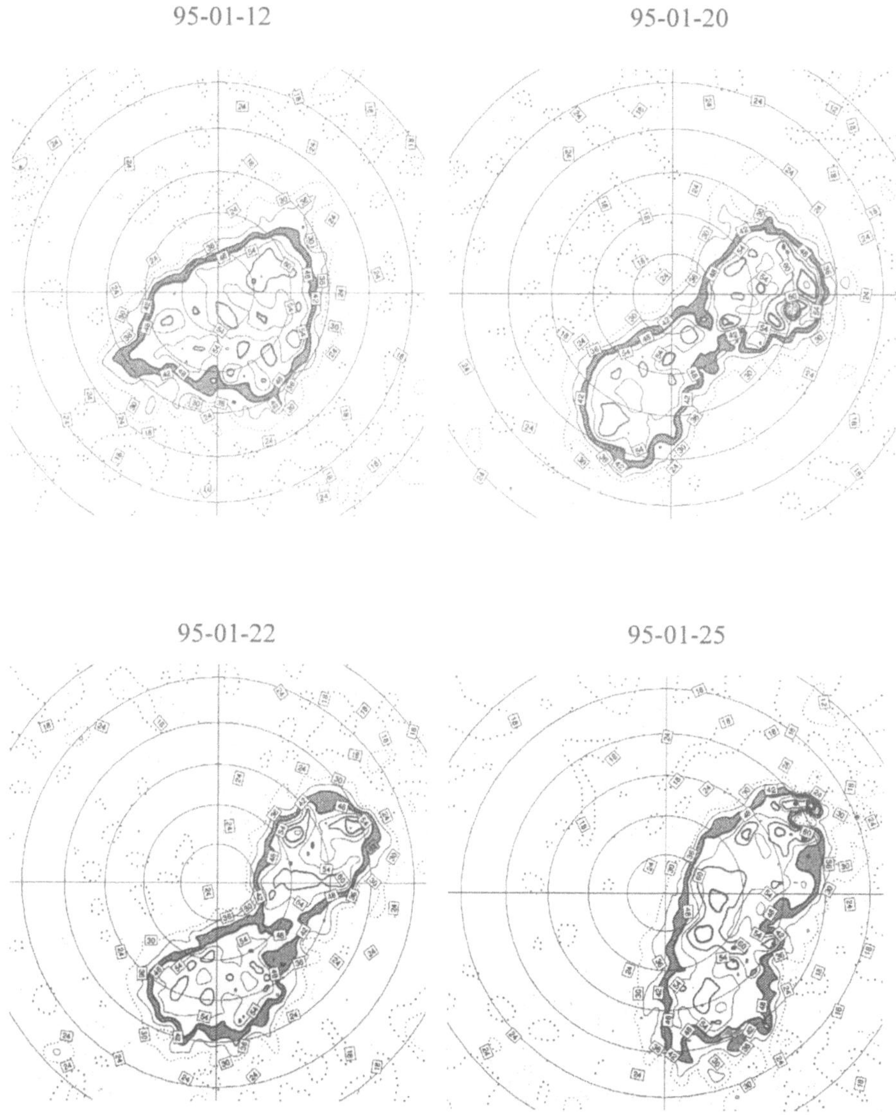

Fig 4. Potential Vorticity maps at 475 K isentropic level. Data produced by DMI and plotted at NILU by isenbw. Circles of latitude are drawn every 10° centred over Europe. Areas between 12 and 48 PV unit have been darkened to approximately show the vortex edge.

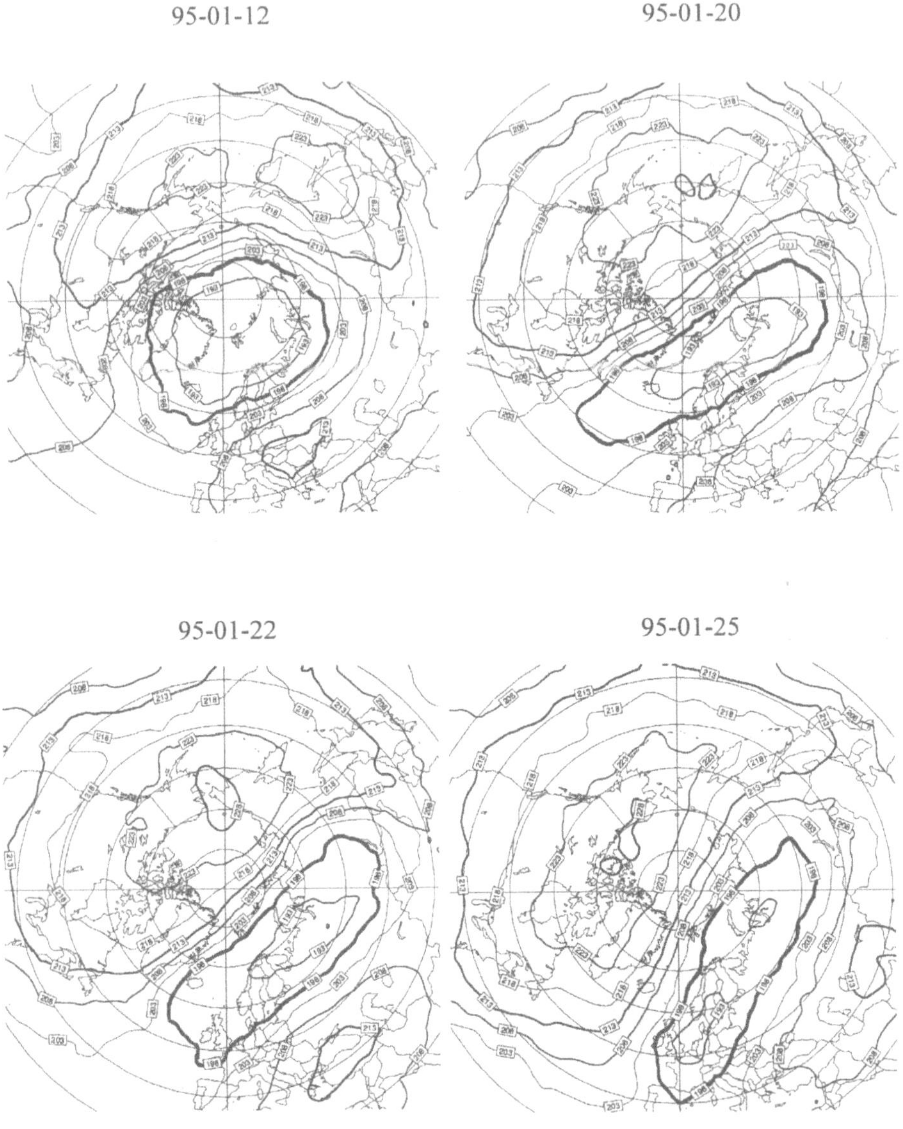

Fig. 5. Temperature polar maps at 475 K isentropic levels. Data produced by DMI and plotted at NILU by isenbw. Circles of latitude are drawn every 10° centred over Europe. The 198 K isotherm has been arbitrarily drawn with a wider line in order to show the cold temperature areas.

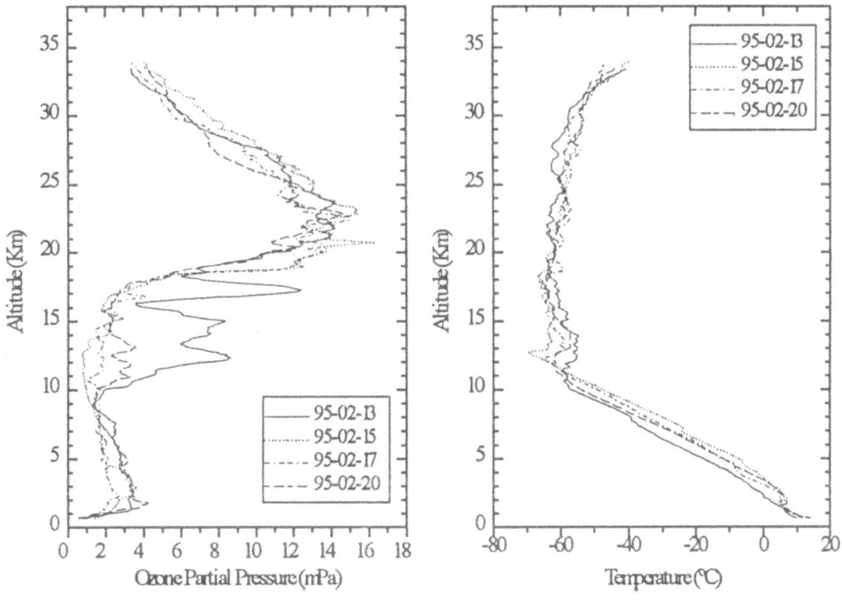

Fig. 6.- Temperature and Ozone profiles from Madrid (40°N, 3°W). (Courtesy from Cisneros, INM).

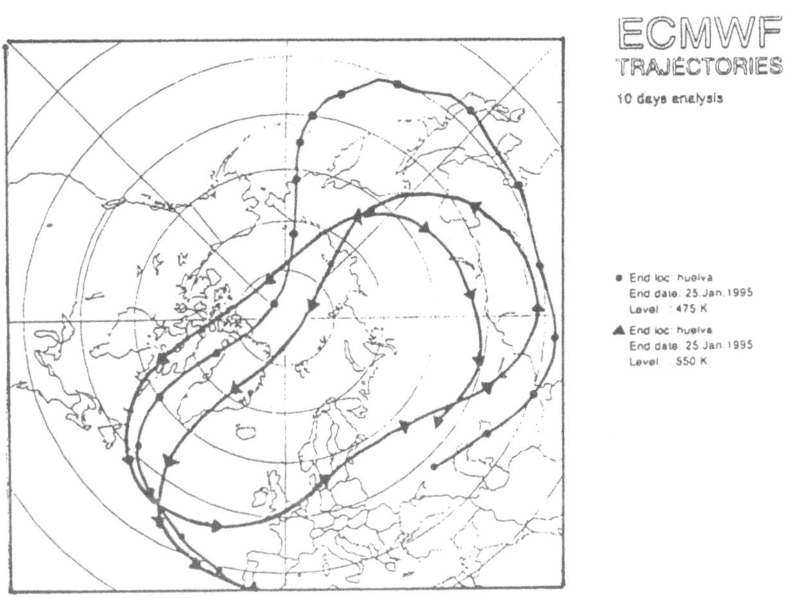

Fig 7.- Ten days back-trajcctories ending at Huelva for 475K and 550 K for day 25.1.95

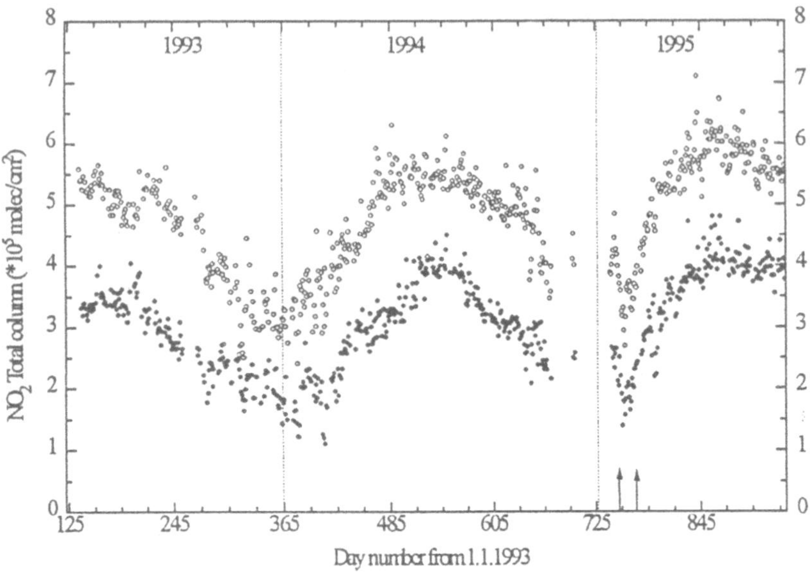

Fig. 8.- NO_2. vertical column over Huelva at sunrise (solid circles) and sunset (open circles) from May 1993 when the instrument was settled. Shown in arrows are the two studied cases. Morning and evening differences are due to photochemical changes in the partition of NO_x .

The case of 15-20 February

Lets move now to the second case studied: the very low values recorded during mid February of about 250 DU (see figure 1). On the contrary of what it is observed in the previous case, during these days nothing noticeable is present in the records of temperature and PV. If any, a slightly decrease in the PV at all levels as compared with the previous days (see figures 2 and 3). Looking at the vorticity maps at 475K in the same way as before, again a distortion of the vortex occurs, but in this case, in the opposite direction.

The vortex elongates in the direction normal to the Greenwich meridian suggesting a forcing from the south (figure 9). The temperature maps at the same level show that the lower stratosphere has warmed up as compared with the January episode. A small area below 198K remains, but always far from our station (figure 10). For these days it is classifier to have a look on the available ozone profiles (figure 11). The first day of the sounding displayed (13.2.95) looks like a profile typical from mid latitudes, with a well developed secondary maximum between the tropopause and the 18 km level. Only two days later (15.2.95) this maximum has completely disappeared and the altitude of the tropopause had rose by 3 km. These ozonesoundings, with no secondary maximum are common at latitudes lower than that of our station. Just to confirm that the origin of this episode was the intrusion of tropical air, we have look at the

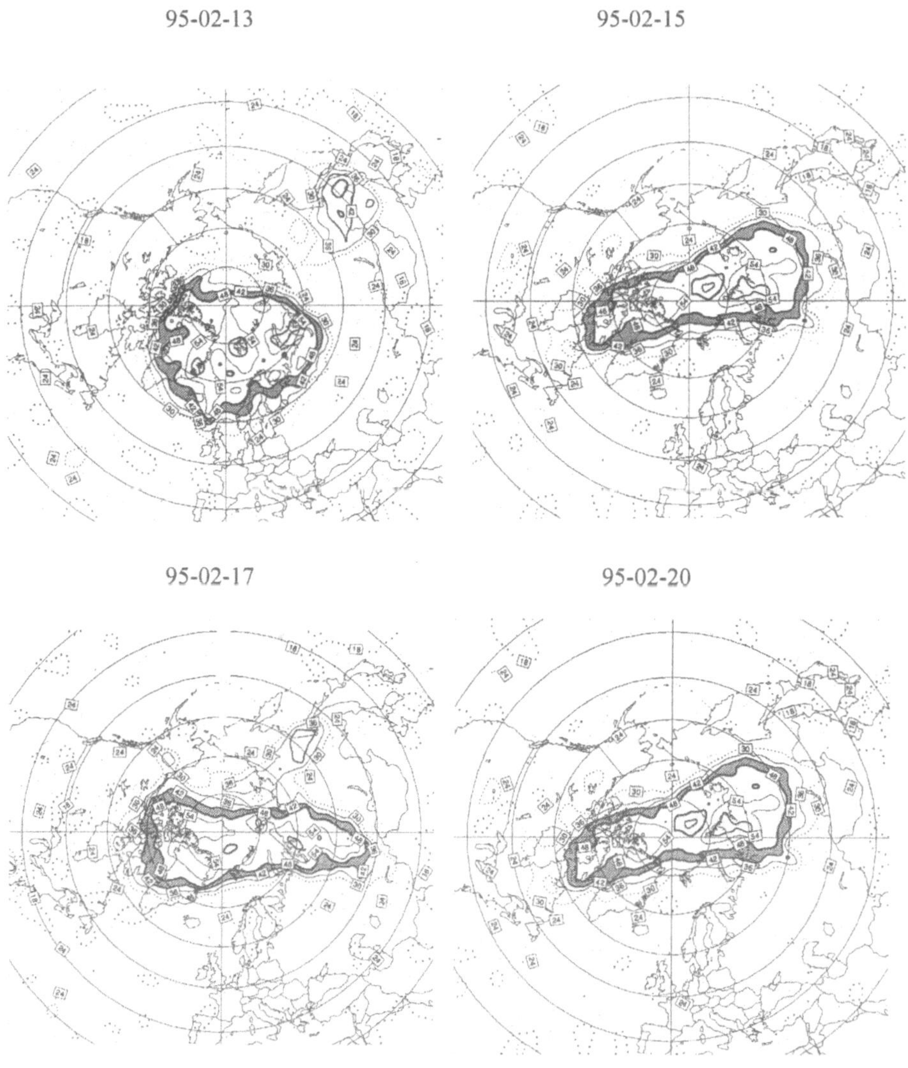

Fig. 9. - Potential Vorticity polar maps at the 475K isentropic level as in figure 4 but for days 13 to 20 of February.

trajectories, but in this case at lower altitudes. Since the station is at the edge of the plots provided by NILU, and trajectories from lower latitudes are out of the map, we have chosen a different way for representation. In figure 12 the history of the airmass for Madrid is represented for the 350K and 380K isentropic levels (aprox. 13 Km and 15 Km) for the days 13.2.95 and 15.2.95, that is before and during the episode. The figure shows how it has been an abrupt change in the aimasses direction reaching the ozonesounding station, and that for 15.2.95 the area in the lower stratosphere was strongly influenced by air flowing from the tropics.

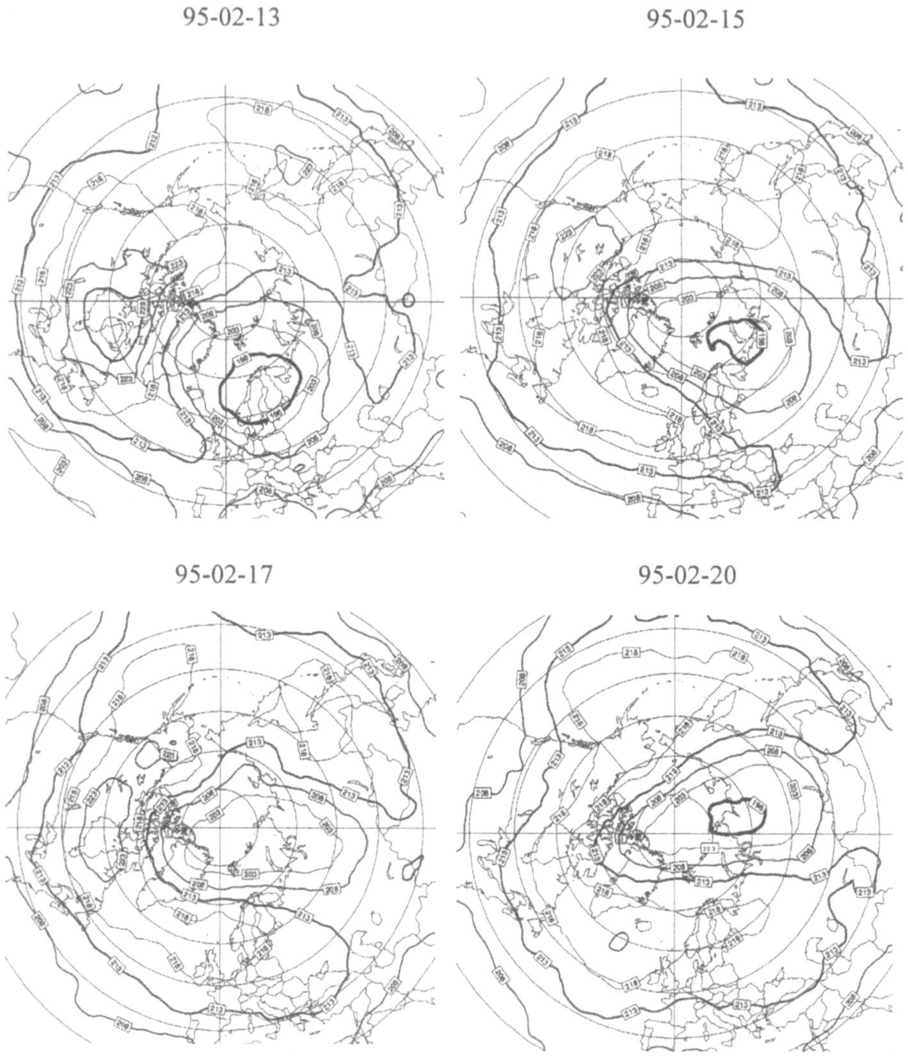

Fig. 10.- Temperature polar maps at the 475K isentropic level for days 13 to 20 of February.

The Madrid (INM) ozonesounding station is at 500 km northward from the El Arenosillo observatory where the Dobson measurements were carried-out. Just to avoid the possibility of an error from considering wrong trajectories, we have drawn the same figure for El Arenosillo (not shown here) and the results are almost identical for the two days considered, that is, the dynamics of both sites for this period was the same, and consequently, it can be stated that the minimum observed was purely dynamical, even thought it took place at a time when chemical reductions at mid latitudes out of the vortex could be expected.

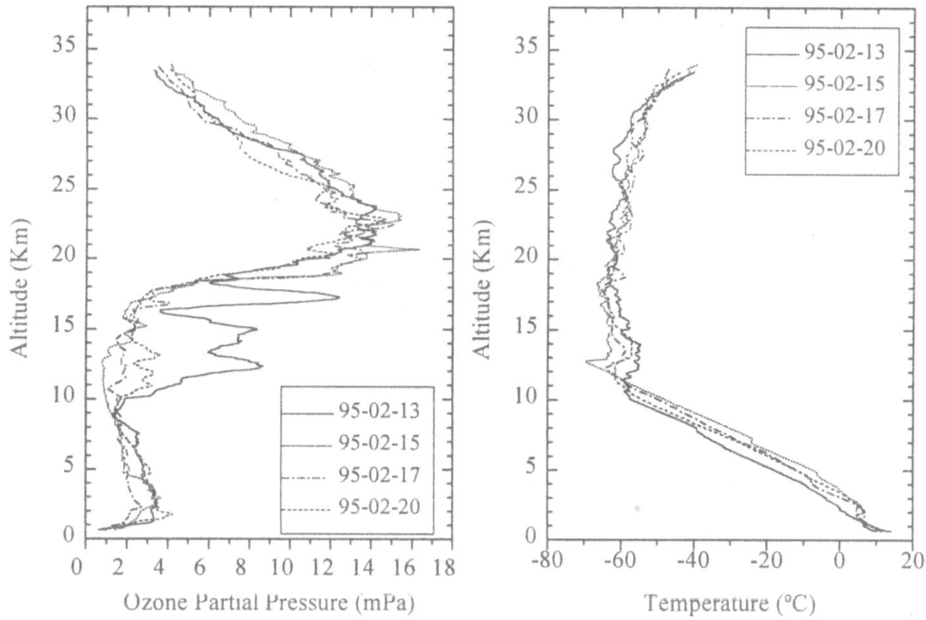

Fig.11.- Ozone and temperature profiles from Madrid (40N, 3W) for day 13 to 20 of February. (Courtesy from Cisneros, INM).

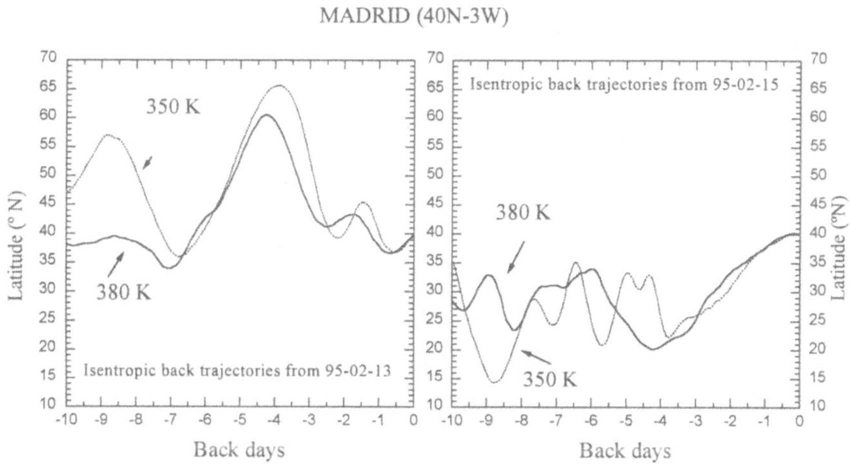

Fig 12.- Ten days back-trajectories ending at Madrid for 350K and 380K for days 13.02.95 and 15.02.95

3 Conclusion

During the winter 1994/95 the absolute total ozone record low for the El Arenosillo station measured by the Dobson spectrophotometer #120, has been reached. In addition, a number of other episodes with unusually low values of ozone have been observed.

Two of the mentioned episodes with a duration of less that one week each one have been analysed with the help of the facilities deployed for the SESAME and related Projects. The result of the study show that during the winter 1994/95 the lower stratosphere has been extremely cold not only at high latitudes, but also over Spain. While the ozone in both of the periods studied reached the same amounts, it has been found that the origin was different nature. The January episode has been found to be related to polar airmasses with very low NO_2 total column and temperatures and high potential voticity. The ozone reductions are observed at almost all altitudes from 16 Km upward being maximum in the 18-21 Km and 24-30 Km.

In the February case the ozone deficiency is located in the 10-18 Km layer. The observed profiles are typical from tropical regions, with high tropopause and no significant change in temperature. In this case, PV was lower than normal and the back-trajectories analyses show low latitude air masses.

Acknowledgements. This work has been partially funded through the SCUVS-II Project, by the EU III Framework Program. We wish to acknowledge Federico Soubrier for Dobson and UV-Vis spectrometer operation at INTA, J.M. Cisneros from INM for making the ozonesoundings from Madrid available for us, McPeters from NASA Goddard Space Flight Centre for the TOMS V6 data in CD rom and the ECMWF, DMI and NILU for provide vorticity and temperature maps and back-trajectories.

References:

Bojkov, R.D., V.E. Fioletov, D.S. Balis, C.S. Zerefos, T.V. Kadygrova, and A.M. Shalamjansky "Further ozone decline during the northern hemisphere winter-spring of 1994-95 and the new record low over Siberia" *Geophys. Res. Lett.*, in press 1995.

Fried A., B. E. Henry, J.G. Calvert and N. Moznrkewich " The reaction probability of N_2O_5 with sulfuric acid aerosol at stratospheric temperatures and compositions. " *J. Geophys. Res.* V99, D2, 3517-3532, 1994.

Mims III, Forrest, W.F. Barnard, A.C. Neuendoffer and G.J. Labow "Unusually low ozone detected over South-Central U.S." EOS V76, March 1995.

Naujokat, B., K. Labitzke, R. Lenschow, B. Rajewski, M. Wiesner, R.C., Wohlfart "The Stratospheric winter 1994/95: A cold winter with a strong minor warming". Bulletin from the Freien Unversität Berlin, 6.9.1995.

Rummukaiken, M., B. Knudsen and P. Von der Gathen " Dynamical diagnostic of the edges of the polar vortices. *Annales Geophys.* V12, 1114-1118, 1994.

THE MIDDLE ATMOSPHERE OVER BULGARIA AT THE TIME OF STRONG GEOMAGNETIC STORM ON OCTOBER 10-th 1988

Tsvetana Gogosheva, Mitko Gogoshev, Vasil Zachariev
Institute of Astronomy, Bulgarian Academy of Sciences,
Stara Zagora - 6000, P.O.Box 442, Bulgaria

ABSTRACT

There is not enough evidence for possible influences of geomagnetic activity on the total ozone content (TOC) and vertical ozone distribution. A broad spectrum of ozone behaviours can be explained by the existence of many intermediate mechanisms between the activity and the ozone density. At the time of strong geomagnetic storm, October 10-18th, 1988, a lot of stratospheric and upper atmospheric observations and ionospheric data were also collected in Bulgaria. The TOC was observed by UV-photometer, and the vertical distribution - by baloon's chemical probe. The stratospheric temperature and the wind were measured by rocket experiments from Rocket Meteorological Station in Ahtopol, Bulgaria. Different neutral upper atmosphere parameters, such as atomic oxygen, mesospheric ozone and nitric oxide density, and the neutral temperature in the mesosphere were measured by means of airglow technics in Stara Zagora observatory. The main results of these investigations show that the TOC before and after the geomagnetic storm shows small variations. A 10% increase of ozone density near the maximum of the layer, in return for the tropospheric decreases was observed. A well pronounced increase of the temperature above 30 km and a cooling of the mesosphere in the same time was clearly seen. This was also accompanied by the increase of the meridional wind speed. We found, by many features, that the effect of a geomagnetic storm on the stratosphere and mesosphere is equivalent to the so called "winter stratospheric warming". The main conclusion is that the geomagnetic storm influence on the ozone layer and on the stratosphere is mainly due to the meridional circulation induced by an auroral precipitation

KEYWORDS: geomagnetic storm, middle atmosphere,ozone.

1. INTRODUCTION.

The investigations of geomagnetic influences on the middle atmosphere (stratosphere, mesosphere and low thermosphere) are very complicate due to the different reasons - [1-5]. As the precipitation radiation controls directly the processes mainly in the polar region, the influences on the middle atmosphere in mid and low latitude is only by transport of masses, energy and impulse. Very important is the fact that the dissipation of energy of the precipitated particles takes place mainly in the mesosphere and low thermosphere, not in the stratosphere - [10]. That is way it can be paid an additional attention to the global stratospheric circulation. In this sense a significant time delay between the particle precipitation processes and middle atmosphere in mid latitude response is evident. This time delay can be, however, some time interpreted not as a inquest of geomagnetic activity, but with different other reasons - secondary precipitation of high energy particles at mid latitudes, solar activity variations, etc. - [6,7,8].

The difficulties to provide these investigations come from the void of experimental methods (in situ measurements) in the middle atmosphere. Upper stratosphere and lower mesosphere are in practice occasionally studied by in situ measurements. The rockets vertical profiles from time to time of some parameters only give a particle, not overlapping information.

NATO ASI Series, Vol. I 53
Atmospheric Ozone Dynamics
Observations in the Mediterranean Region
Edited by Costas Varotsos

That is way in our investigation we try to use the data from different methods - remote sensing from ground based stations, balloon probes, rocket measurements etc.

The main purpose of this paper is to study the behaviour of middle atmosphere over Bulgaria (as a part of Mediterranean region) at the time of comparatively strong (for 1988) geomagnetic storm of October 10th 1988.

2. DATA

In this paper we use the following data:

2.1. GROUND-BASED REMOTE SENSING OF THE MESOSPHERE AND LOW THERMOSPHERE SITUATED IN STARA ZAGORA OBSERVATORY (φ=42,5N, 25,5E), BULGARIA.

By air-glow technics (photometers with narrow interference filters) we investigate some of very important processes in the mesosphere and lower thermosphere as well:

A) The density of atomic oxygen at the heights of the maximum of the layer - 97 ± 10 km.

These data come from the variations of intensity of airglow emission at wavelength 5577 A.

B) The variations of the density of upper atmosphere near the mesopause - 82 ± 10 km.

These variations are obtained by means of measurements of total intensity of OH(8-3) band in the near infra-red part of the spectrum of air-glow. Two spectral filters have been used. The first one has been centered on the R-branch of OH(8-3) near to λ =7240 A, and the next one - at the Q-branch, close to λ =7270 A. By measuring the intensity of those branches we receive the total intensity of the band, throughout the technics, described in [9]. As it is known, the OH radicals near the mesopause are a production of following reaction:

$$O_3 + H \dashrightarrow OH(v>6) + O_2 \qquad (1)$$

In this case, every fluctuation in the density in both components - O_3 and H, immediately have to reflect in OH production. And if OH is excited to the high therms (v>6), this immediately will reflect to OH(8-3) emissions.

C) The densities of NO in the region 95 ± 10 km.

This is also from airglow measurements of the spectral background near λ=7200 A. As it is known, the intensity of this spectral background in proportional to:

$$i_f = C[NO]/[O] \qquad (2)$$

where i_f is intensity in R/A (R - Raleigh), C - is a constant, [O] is extracted by the above mentioned method and the concentration of NO is the investigated parameter. It must be taken into account that NO is very important parameter in the lower thermosphere, especially in the polar region, where at the time of particles precipitation it can be produced in the big quantity following the reactions:

$$e(precip.) + N_2 \dashrightarrow N_2^+ + e + e \qquad (3)$$

$$N_2^+ + e \dashrightarrow NO(^2D) + N \qquad (4)$$

$$N(^2D) + O_2 \dashrightarrow NO + O \qquad (5)$$

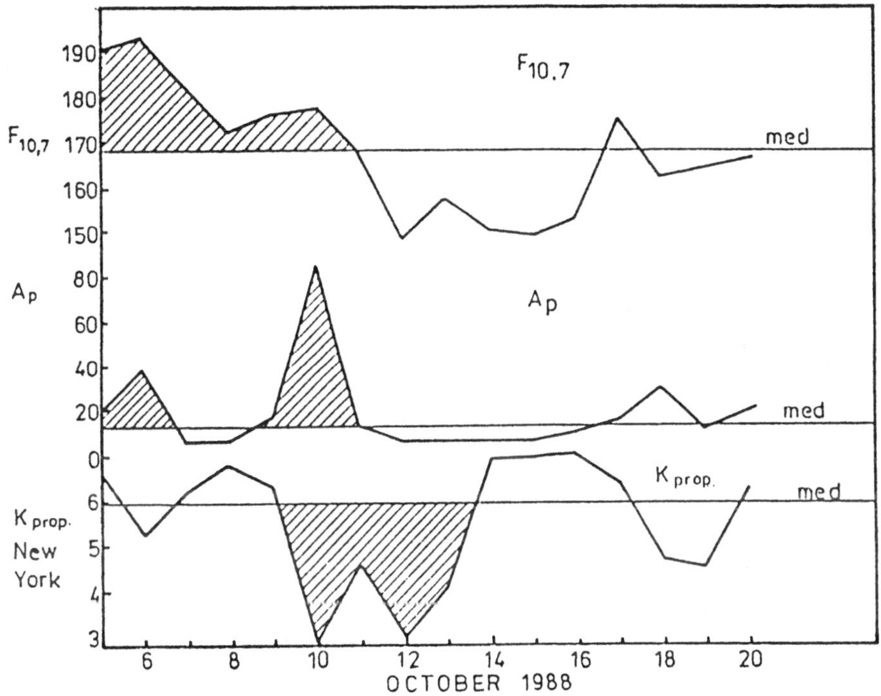

Fig.1. The geomagnetic situation in the first half of October 1988. The upper curve gives the $F_{10,7}$ behaviour. The monthly median is shown. The shaded area expresses the extra-activity. The middle curve is the A_p-index, and at the bottom is the radiopropagation K-index.

Many time, after the strong precipitation, the NO density can increases in few order of magnitude, and can significantly change the photochemistry at a level of 95+10 km.

D) The density of mesospheric ozone.

In the atmosphere of prevailing atomic oxygen density (between 90 and 110 km.) the ozone can be formed as follows:

$$O + O + O \dashrightarrow O_3 \quad (6)$$

Obviously this reaction determined the second (mesospheric) peak of ozone density.

E) Determination of the neutral temperature near to mesopause region (82+10 km.)

At the time of OH formation (see reaction (1)) the excitation of R and Q branches depends on the so called rotational temperature. That is way, by measuring the ratio between the intensities of these branches we can calculate this temperature. If we take into consideration the comparatively long time of methastable R and Q branches, we can assume, with high accuracy, that the rotational temperature will be equal to the neutral temperature around the mesopause.

All this technics for estimation of the above mentioned mesospheric and low thermospheric parameters are in details described in [10,11,12].

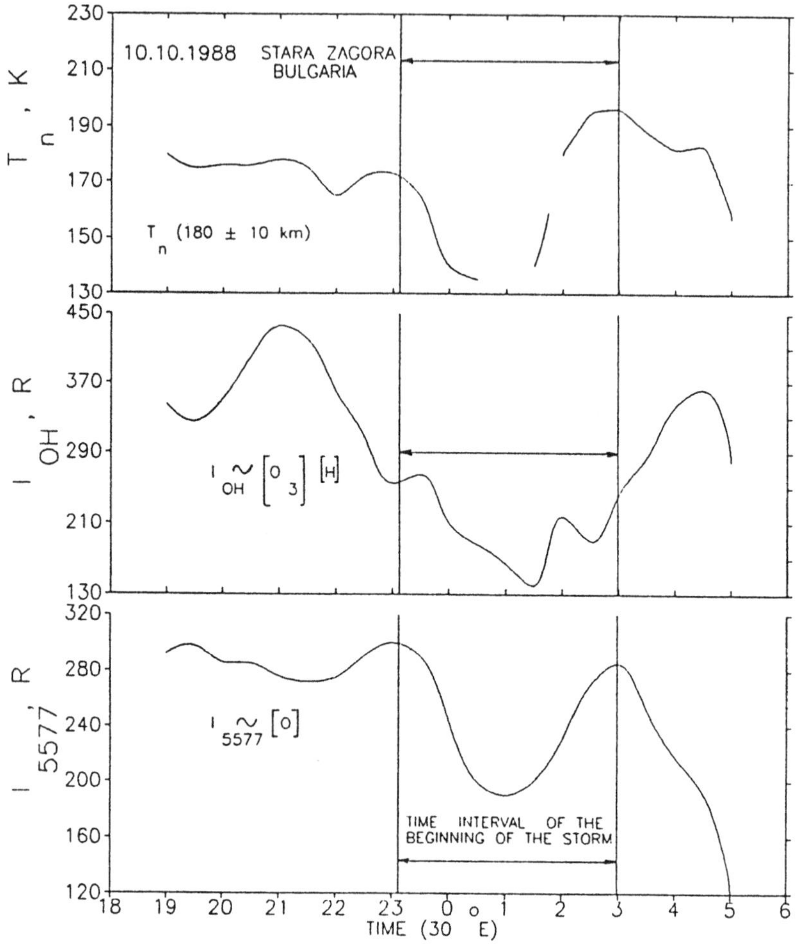

Fig.2. Time interval of storm beginning signature at mid latitude (Stara Zagora, Bulgaria, geographic latitude - 42°)

2.2. BALLOON OZONOSONDE DATA

A chemical balloon ozonosonde has been used for measuring the ozone distribution in the troposphere and stratosphere. Two successful flights give the distribution of this atmospheric component before and after the storm. The balloon ozonosonde station is situated in Sofia (φ =42°N).

2.3. ROCKET DATA FOR THE UPPER STRATOSPHERE AND MESOSPHERE

In the rocket sounding station Ahtopol, Bulgaria, situated on the Black seaside (φ =41°N), the data for temperature, density and wind have been obtained.

2.4. BALLOON SOUNDING UP TO 30 km. OF TEMPERATURE HUMIDITY AND WIND-METEOROLOGICAL STATION SOFIA (φ =42°N)

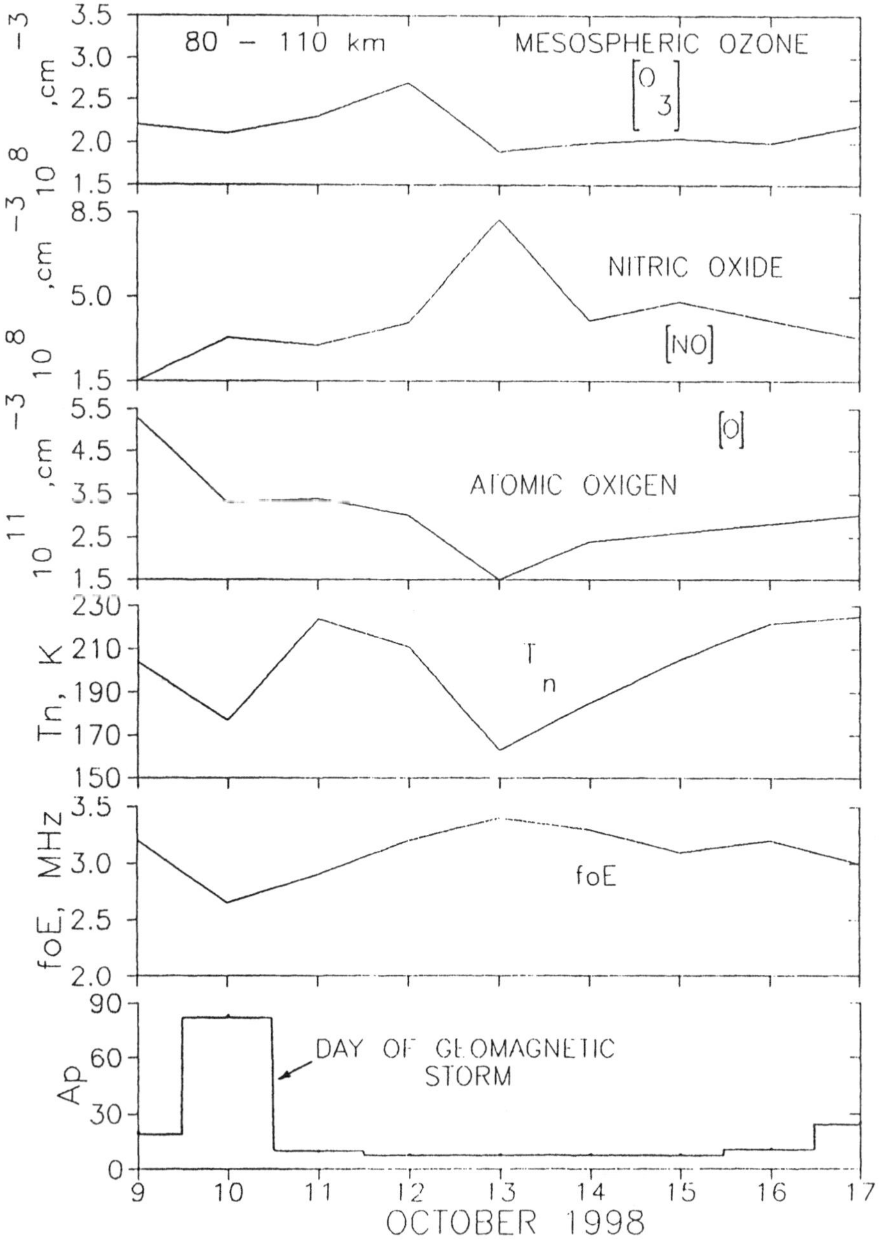

Fig.3. The main effect of geomagnetic storm in the mid latitude middle atmosphere has been observed on the third day after the beginning of the storm. The decreases of mesospheric ozone (upper panel) follows the depletion of atomic oxygen density (panel 3). In the same night a strong bulge of nitric oxide density has been seen. *NO* has been transported from auroral region by the air circulation. The infra red radiation of *NO* provokes the temperature minimum. At the same time the critical frequency of E-ionospheric layer increases (following *NO* density). All these lower thermospheric and mesospheric processes are very complicatedly related with stratospheric circulation, and temperature behaviour.

2.5 IONOSPHERIC DATA for middle and low ionosphere (absorption, critical frequency) - vertical ionospheric sounding and radioabsorption (A_3 method) data have been used. All they come from ionospheric station in Sofia (φ =42°N).

3. DATA ANALYSIS

Fig.1 gives a good illustration for the solar and geomagnetic activity for the period from 5th to 20th October 1988. The upper curve is the behaviour of solar index $F_{10.7}$. The shadowed area is the excess of solar activity, that means - a comparatively high solar UV-irradiance. The middle curve represents the disturbed days by means of daily index A_p. As it is seen - on October 10th - one of the strongest geomagnetic storm for 1988 has been observed. Only as an example of poststorm effect we show the so called "propagation index" - from "Solar Geophysical Data" Bulletin. As it is seen a few days after the storm, due to the increased low ionospheric absorption, this index in below the monthly norm.

Consequently, after the storm, for a few days many of the parameters, as we shall see later, remain in the disturbed conditions.

By the analyses of the different magnetograms from the stations, around the polar oval, it is seen that the storm began with SC (sudden commencement) at 02:30 UT on October 10th. The main phase of the storm began at about 10 UT, and the maximal declination of about 170 nT of H-component of the magnetic field, was reached at about 20 UT in the same day. The recovery phase of the storm follows after that.

From the Fig.2 the reactions of some parameters of the middle atmosphere can be seen. The upper curve shows the mesopause neutral temperature, the middle curve - the intensity of OH(8-3), which corresponds to the density at the height of mesopause, and the lower curve represents the intensity of the green oxygen line 5577 A, representing the O-density fluctuation. Around the local midnight (about 22 h UT) in all curves a well pronounced dip minimum has been observed.

How can we interpret the behaviour of all these parameters? 10-12 hours after the maximum of the main phase of the geomagnetic storm a packet of acoustic and internal gravity waves, launched from the polar oval and travelling with a speed of several hundred meters per second, reached the mid latitudes.

Evidently, here is a case of nondisspative, nonadiabatic processes, but only wavelike phenomena. The problem of transportation of masses and energy we have to discuss later.

Let us describe Fig.3. This figure has been obtained by the following way. Because we can study (by air glow methods) only the behaviour of night-time atmospheric processes, we can calculate from our observations a "mid-night" value of each parameter. This value is the mean value of the observed parameters, averaged for 4-hours period around midnight. We assume that these averaged parameters represents the state of the night upper atmosphere from night to night. The upper panel of

Fig.3 in this manner represents the behaviour of mesospheric O_3 for the nights in the period from October 9th (before the storm), until October 17th. The second panel is the NO-behaviour. The third - O density, the forth - T_n, the fifth - the ionospheric radioabsorption at 7,67 MHz (absorption in D-region), the sixth curve - critical frequency for E-region, and the shaded area below - A_p geomagnetic index. Everywhere the October median as a line is shown. What we can see from this picture is the following. Almost all parameters mark the day of the geomagnetic storm - October the 10th. But the main reaction of these parameters has been on the third day after the storm - on October 13th. This third night has been characterized by the:

a) decreasing the temperature at mesopause level;

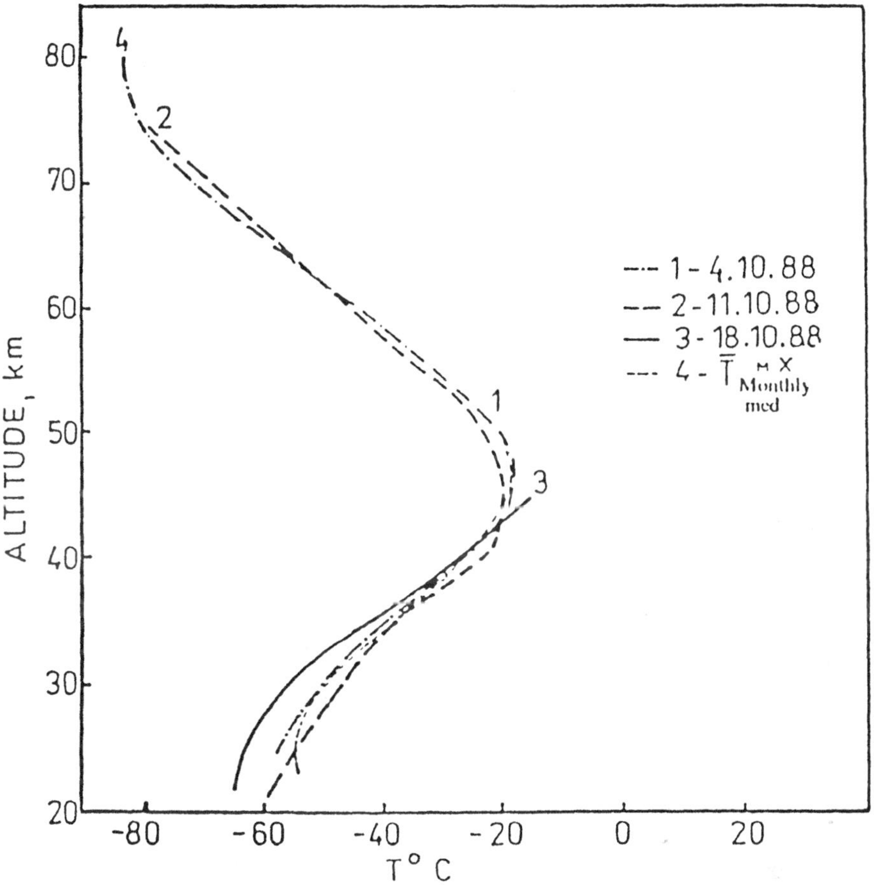

Fig.4. The vertical profiles of the temperature before and after the geomagnetic storm. A tendency of strong increasing after the storm (curve 3) is seen.

b) decreasing the density of O and O_3;

c) increasing of NO;

d) increasing the absorption in D-region and of the critical frequency in E-region.

Here we like to denote that the deviation of these parameters from the median are much more than 1σ and in every case can be assumed as a real declination.

The interpretation will be done latter.

In the following days some fluctuations of investigated parameters have been also seen, but due to the lower amplitude they cannot be interpreted.

On the Fig.4 the vertical profiles of the temperature in the stratosphere are shown. The dashed line denotes the October median. One curve represent the temperature profile - before storm - on October 10th, marked by dashes, and two others - after the storm. Unfortunately, very important curve - on October 18th, after the storm, due to the technical reasons, has been interrupted on a height of 45 km. This last profile shows decreasing of temperature in the low stratosphere and the start of increasing in the upper stratosphere.

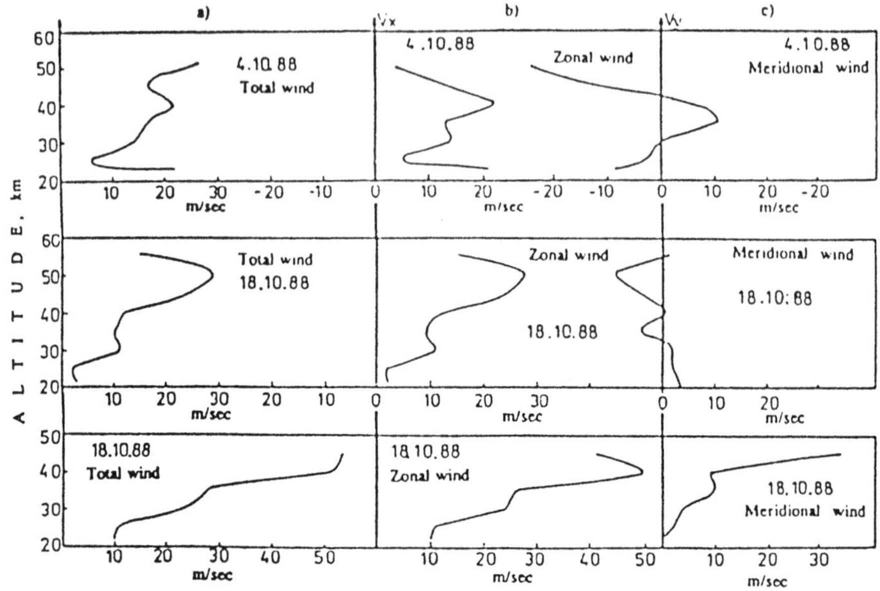

Fig.5. The behaviour of the total a), zonal b) and meridional c) winds, observed by vertical rocket experiments are shown. As it is seen from the bottom panel of column a) the total wind reached after the storm a maximal value of more than 50 m/s. Both components - zonal and meridional are increased.

Very important is the behaviour of the stratospheric winds, shown on Fig.5. For three days data are presented - October 4th the first row, October 11th the second and October 18th. The first column is the total wind, the second - zonal one and the third - meridional component.

As it is seen as a general conclusion - a week after the storm the absolute value of the total wind reached in the stratosphere more than 50 m/s. Up to 40 km this is due mainly to zonal component, but the meridional component is prevailing above this height. Without doubt, the stratospheric winds a week after the storm have been characterized by strong increasing of their velocities, related as we believed with the strong energy deposition in the polar region.

If we see a decreasing of mesospheric ozone as an inquest of low oxygen density, very complicate is the behaviour of the stratospheric ozone. On Fig.6 we show the vertical profiles of ozone and the corresponding profiles of low stratospheric winds. We suppose the deviation of ozone to be related with the deviation of winds, measured at the same time. The most important feature of ozone vertical profile after the storm (on October 19th) is the increasing of stratospheric part and decreasing of tropospheric ozone. The deviations of the wind are in the opposite direction.

We can conclude, only from the analysis of the last figure, that increased stratospheric winds leads in some manner to the increasing of stratospheric ozone and to the decreasing of tropospheric part of this component.

And finally we investigate the behaviour of the circumpolar stratospheric vortex ut to a level of 10 hPa. At the time of the storm and a few weeks later a significant part of the northern hemisphere has been occupied by this vortex.

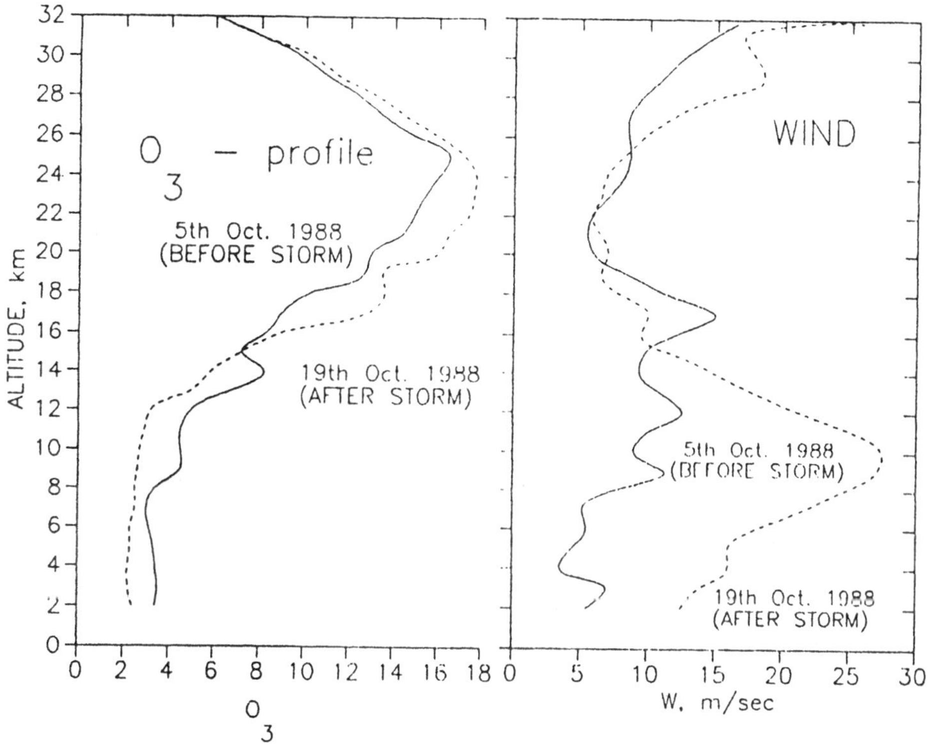

Fig.6. As a consequence of the increases of the wind in the stratosphere we can interpret the observed access of stratospheric ozone. At the same time, after the geomagnetic storm decreasing of the tropospheric ozone is seen. The TOC after the storm remains almost the same. No satisfied physical explanation for the whole phenomena.

We found that the almost sustaining circulation in the vortex before the storm, was strongly influenced by the polar region heating after the storm. The zonal, but mainly the meridional components of the wind increased. By investigation the geopotential height H=3104 dkm at a level of 10 hPa we see the deformation of his shape and size after the storm in relation, as we think, with the increasing of induced circulation.

DISCUSSION

A lot of experimental facts at the time before and after this geomagnetic storm have been investigated. It is very complicate to relate the data in general picture without some contradictions. Let us summaries briefly the facts:

1. In the lower thermosphere and mesosphere:

1.1. In the day of the storm and three days after that we have a decreasing of the neutral temperature and the density of oxygen and mesospheric ozone content.

1.2. Increasing of the density of NO, ionospheric absorption and critical frequency of E-ionospheric layer.

2. In the stratosphere:

2.1. The existing of strong circumpolar stratospheric vortex.

2.2. A decreasing of geopotentioal height at 10hPa level.

2.3. A general increasing of the velocity of the winds and especially - of the meridional component as well.

2.4. An increasing of the stratospheric ozone concentration.

2.5. An increasing of the stratospheric temperature (above 30 km).

Let us try to summarized all this facts.

The absolute start of all these phenomena is due to the particle precipitation in the polar region as a inquest of a geomagnetic storm. The auroral heating at the height of 100+10 km on October 10th, produces a lot of secondary processes. It induces the strong vertical and horizontal circulation as well. A packet of internal gravity and acoustic waves, generated by the heating is launched to the mid and low latitudes. At the same day the disturbed they mid and low latitude mesosphere and low thermosphere. The effects of these travelling disturbances are seen as fluctuations in middle atmosphere density and temperature.

The main mesospheric and upper stratospheric circulation is influenced a few days later. As we mentioned before, after the precipitation, around the 100 km level, the air masses, reached by NO, start to circulate in the direction of mid and low latitude. The composition of the masses is as follows: reached by NO and N_2 and purred by O. The effect was observed at least 25 years ago by the rocket and satellite data.

By means of this induced circulation it is evidently how three days after the storm, the mesosphere/low thermosphere density goes down. The decreasing of the mesospheric temperature is an effect of NO-infrared radiative features. As it is known, the strong emitting of NO can produce a significant cooling of the upper atmosphere.

The air masses reached by NO, can lead as a consequences of photoionization to a significant increases of electron density in D and E-ionospheric regions. The observed radioabsorption there and increasing of E-region critical frequency is an evident effect.

No doubt that the increased stratospheric winds are an inquiry of mesospheric heating in the polar region. But the concrete physical mechanism is not evident. We believe the stratospheric ozone surplus in mid latitudes and related stratospheric heating is an effect of ozone transportation from the other regions. The detail picture, however, is not clear.

5. CONCLUDING REMARKS

Without doubt, the geomagnetic storms, heated by particle precipitation the polar regions, and influence the mid and low latitude middle atmosphere. The general way in through the induced mesospheric and stratospheric circulation.

The mesospheric air masses with the changed composition (reached by NO and N_2 and pure by O), a few days after the storm, come to mid latitudes, provoking the significant disturbances. They change the temperature, density and composition.

An induced by the above effects stratospheric circulation (increasing of stratospheric winds) leads to midlatitude stratospheric ozone surplus and to related with this stratospheric heating.

By comparison of these data with other similar disturbances in the middle atmosphere we found the same behaviour of the stratosphere and mesosphere at the time of so called "winter stratospheric heating". No physical explanation of this similarity.

This paper is supported by Grant NZ-511/1995 of National Scientific Foundation of Bulgaria.

REFERENCES

1. Brasseur G.S., Solomon S. Aeronomy of the Middle Atmosphere. Reidel Publ. Company Dodrecht, 1984.

2. Danilov A., E.S. Kazimirovsky, G.V. Vergasova, G.A. Hachinyan. The Meteorological Effects in the Ionosphere. Monog. (in Russian), Gidrometeoizdat, 1987.

3. Mitra A.P. Ionospheric Effects of Solar Flares, Reidle Publ. Comp. Dodrecht, 1974.

4. Kazimirovsky E, V.D. Kokourov. Ionospheric.Dynamics. (in Russian). Nauka, 1979.

5. Nesterov G. The Absorption and Morphology in the Lower Ionosphere. BAN, Sofia, 1986.

6. Stolarsky R.S. J. Atm. Terr. Phys., 38, 1976, 963

7. Stolarsky R.S., P.B. Hays, R.G. Roble. J. Geoph. Res., 80, 1975, 2266.

8. Killen T.L., P.B. Hays, N.W. Spencer, L.E. Wharton. Adv. Sp. Res., 2, 1983, 133.

9. Chamberlain J. Physics of the Aurora and Airglow. Acad. Press New York, London 1961

10. Fishkova L.M. The Night Airglow on the Earth Midlatitude Upper Atmosphere. 1983, Tbilissi.

11. Gogosheva Ts.N., Gogoshev M.M. Bulgarian Geophys. Journal, vol. 16, 1990, N3, 37.

12. Gogosheva Ts., M. Gogoshev, K. Serafimov. On a Self Consistent Method for Determining Some Smoll Components in the Lower Thermosphere and in the Mesosphere. Bulg. Geophys. Journal, vol. 19, 1993, 38.

VARIATIONS OF STRATOSPHERIC OZONE AT THE TIME OF EXTREMELY HIGH SOLAR ACTIVITY

Mitko Gogoshev, Boyan Petkov, Tsvetana Gogosheva
Institute of Astronomy and National Observatory
Bulgarian Academy of Sciences
Stara Zagora-6000, P.O.Box 442, Bulgaria

ABSTRACT

Many investigators have found the 5-7% trend, related with 27-day and 11-year solar activity cycles by means of the analyses of the Total Ozone Content (TOC) data. In this paper we have investigated the possible TOC variations at the time of extremely high solar activity when $F_{10,7}$ is about 350. Such unobserved values could be reached (in the future or in the past) at the time of supercenturial solar cycle. From the climatic point of view this high activity could influence the climate through the stratospheric ozone. For this purpose we have constructed a model of solar flux in the region 1216-3300 A, based on the ratio of L_{α}(max/min) equal to 2 given by Lean, 1982 (ref. in WMO, 1985). Mainly experimental data has been used. The variation of stratospheric ozone has been calculated by means of a stationary photochemical model. The maximal value of $F_{10,7}$ has been taken as 350. The main results from the modelling are: 1) 5% increase of TOC from minimum to maximum solar activity ($F_{10,7}$ 100-200) is due basically to the photochemistry; 2) At $F_{10,7}$ equal to 350, the calculated photochemical increase of TOC is about 9%; 3) In all diapason of data for $F_{10,7}$ (from 100 to 350) the increase of TOC follows the law $\Delta TOC \sim (\Delta F_{200})^{1/2}$, where ΔF_{200} is the flux of wavelength of 2000 A; 4) The existing reverse negative interrelations between the TOC and temperature could only decrease the ozone growth. The main conclusion is that the extremely high solar activity cannot dramatically influence the behavior of stratospheric ozone and is not compatible with anthropogenious factors.

KEYWORDS: stratospheric ozone, UV-B radiation, solar activity, modelling.

1. INTRODUCTION

By applying cross spectral analysis to daily and monthly total ozone records, after removal of well known natural contributors as volcanic and QBO signals, Zerefos et all, 1995 found an 1-2% amplitude of solar activity influence. Almost similar results has been shown in Fleming et al (1994), who found a 3,5% increases from minimum to maximum. But it is necessary to stress on the fact, that all the investigators usually have used $F_{10,7}$ index as a main indicator of solar activity. In the context in this approximation a full correspondence of UV-variations to $F_{10,7}$ changes has been used. But Hoegy at al, 1993, have reported, based on the Pioner-Venus space probe data, an 20% excess of total EUV radiation over $F_{10,7}$ three-month average values. The authors cannot discriminate different parts of ultraviolet radiation in their integral EUV-flux, as it has been reported, due to the technical reasons. But we can interpret these data as follows. There is no doubt that $F_{10,7}$ emission corresponds adequately to those part of solar EUV radiation, which is closely related with the active regions of solar activity - flares, spots, protuberances, spicules etc. It is well known at the same time that a part of UV radiation, emitted by the chromosphere, is due to the dissipation of the energy of internal gravity and acoustic waves, generated in the photosphere, and some time - out of the active

NATO ASI Series, Vol. I 53
Atmospheric Ozone Dynamics
Observations in the Mediterranean Region
Edited by Costas Varotsos

regions. Hoegy at al, 1993, said for this part "radiation from background region". Finally, we can assume an excess of UV-radiation even in the part of UV spectrum from 1000 to 3000 A prevailing the $F_{10,7}$ intensity with about 20%, not taken into account until now in the analysis of total ozone content fluctuations.

Consequently, the solar activity contributions of 1-3 per cents in the ozone changes, usually accepted now by scientific community, can be underestimated in the light of observed in the 22 solar cycle overflow in UV radiation and unpredictable possibly high radiation in the next cycle as well.

2. EMPIRICAL MODEL OF SOLAR SPECTRUM (1760 - 3200 A) VARIATIONS

In order to reconstruct the solar UV spectrum at different level of $F_{10,7}$, we investigate the behaviour of L_α . This frequently measured line shows a wide amplitude of fluctuations from minimum to maximum of 11-year cycle. We neglect such strong enhancement as 6 time, reported by some authors, but 2 time amplitude (Lean, 1991) seems to be more reasonable. In the same paper, an amplitude of 9% for the whole spectrum for 2000 to 3000 A also given. But if we take into account, that at wavelength 3000 A the possible increases is only 2%, for the wavelength 2000 A we can assume an increases from minimum to maximum of 2000 A equal to 25%. Such comparatively high value has been also taken by Lean, 1984 (ref. in WMO, 1985). We believe this magnitude of fluctuation of 25% would be more reasonable in the light of observed EUV overflow radiation in XXII-cycle , as it was reported above. These 25% at 2000 A are the reference point in creation of our models.

On Fig 1 on the lower panel is shown the reference spectrum at $F_{10,7}=150$ (moderate activity) as it has been accepted in WMO-85 report, and on the top panel - the corresponding increases in per cents for growing level of $F_{10,7}$. On Fig.2 the calculated spectra from $F_{10,7}=150$ to $F_{10,7}=350$ in the same spectral range are given. The convergence point (no variations of the intensity of UV-radiation according $F_{10,7}$ fluctuation) has been taken at λ =3200 A. The highest deviation in this spectral range is those at λ=1750 A.

3. PHOTOCHEMICAL MODELING OF OZONE DISTRIBUTION AND TOTAL OZONE CONTENT (TOC) AT DIFFERENT LEVELS OF SOLAR ACTIVITIES

The first photochemical theory concerning the stratospheric ozone distribution was made by Chapman (1930). Since then there has been a lot of investigators who have developed this theory (Wulf and Deming, 1936; Craig, 1950,; Dutsch, 1956).

According to the photochemical theory the forming of the stratospheric ozone is mainly due to the processes as photodissociation of oxygen molecule and following recombination of atomic and molecular oxygen (London and Prabhakara, 1966). The basically photochemical reactions are:

$$O_2 + h\nu \dashrightarrow O + O$$

(1)

$$O + O + M \xrightarrow{k_1} O_2 + M$$

$$O + O_2 + M \xrightarrow{k_2} O_3 + M$$

$$O_3 + h\nu \dashrightarrow O_2 + O$$

$$O + O_3 \xrightarrow{k_3} 2O_2$$

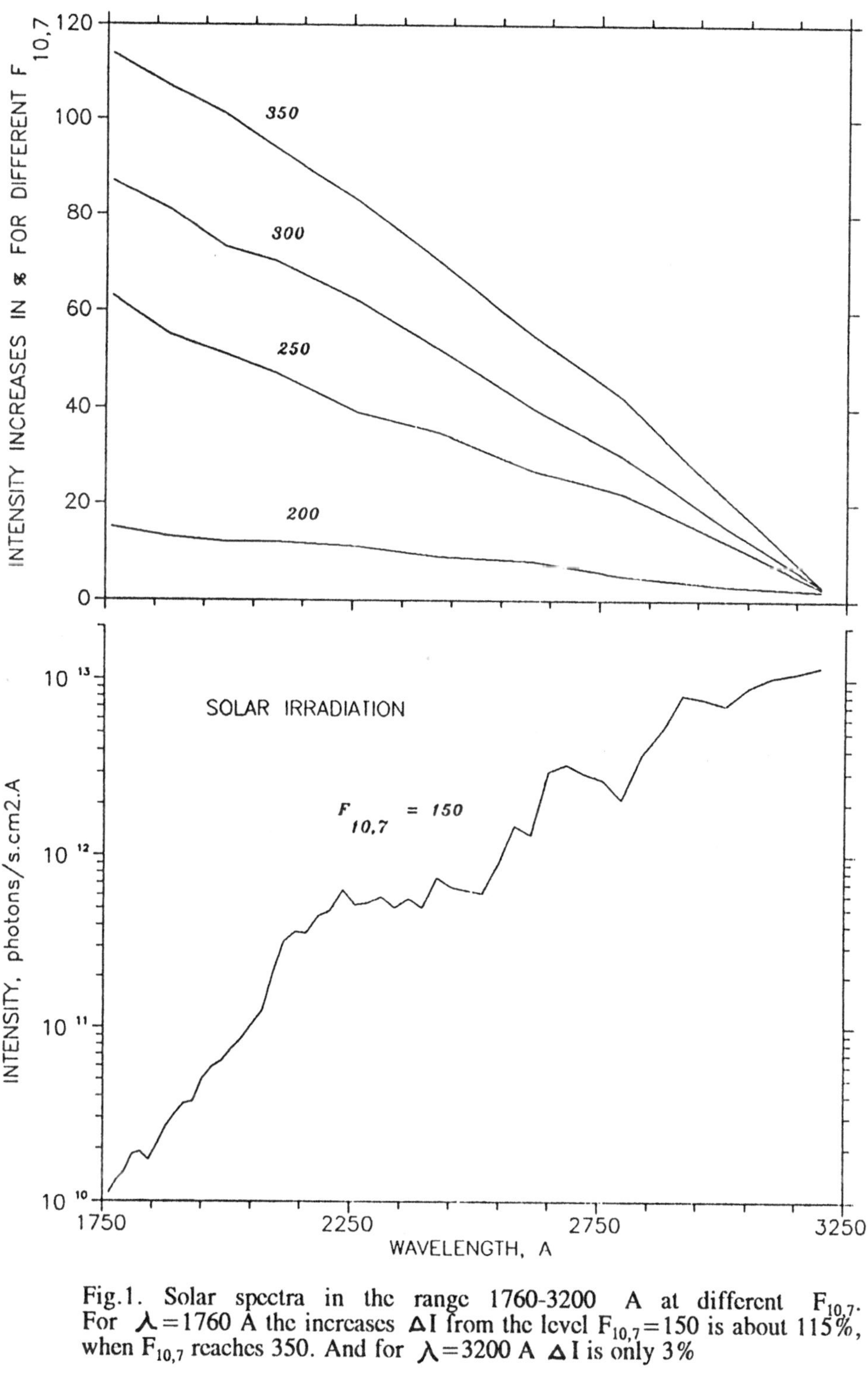

Fig.1. Solar spectra in the range 1760-3200 A at different $F_{10,7}$. For $\lambda=1760$ A the increases ΔI from the level $F_{10,7}=150$ is about 115%, when $F_{10,7}$ reaches 350. And for $\lambda=3200$ A ΔI is only 3%

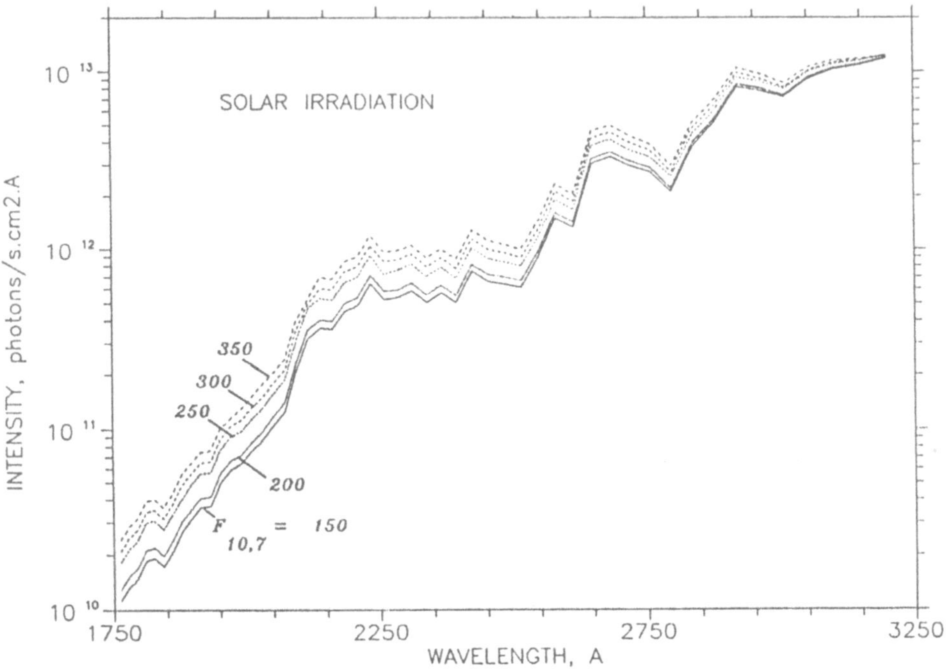

Fig.2. Model of the solar spectrum in the range 1752- 3200 A as a function of $F_{10,7}$.

where $h\nu$ is a quantum of energy necessary for dissociation of oxygen or ozone, k_1, k_2, k_3 are coefficients defining the speed of reactions, M is the third partner.

This model leads to the following expression about the ozone concentration n_3 at given altitude (London and Prabhakara, 1966):

(2) $$n_3 = 2.25 n_2^2 [(f_2/f_3) k n_3]^{1/2}$$

where n_2 is a molecular oxygen concentration at this altitude; f_2, f_3 (number of solar radiation quanta absorbed by O_2 and O_3) and k are given as

(3) $$f_2 = \int_\lambda \sigma_2 q_{\infty\lambda} \exp(-\sigma_2 N_2 - \sigma_3 N_3)\, d\lambda$$
$$f_3 = \int_\lambda \sigma_3 q_{\infty\lambda} \exp(-\sigma_2 N_2 - \sigma_3 N_3)\, d\lambda$$
$$k = k_2/3k_3 = 1{,}9.10^{-23} \exp(2282/T)\ \mathrm{cm^3/molecules}$$

where σ_2 and σ_3 are absorption cross sections of oxygen and ozone, $q_{\infty\lambda}$ is the intensity of solar radiation at the border of the atmosphere; N_2 and N_3 are the column density of molecules of oxygen and ozone above given level at solar direction; T is the temperature. The total ozone content (TOC) has been calculated as

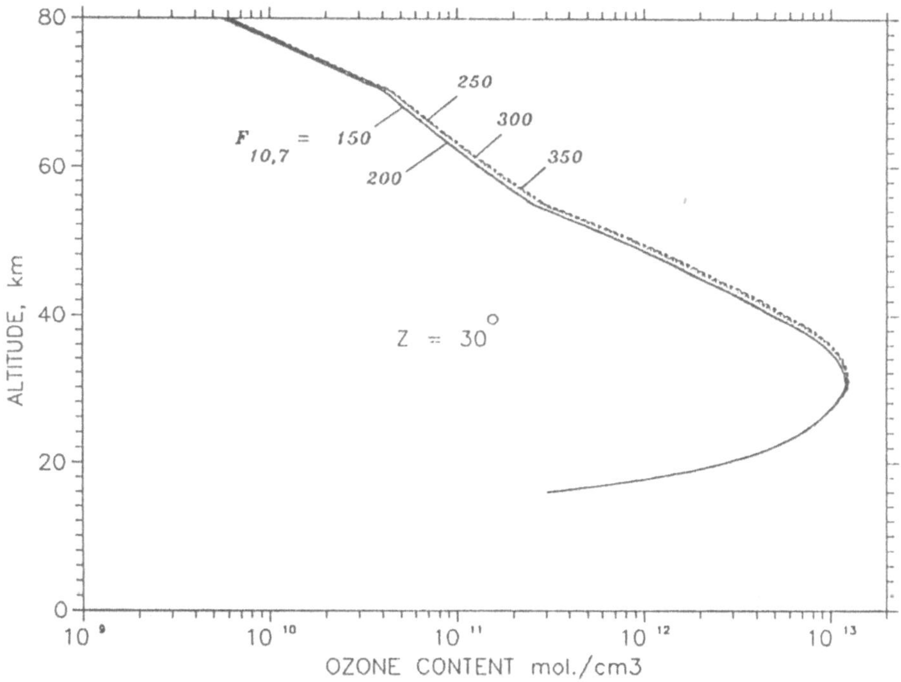

Fig.3. The theoretical profiles of ozone production as a function only of UV-increases. As it is seen comparatively higher production is observed in the upper part of the profiles.

(4) $$TOC = \int_V n_3 dv$$

where V is the volume of an atmospheric column with an unit section.

This simple photochemical model has been used for estimating the solar activity influence on the ozone contents. The above explained solar spectrum intensity, in the wavelength region 1760 - 3200 A responsible for producing of the ozone has been used. The cross sections σ_2 , σ_3 has been taken from WMO-85 report, while the temperature profile and concentration n_2 from Banks and Kockarts G, 1973].

4. MAIN RESULTS

On Fig.3 we show different theoretical profiles, calculated by above mentioned model, for solar zenith angle (SZA) of 30 degrees and for solar activities from $F_{10,7}=150$ to $F_{10,7}=350$

As it is seen the photochemical excess, produced by increasing solar activity, reflects mainly on the region of maximum and above of the emitting layer.

The calculated TOC at SZA of 30 deg show that the photochemical increases for $F_{10,7}=250$ (a normal level for the maximum of 11-year cycle) is about 5%. Here we would like to stress that these are results only by the photochemistry. In the simple photochemical model, used above, no others processes leading to the ozone dynamics are taken into account. In this sense the value of 5% is more than twice grater in comparison with the values of few per cents received by other authors (see

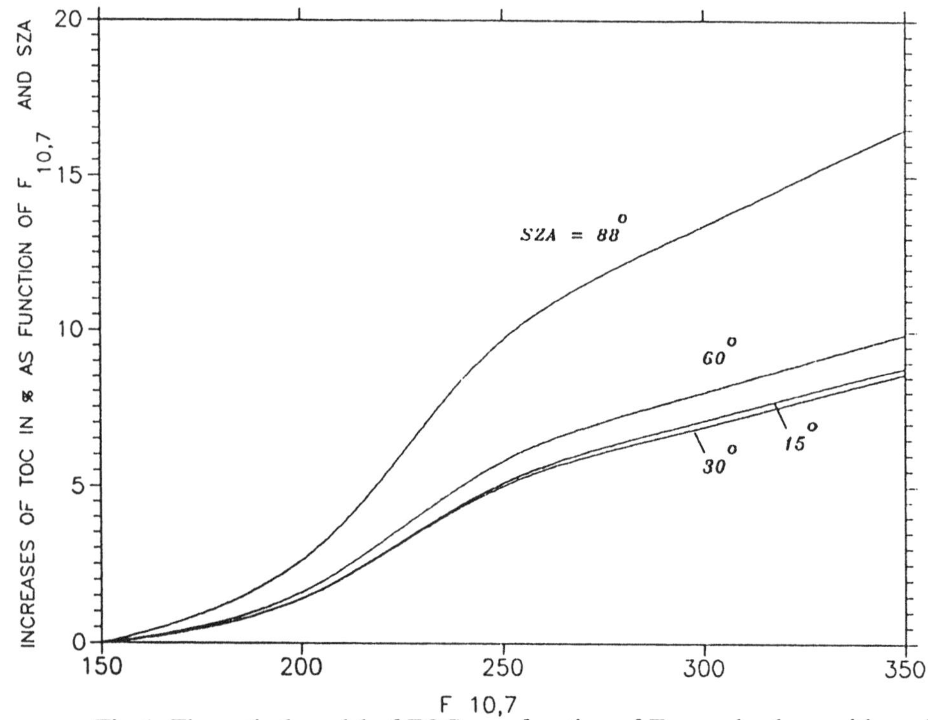

Fig.4. Theoretical model of TOC as a function of $F_{10,7}$ and solar zenith angle (SZA). As it is seen in the range of $F_{10,7}$ from 150 to 250 we have a comparatively sharp increases of TOC. After $F_{10,7}=250$ the curves show a saturation of TOC. This fact can be interpreted as a falling of the dependence of TOC on solar activity

introduction).

For the solar activity levels of $F_{10,7}=300$ and $F_{10,7}=350$ (extremely high solar activity.

activities), the photochemical excess riches the values of 7% and 9% consequently. 30 degrees of SZA is chosen as a typical for mid latitudes in summer time.

On Fig.4 the dependence of TOC increases (a photochemical excess) as function of both - solar activities and SZA, is given. This excess is higher at big SZA. But we can pay attention to an interesting fact. This is the tendency for saturation of the curves on $F_{10,7}$. That means that even such high level of solar activity as $F_{10,7}=350$ or higher cannot influence significantly the variations of stratospheric ozone.

Let us now investigate the dependance of photochemical excess, as function of increasing of solar irradiation (following the solar activity). We found that solar flux at a wavelength of 2000 A is very important for both - photolysis of stratospheric ozone and oxygen. That is way we can use this line as an indicator of possible changes in ozone photochemistry. From the results, obtained in these investigations we found the following relationship:

(5) $$\Delta\, TOC_{ph} \sim (\Delta\, F_{200})^{1/2} \text{ (in per cents)}$$

ΔTOC_{ph} means the excess of total ozone production by the photolysis of stratospheric oxygen. ΔF_{200} is increases (in per cent) of solar irradiation at wavelength 2000 A.

The above approximation is useful for physical estimations of ozone fluctuations due to solar activity.

5. DISCUSSION

We calculate the photochemical excess in ozone production by constructing the solar spectrum in the range 1760-3200 A at different level of solar activity. Two crucial points at the reconstructing the solar irradiation exist in this investigations. They are a 200 per cents increasing of L and 25 per cents increasing of 2000 A from minimum to maximum in the 11-year variations. These assumptions follows the Lean 1982, 1984 (ref. in WMO, 1985) models. They are very "optimistic" one in comparison with more moderate 60 per cent for L_α and 9 per cent for λ=2000 A observed by Solar Mesosphere Explorer Satellite in the period from 1982 to 1988 (Kondratyev and Varotsos, 1995). A most of authors (Chandra, 1991; Chandra and McPeters, 1994) usually have used these low variations of solar radiation and as a results they have obtained only a few per cent solar activity variations in the stratospheric ozone.

But the Pioner-Venus EUV integral flux in the maximum of XXII solar cycle (1990-1992) shows a significant excess of about 20% in total EUV-radiation. As we mentioned above this extra flux can be attributed also to the long-wave solar UV radiation and especially to the range of very important for ozone production wavelengths. In this sense the accepted by us increases of 200 per cent for L and 25 per cents for λ =2000 A seems to be not unbelievable.

Therefor an increases of TOC, as a results of increased photoproduction, in the range of 5 per cent from minimum to maximum of 11-year solar cycle seems to be very reasonable. Also an excess of 9 per cents total ozone increases for never observed, but only theoretically postulated $F_{10,7}$=350 value looks reasonable.

An important empirical dependance, given above, and denoted (5), gives a practical possibility to estimate the TOC changes as function of λ =2000 A solar irradiation excess. Shortly, it can be explained as a square root dependance of TOC from λ =2000 A solar flux. By the way, a few per cent increases of TOC from minimum to maximum also perfectly confirms this relationship, taking into account a 9 per cent increases of λ =2000 A.

6. CONCLUDING REMARKS

An extraflux of solar radiation in the range of 1760 A to 3200 A, responsible for ozone photoproduction is discused. Such excess of solar EUV-flux was observed at the time of unusual maximum of solar cycle N22 (1990-1992) by Pioner-Venus space probe. These observations give us the possibility to define an increasing of solar flux at the upper limit of the Earth atmosphere as 25% for λ=2000 A and 3% for λ=3200 A. In the case of extremely high solar activity these increases look quite resalable.

By the simple photochemical model we calculate the ozone vertical profiles and TOC as well. An extra volume of 5% from maximum to minimum is calculated. At the time of extra activity ($F_{10,7}$=350) this excess reaches a value of 9%. These results can be expressed as a square root dependance of TOC from solar irradiation of 2000 A increases.

A very important consequences from our investigations are the results showing a saturation of TOC increases at the time of very high solar activities. This only confirm that solar activity cannot be accepted as a competitor in the dramatic ozone depletion, observed in the last decade.

We believe our results are not in contradiction with the solar activity variations in stratospheric ozone, observed by other authors.

This paper is supported by the Grant NZ-511/1995 with the National Scientific Foundation of Bulgaria.

REFERENCES

Banks P.M. and G. Kockarts. Aeronomy, part A, Academyc press, New York and London, 1973.

Chandra S. The solar UV related changes in total ozone from a solar rotation to a solar cycle, Geoph. Res. Lett., 18, 1991, 837.

Chandra S.and R.D. McPeters. The solar cycle variation of ozone in the stratosphere inferred from the Nimbus-7 and NOAA-11 satellites. JGR, 100 (1995) in press.

Chapman S.A. A theory of upper atmospheric ozone. Mem. Roy. Meteorol. Soc., 3, 1930, 103.

Craig R.A. The observation and photochemistry of atmospheric ozone and their meteorological significance. Meteorol. Monographs, 2, 1950, 50.

Dutsch H.U. Das atmospharische ozon als indikator fur stromungen in der stratosphare. Arch. Meteorol. Geophys. u. Bioklimat. A, 9, 1956, 87.

Fleming E.L., S. Chandra, C.H. Jackman, D.B. Considine, A.R. Douglas. The middle atmospheric response to short and long term solar UV variations: Analysis of observations and 2D model results. J. Atm. and Terr. Phys., 1994 (in press)

Hoegy W.R., W.D. Pesnell, T.N. Woods, G.J. Rottman. How active was solar cycle 22? Geoph. Res. Lett., vol. 20, 1993, 1335.

Kondratiev K.Ya., C.A. Varotsos. Atmospheric ozone variability in the context of global change. Int. J. Remote Sensing, vol. 16, 10, 1995,1851.

London U., K. Prabhakara. Absorbing processes in stratosphere and mesosphere.In the book "Ozone in the earth atmosphere", S.Peterburg, 1966 (in Russian).

WMO Report. Atmospheric ozone. Assesment, vol.1, 1985.

Wulf O.R., L.S. Deming. The theoretical calculation of the distributions of photochemically-formed ozone in the atmosphere. Terr. Mag., 41, 1936, 299.

Zerefos C.S., K. Tourpali, V. Fioletov. Future studies on solar activity - ozone relationship. Paper, presented on International Conference on Ozone in lower stratosphere. Halkidiki, Greece, 15-20 May 1995.

Total Ozone Variations and Meteorological Processes

Kjell Henriksen[1] and Valentin Roldugin[2]

[1] The Auroral Observatory, University of Tromsø, N–9037 Tromsø, Norway
[2] Polar Geophysical Institute, Apatity, Murmansk Region, 184200, Russia

Keywords. Ozone distribution, vertical transport, cooling and warming of the lower stratosphere

Introduction

Ozone is a relatively stable trace gas with highest density in the lower stratosphere. As the air moves around, the ozone is transported, and the distribution changes especially when vertical mixing takes place, since the steepest ozone gradients exist in the vertical direction.

The basic production mechanism for ozone in the atmosphere was ruled out by Chapman (1930) indicating also some loss processes. It seems that the loss processes quoted by Chapman (1930) are insufficient to balance the natural production rate (Johnston, 1975), and presently much research is concentrated on the part played by several photochemical and chemical reactions destroying stratospheric ozone. Almost from the initiation of the ozone research at Oxford in 1920's it has been evident that tropospheric weather has a significant influence on stratospheric ozone distribution (Dobson, 1968).

Presently the Antarctic ozone hole is a well-known phenomenon and the related ozone depletion can be closely related to low stratospheric temperatures and polar stratospheric clouds (Wayne, 1991). Ozone depletion related to stratospheric clouds was clearly notified in Norway during the 1940's. Then Langlo (1952) describes 20 observations of mother of pearl clouds and found on the average an ozone decrease of 27 D.U. on the days with mother of pearl clouds. It was also experienced that this average value was much dominated by a few large negative deviations from the annual mean curve, and it is likely that polar stratospheric clouds and mother of pearl clouds are closely related and appears as a consequence of low stratospheric temperature.

More recently Rabbe and Larsen (1992, 1995) have studied the effect of horizontal and vertical motion in the ozone layer, and below their phenomenological model is reproduced, see Fig. 1. Ascending motion through the tropopause will bring ozone-poor air into the stratosphere, and thereby attenuating the ozone

NATO ASI Series, Vol. I 53
Atmospheric Ozone Dynamics
Observations in the Mediterranean Region
Edited by Costas Varotsos

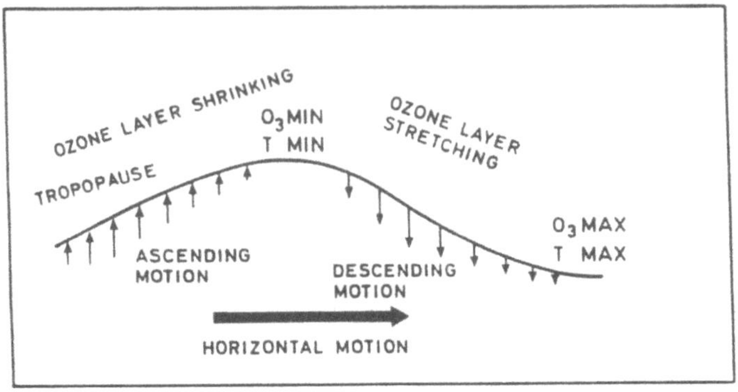

Fig. 1. Phenomenological model, relating vertical air motion to variations in total ozone (Rabbe and Larsen, 1992)

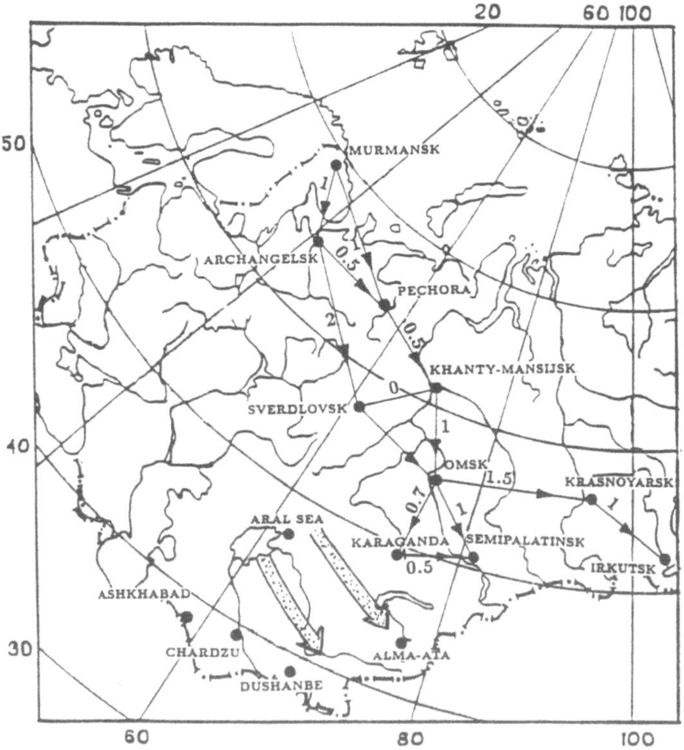

Fig. 2. Map of ozonometric stations, used in this study. The wide dotted arrows show direction of the ozone movement, observed by the Middle-Asian stations. The lines between the stations mean that the cross-correlations of these pairs have distinct maxima. The direction of movement is indicated by the arrows and the associated numbers give the time delay between the stations in units of days

layer. On the other hand, descending air from the upper stratosphere will bring ozone-rich air into the ozone layer, enhancing the density of the ozone layer. The vertical motions are related to horizontal advection, and therefore tropospheric weather will to some extent always have some effect on the ozone layer.

Correlation Studies of Total Ozone Variations

Using data from several of the 56 ozone stations in the former USSR from 1978, 1979, and 1986, horizontal movements in the ozone layer were quantified through cross-correlation analysis (Roldugin and Henriksen, 1995). In Fig. 2 the stations used in this analysis are given, and disturbances are found to shift between the stations in the direction of the arrows. The numbers give the temporal shift in days. The weather pattern of this area shows a similar drift pattern as the ozone drift in Fig. 2.

In Fig. 3 the ozone records from most of the stations in 1986 are shown. Superposed on the clear seasonal variation, amplitude disturbances of a few tens of DU appear. These records are treated by cross-correlation analysis, and periodic disturbances of 15–25 days duration are visualized.

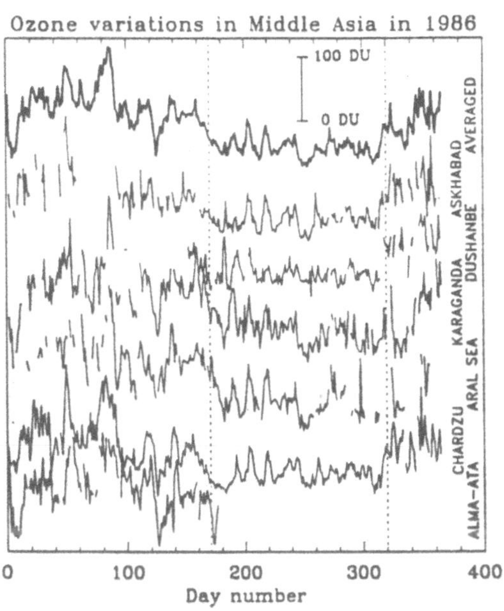

Fig. 3. Total ozone variations at six Middle-Asian stations in 1986 and their averaged curve. The vertical dashed lines mark the interval used for detailed cross-correlation analysis

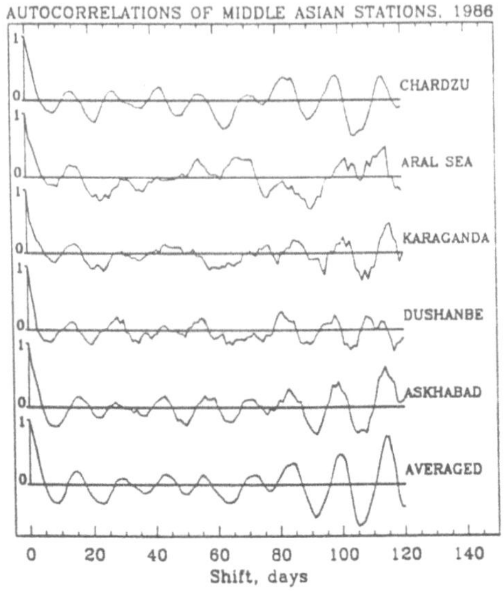

Fig. 4. Autocorrelation functions of the total ozone variations for six Middle-Asian stations and of their averaged value. The period of the treated data is indicated by the dashed vertical lines in Fig. 3

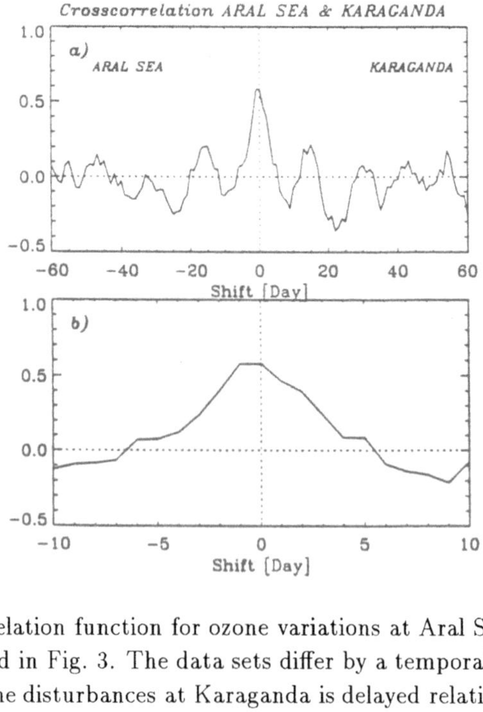

Fig. 5. Cross-correlation function for ozone variations at Aral Sea and Karaganda for the period indicated in Fig. 3. The data sets differ by a temporal shift of 0.5 day, indicating how much the disturbances at Karaganda is delayed relative to the disturbances at Aral Sea

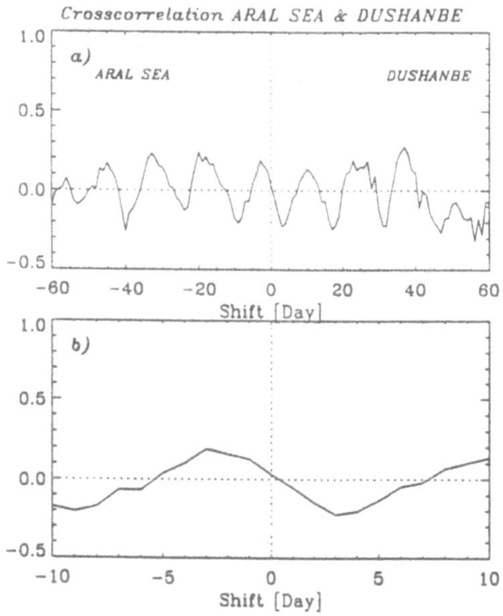

Fig. 6. Similar analysis as shown in Fig. 5, using data from Aral Sea and Dushanbe, showing a delay of about 3 days for disturbances at Dushanbe relative to the disturbances at Aral Sea

The temporal shift of disturbances arriving at the different stations as shown in Fig. 2 is obtained by cross-correlation analysis. By autocorrelation analysis five different records and their averaged value periods of 15–25 days in total ozone appear, see Fig. 4. In Fig. 5 the results of the cross-correlation analysis of the records from Aral Sea and Karaganda is shown and a shift of 0.5 day between Aral Sea and Karaganda appears. The disturbances transfer from Aral Sea towards Karaganda. Similar analyses are carried out for all the station pairs connected by lines and arrows in Fig. 2, indicating the speed and direction of horizontal movements in the ozone layer.

Vertical Movements

In order to study correlations between vertical air movements and variations in total ozone, ozone data from three Soviet stations in Middle Asia were used (Henriksen and Roldugin, 1995). The stations are Aral Sea (46.8°N, 61.7°E), Chardzu (39.1°N, 58.3°E), and Ashkhabad (38.0°N, 58.3°E). The daily ozone values from these stations were averaged for each day, and this mean value is representative for this area. This data set was compared with heights and temperatures of the 100 and 500 hPa levels at the centre of the triangle (43°N,

60° E) on the southern shore of Aral Sea, Fig. 7. The meteorological data were taken from the daily weather maps of the European Meteorological Bulletin, issued by Deutscher Wetterdienst.

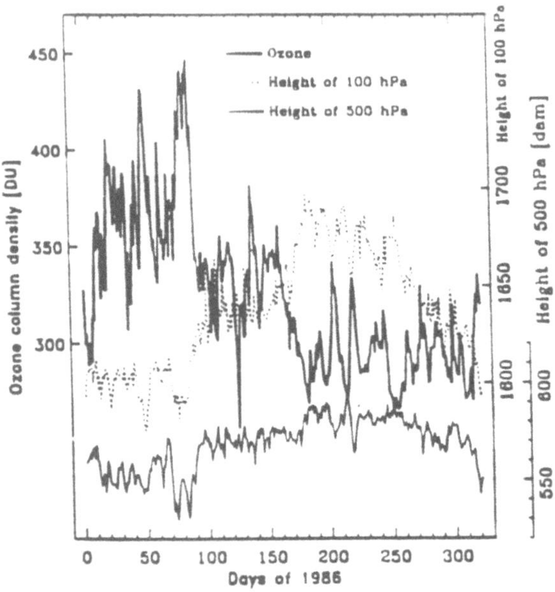

Fig. 7. Time variations of total column ozone, averaged for three ozone stations around Aral Sea. Heights of the 100 and 500 hPa levels for the region of Aral Sea in 1986 are also shown, and heights are given in decameters [dam]

When the seasonal variations in the ozone data are eliminated the shorter variations of 15 to 25 days remain, see Fig. 8. An eye inspection of the variations indicates anticorrelation between the height and ozone variations.

Simultaneous variations in total ozone and temperature at the 100 hPa level are shown in Fig. 9, and by eye inspection some correlation between these variations seems to exist. Cross-correlation analysis between total ozone variations and temperature variations at the 100 and 500 hPa levels are carried out both with deseasonalized ozone data and non-reduced ozone data.

Cross-correlation analysis between temperature and ozone variations may reveal and any temporal shift between these variations, and the result of this analysis is shown in Fig. 10, showing no shift at all. It must be realized that the sampling period of these data is one day, giving the threshold of temporal resolution in this analysis.

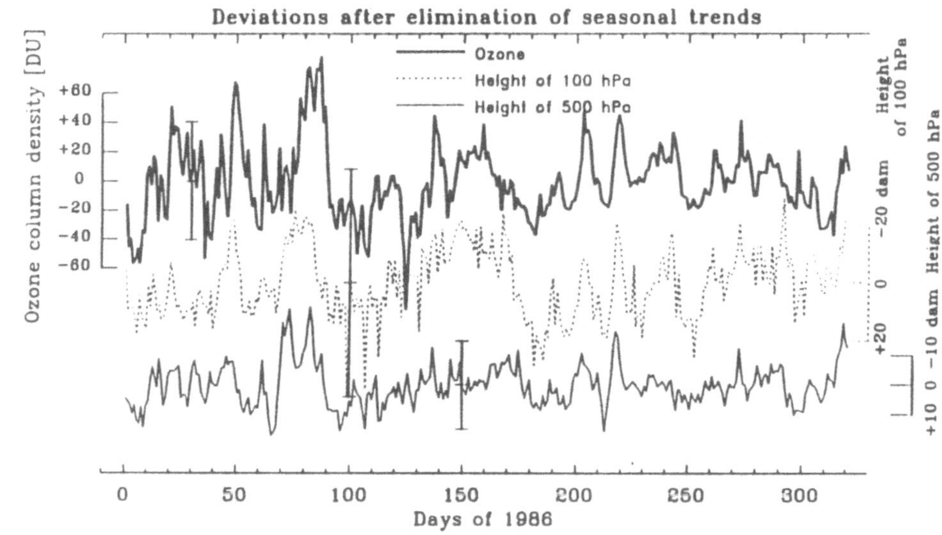

Fig. 8. The same records as in Fig. 7 after the elimination of the seasonal variation in total ozone and heights

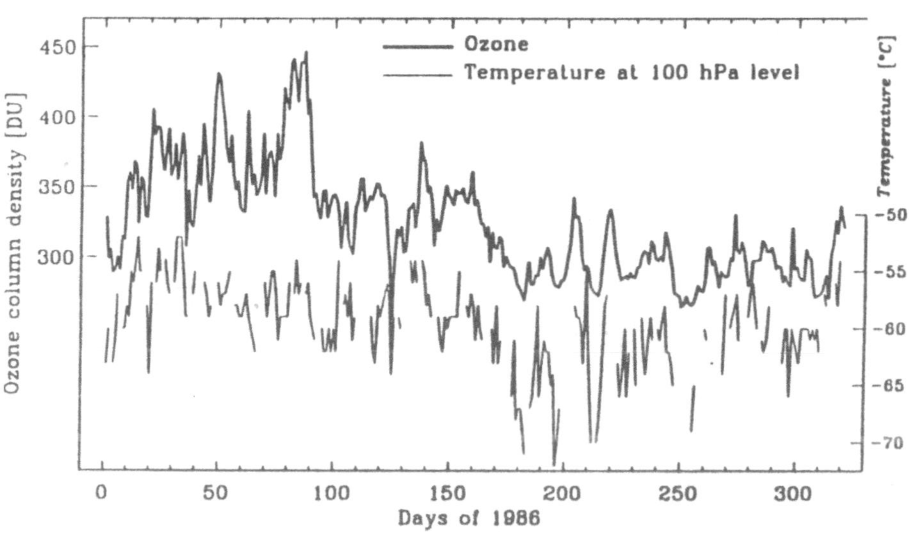

Fig. 9. Temporal variations of the total ozone and temperature at the 100 hPa level in Middle Asia, using total data sets

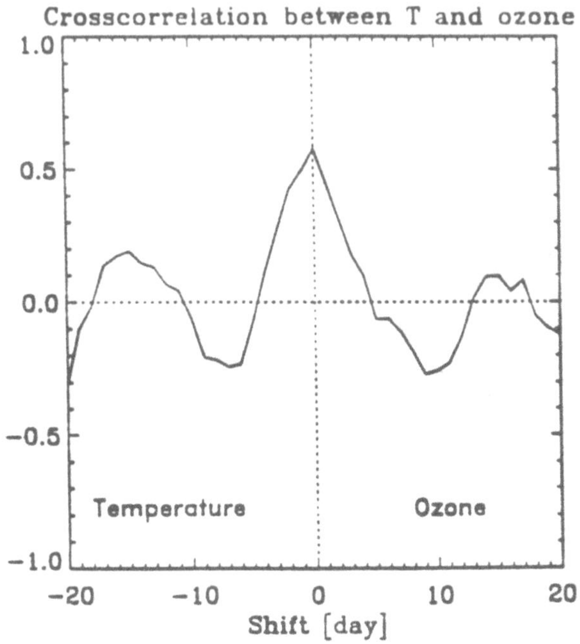

Fig. 10. Cross-correlation curve for total ozone and temperature variations in Middle Asia from 19 June to 16 November, 1986, using total data sets

Table 1. Correlation coefficients for variations of total ozone content, and heights and temperatures at the 100 and 500 hPa levels. Values in up-right part of the table (bold) are calculated for the complete data, without subtraction of slow seasonal variations, and the numbers in down-left part are for the fast variations only

	Ozone	Height of 100 hPa	Height of 500 hPa	Temperature at 100 hPa	Temperature at 500 hPa
Ozone	*1.00*	**−0.70**	**−0.75**	**0.61**	**−0.75**
Height of 100 hPa	−0.49	*1.00*	**0.90**	**−0.63**	**0.91**
Height of 500 hPa	−0.53	0.63	*1.00*	**−0.66**	**0.93**
Temperature at 100 hPa	0.39	−0.39	−0.42	*1.00*	**−0.62**
Temperature at 500 hPa	−0.57	0.64	0.77	−0.34	*1.00*

The results of the cross-correlation analysis are given in Table 1. It shows higher correlation coefficients when the full ozone data is used than the de-seasonalized data.

Concluding Remarks

The cross-correlation analysis of the ozone data from the former Soviet Union indicates a pattern in the horizontal movements in the ozone layer, and this pattern is similar to the weather pattern found in this region. This similarity gives an indication that the drifts in the ozone layer and stratosphere is somehow powered by tropospheric motions, since there is much more energy and power involved in tropospheric than in stratospheric movements.

Vertical movements are inferred from height variations of the 100 and 500 hPa levels. Table 1 shows that downward motion results in decreasing ozone and vice versa, in agreement with the phenomenological model of Rabbe and Larsen (1992). At 100 hPa there is a positive correlation between temperature and ozone variation, indicating an ozone increase when high temperature, ozone-rich air from the upper stratosphere penetrates into the lower stratosphere. On the other hand, when low-temperature, ozone-poor tropospheric air penetrates into the ozone layer, an anticorrelation between the total ozone and the temperature at 100 hPa is expected. However, the correlation at the 500 hPa level is opposite since downward motion bring cold, but relatively ozone-rich tropopause air into the middle troposphere, the 500 hPa level. Similarly and upward motion at the 500 hPa level brings higher temperature air, but relatively ozone-poor air from the lower troposphere into the middle troposphere.

The near simultaneity between ozone and temperature variations indicates that the fast variations with duration between 15 and 25 days are generated through dynamical motions. On the other hand, the meteorological and ozone data used in this study do not give any indication that the ozone variations are related to chemical processes, but strongly evidence the meteorological control of fast ozone variations.

References

Chapman, S., Theory of upper-atmospheric ozone, *Memoirs of the Royal Meteorological Society*, *3*, 103–125, 1975.

Dobson, G. M. B., Forty years' research on atmospheric ozone at Oxford: a history, *Applied Optics*, *7*, 387–405, 1969.

Henriksen, K. and V. Roldugin, Total ozone variations in Middle Asia and dynamic meteorological processes in the atmosphere, accepted for publication in *Geophys. Res. Lett.*, 1995.

Johnston, H. S., Global ozone balance in the natural stratosphere, *Rev. of Geophys. and Space Phys.*, *13*, 637–649, 1975.

Langlo, K., On the amount of atmospheric ozone and its relation to meteorological conditions, *Geofysiske Publikasjoner*, *Vol. XVIII*, No. 6, 1952.

Rabbe, Å. and S. H. H. Larsen, Ozone variations in the Northern Hemisphere due to dynamic processes in the atmosphere, *J. Atm. Terr. Phys.*, *54*, 1107–1112, 1992.

Rabbe, Å. and S. H. H. Larsen, On the low ozone values over Scandinavia during the winter of 1991–1992, *J. Atm. Terr. Phys.*, *57*, 367–373, 1995.

Roldugin, V. and K. Henriksen, Horizontal movements in the ozone layer, accepted for publication in *J. Atm. terr. Phys.*, 1995.

Wayne, R. P., *Chemistry of Atmospheres,* Oxford Science Publications, Clarendon Press, Oxford, 1991.

Decrease in Surface Ozone Related to Car Exhaust at High Latitudes

Kjell Henriksen and Sriharan Sivasingarajah

The Auroral Observatory, University of Tromsø, N-9037 Tromsø, Norway

Keywords. Nitric oxides, chemical depletion, solar spectral UV irradiances

It is well-known that car exhaust contains high densities of NO_x as a consequence of chemical destruction. No ozone can be measured inside car tunnels. Ozone is mainly destroyed by the reaction

$$NO + O_3 \longrightarrow NO_2 + O_2$$

However, NO_2 photolysis followed by $O + O_2$ combination is the only known tropospheric source of O_3 (Wayne, 1991). The photolysis takes place for $\lambda \leq$ 400 nm.

At mid and low latitudes local ozone production from NO_2 photolysis is a significant source and mixing ratios of 100–200 ppb can be measured (Colbeck and Harrison, 1985). That is one order of magnitude higher than the mixing ratio normally 20–30 ppb found in the free troposphere.

Enhanced ozone mixing ratio is generally occurring in metropols like Athens and Los Angeles, in the neighbourhood of heavy traffic. At these sites the solar irradiance is normally sufficient to promote the necessary photolysis rate of NO_2.

At high latitudes, however, local ozone production is rather hard to find (Theodorsen et al., 1995), and there only significant depletions of surface ozone can be measured in connection with high concentrations of NO_x.

In Fig. 1 typical records of surface ozone are shown for midsummer and early fall 1995. The records are obtained in Tromsø (70°N) at a measuring site 0.5 km away from the nearest highway and 3 km outside the city centre. The ozonometer is the well-known American instrument Monitor-Labs 8810/M, sampling air at a height of 2 m above the ground. The background ozone mixing ratio is 20–30 ppb, and characteristic ozone decreases can occur at any time of the day. These decreases are most likely generated by high density of NO_x (Theodorsen et al., 1995). In Tromsø the solar UV irradiance may be insufficient to generate significant ozone production, and therefore the chemical loss of O_3 in interaction with NO_x is dominating, producing the characteristic depletions.

NATO ASI Series, Vol. I 53
Atmospheric Ozone Dynamics
Observations in the Mediterranean Region
Edited by Costas Varotsos

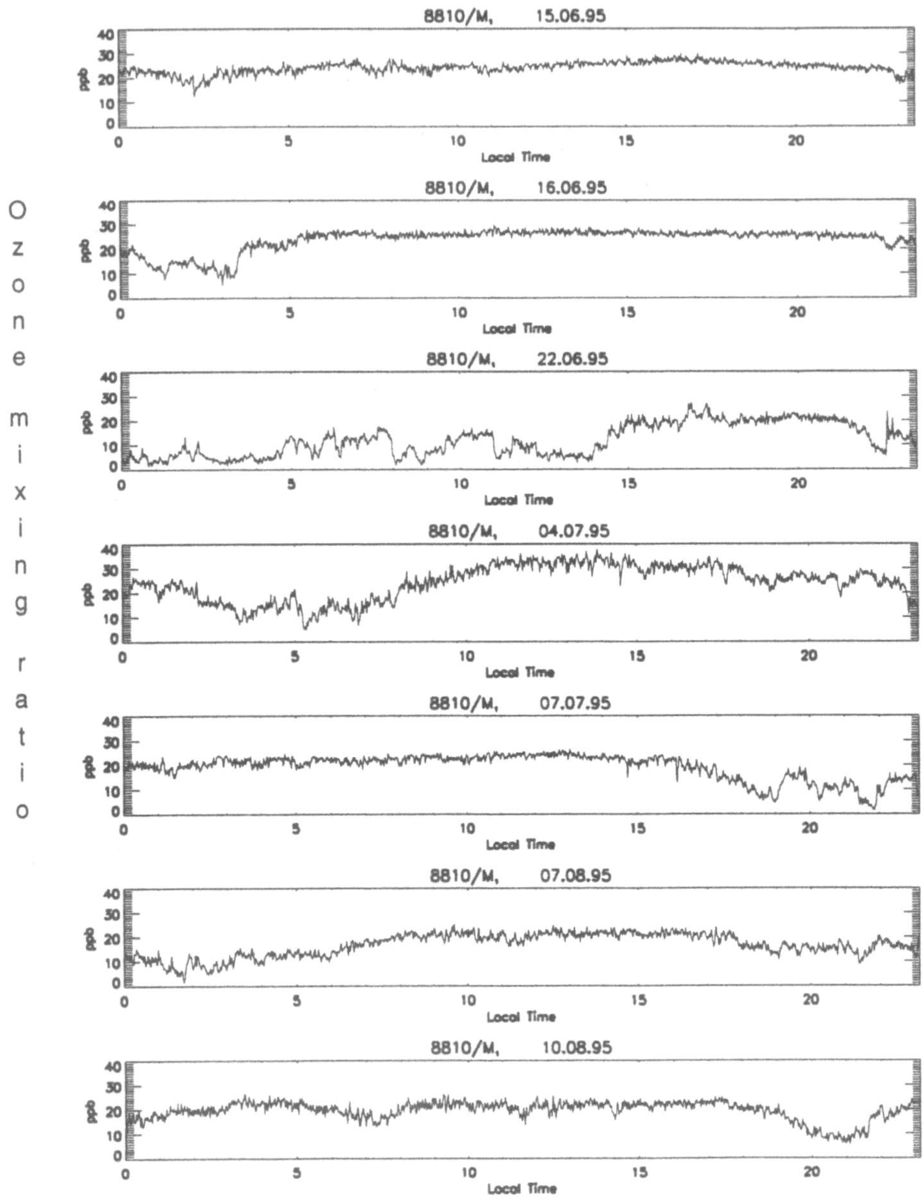

Fig. 1. Records of surface ozone obtained at Tromsø using the ozonometer Monitor Labs 8810/M. The sampling time is 20 sec, and thereby ozone decreases lasting only a few minutes can be visualized. These records are obtained when the sun is highest, but even then no diurnal variation or local ozone production can be seen. The spectral solar UV irradiances are probably too low to generate ozone production via NO_2 photolysis.

The different behaviour of surface ozone at high and low latitudes in connection with enhanced densities of NO_x can be explained from the difference in UV spectral irradiances. It is a future task to measure and quantify the relations, making simultaneous spectral measurements, and in situ measurements of NO, NO_2, and O_3.

References:

Colbeck, I. and R. M. Harrison, The photochemical pollution episode of 5-6 July 1983 in North-West England, *Atmospheric Environment, 19,* 1921-1929, 1985.

Theodorsen, A., K. Henriksen, S. Bersaas, H. Oernes, V. Sirota, and A. Vasiliev, Surface ozone at high latitudes, accepted for publication in *Geophys. Res. Lett.,* Nobember 1995.

Wayne, R. P., *Chemistry of Atmospheres,* Oxford Science Publications, 1991.

The Portuguese Ozone and UV-B Monitoring Stations

Diamantino V. Henriques[1]

[1] Institute for Meteorology, Rua C do Aeroporto, 1700 Lisboa, PORTUGAL

Abstract. The first total ozone measurements in Portugal were carried out in Santa Maria Island (Azores) in 1951 with the Dobson spectrophotometer #13 borrowed by the IOC (International Ozone Commission) in the scope of the IGY (International Geophysical Year) activities. After that, regular total and Umkehr ozone measurements have been made in Lisbon since 1967 with the same instrument. Recently, three stations equipped with automated Brewer spectrophotometers were installed at Funchal (Madeira Is.) in 1989, Angra do Heroismo (Terceira Is. - Azores) in 1992 and Penhas Douradas (Portugal Mainland) in 1994. The Brewer instruments are equipped with UV-B option and perform daily measurements of global UV-B radiation. The actual monitoring program, some results and comparisons between these stations are shown.

Keywords. Total ozone measurements, UV-B measurements

1 Historical review

In 1950, the International Ozone Commission (IOC) asked to the portuguese National Meteorological Service (SMN - Serviço Meteorológico Nacional) for a collaboration in an observational program at Azores Is. In this way, Mr. L. Neves, one of the former meteorologists of the SMN, went to Oxford in order to learn how to operate the Dobson ozone spectrophotometer with Sir Charles Normand. After this short training period, Mr. Neves carried out the first total ozone measurements at Santa Maria Island with the Dobson spectrophotometer #13. However, due to technical reasons, the instrument was moved to Lisbon in 1958, were is operated until now. The total ozone observations at Lisbon began in 1960 with Prof. Mário Figueira who developed the most part of the scientific research on the Ozone Layer field in Portugal until 1991. The first years of operation had some technical problems and some irregularities can be seen in the record until 1975. The regular Umkehr observations started in 1967, and later in 1978, some ozone soundings were made using Brewer-Mast ozone sondes. Continuous surface ozone measurements were also started in Lisbon and in other locations (Castelo Branco and Beja) using Max Planck instruments and then KWS 775 ozone-meters based in the KI method.

NATO ASI Series, Vol. I 53
Atmospheric Ozone Dynamics
Observations in the Mediterranean Region
Edited by Costas Varotsos

Later in 1987, the portuguese National Institute for Meteorology and Geophysics (INMG - Instituto Nacional de Meteorologia e Geofísica), who replaced the first SMN, and the portuguese National Agency for Scientific Research and Technology (JNICT - Junta Nacional de Investigação Científica e Tecnológica) supported the purchase and installation of two Brewer ozone spectrophotometers at Funchal (Madeira Is.) in 1989, and at Angra do Heroismo (Terceira Is. - Azores) in 1992. These instruments are automated and performs total ozone and SO_2, Umkehr and UV-B measurements. Surface ozone photometric UV analyzers were also purchased and installed in both places.

Also during 1983 to 1992, continuosly measurements of erithemally effective UV-B global radiation was performed by Mr. C. Gonçalves et al. using a Robertson-Berger detector. However, due to calibration problems, the data needs to be revised.

More recently in 1994, the portuguese Institute for Meteorology (IM - Instituto de Meteorologia), who replaced the last INMG, supported the purchase of a third Brewer spectrophotometer, that was installed at the Penhas Douradas meteorological Observatory in Portugal Mainland, as well as a surface ozone photometric UV analyzer.

All these instruments are located in meteorological observatories and are daily operated by the meteorological observer personnel of each one. The data is sent to the IM headquarters in a monthly basis, and then it is validated, processed and sent to the World Ozone Data Center. Figure 1.1 shows the geographical location of the four portuguese ozone stations.

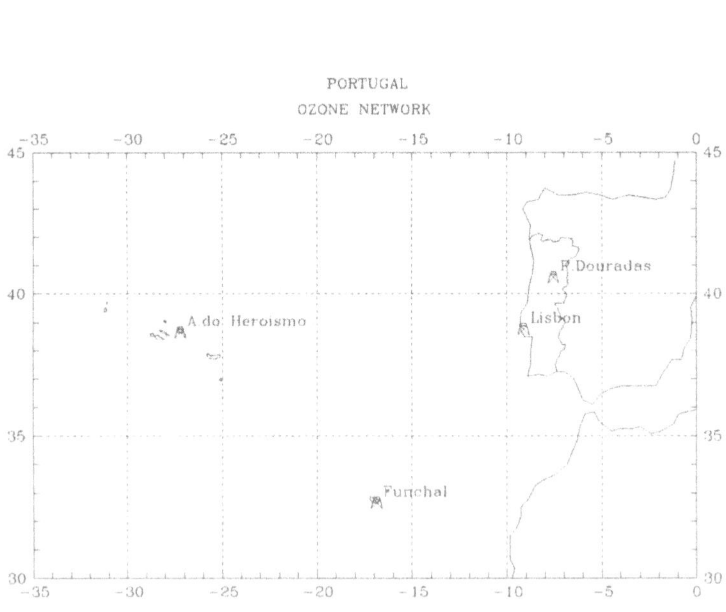

Figure 1.1. Geographical location of the portuguese Ozone/UV-B stations.

2 Objectives

The objectives of these stations are the following:

- To understand the spatial and temporal variability of the atmospheric ozone and UV-B radiation fields over the North Atlantic and West Europe regions.
- To complement the existing satellite data with high quality ground based measurements.
- Implementation of the results in numerical / statistical models for daily forecasting of total ozone and UV-B radiation in Portugal.
- Improvement of the radiative transfer models for UV-B in Portugal.

To match these objectives, a great effort is made by the IM in order to purchase, to install, and to operate the instruments for long term periods. At the same time, some of the records are processed, validated and stored and can be available for scientific research. However, there is a lot of data that needs to be processed, and/or re-evaulated according changes ocurred in the calibration levels of the instruments or in the absorption coefficients. Unafortunately, due to personnel limitations, only few graduated people (1-2) are actually involved in this work and for this reason an international cooperation is needed in order to match these objectives.

3 Location of the stations and the actual observing program

All stations are equipped with ozone spectrophotometers and other ancillary instruments, like surface ozone analyzers, carbon dioxide analyzers, and some of them also with high volume samplers (suspended matter) and automated precipitation collectors, integrated in the BAPMoN program. Table 3.1 shows the actual O_3/UV-B portuguese network characterization.

4 Some results

The ozone stations are equipped also with personal computers that can be used to punch data from manual instruments or to store the data collected from automated instruments and data loggers. Then, the data is saved in floppy disks and sent by mail to the IM headquarters (Lisbon) in a monthly basis. After that, the data is stored properly and processed. The processing consists in a simple control quality program to find and eventually correct some wrong data that can occur. In addition, this processing performs statistical computations for hourly, daily and monthly scales. The following section shows some of the daily results in graphical form.

Table 3.1

Station	Lisbon	Penhas Douradas	Funchal	A. do Heroismo
WMO Nr.	08579	08568	08522	08511
GO_3OS Nr.	082	288	287	327
Latitude	38.77° N	40.42° N	32.64° N	38.66° N
Longitude	9.15° W	7.55° N	16.89° W	27.22° W
Altitude	105 m	1380 m	49 m	74 m
Site type	Portugal mainland: coastal region, moderately polluted site close to the international airport and to the Tejo river.	Portugal mainland: Mountain region far from pollution sources and with near clean air conditions.	Madeira Is.: Moderately polluted site at the surroundings of the town very close to the sea shore (100 m).	Terceira Is. (Azores): Lightly polluted site at the surroundings of the town , close to sea shore (1 km).
Instrumentation	Dobson #13 (IOC) since June 1960; Brewer MARK II #047, since July 1989 to June 1992.; Robertson Berger sunburn UV meter, since 1983 to 1992.	Brewer MARK II #102, since October 1994.	Brewer MARK II #048, since October 1989.	Brewer MARK II #047, since July 1992.
Observing program	Daily AD/DS[1] measurements at μ^2 = 3.0, 2.5, 2.0, 1.5 and at noon; Daily Umkehr observation (mainly p.m.)	Daily DS measurements at $Z^3<$ 79.5°; Daily ZS[4] measurements at Z< 69.5°; Daily Umkehr observations (a.m. and p.m.); Daily Global UV-B spectral irradiance measurements at Z< 69.5°.	Daily DS measurements at Z< 79.5°; Daily ZS measurements at Z< 69.5°; Daily Umkehr observations (a.m. and p.m.); Daily Global UV-B spectral irradiance measurements at Z< 69.5°.	Daily DS measurements at Z< 79.5°; Daily ZS measurements at Z< 69.5°; Daily Umkehr observations (a.m. and p.m.); Daily Global UV-B spectral irradiance measurements at Z< 69.5°.

[1] AD/DS: Measurement performed with the operational pair of wavelengths A and D; DS : Direct Sun measurement.

[2] μ: Optical mass.

[3] Z: Solar zenith angle.

[4] ZS: Zenith Sky measurement.

4.1 Total ozone

The Figure 4.1.1 shows the daily total ozone averages (dots) and 365 days moving averages (solid line) for Lisbon station, obtained with the Dobson #13 spectrophotometer from AD/DS measurements. It should be noted that this data needs to be re-evaluated, mainly before 1972, however it can be seen the annual variation and a marked negative trend from 1972.

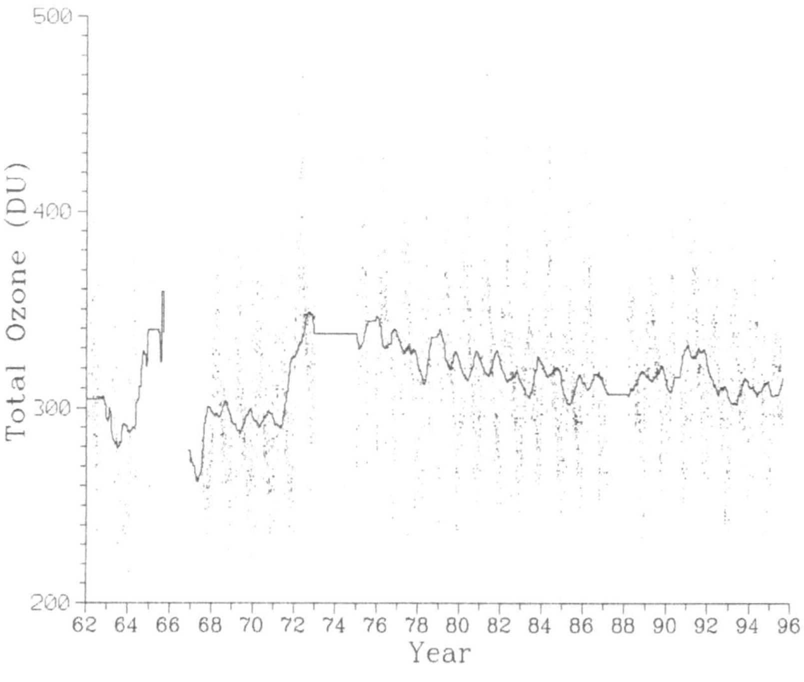

Figure 4.1.1. Total ozone at Lisbon station, measured with Dobson #13. Dots: daily averages of AD/DS measurements; solid line: 365 days moving averages.

In the same way, Figures 4.1.2 - 4 shows the daily DS results for Funchal, Angra do Heroismo and Penhas Douradas stations respectively, obtained with Brewer spectrophotometers.

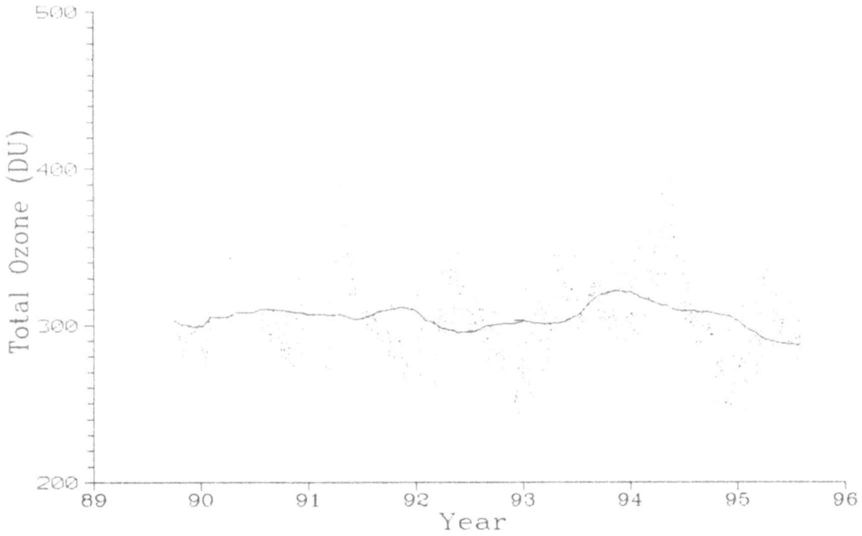

Figure 4.1.2. Total ozone at Funchal station, measured with Brewer #048. Dots: daily averages of DS measurements; solid line: 365 days moving averages.

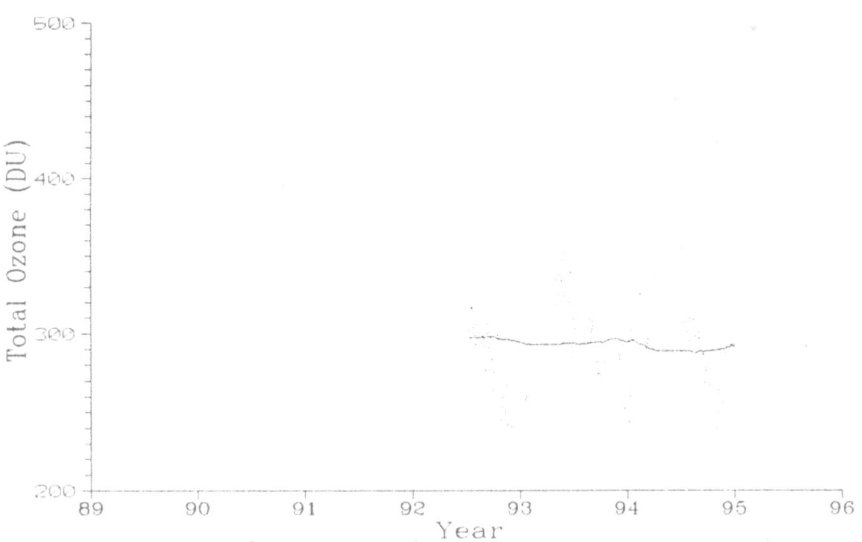

Figure 4.1.3. Total ozone at Angra do Heroismo station, measured with Brewer #047. Dots: daily averages of DS measurements; solid line: 365 days moving averages.

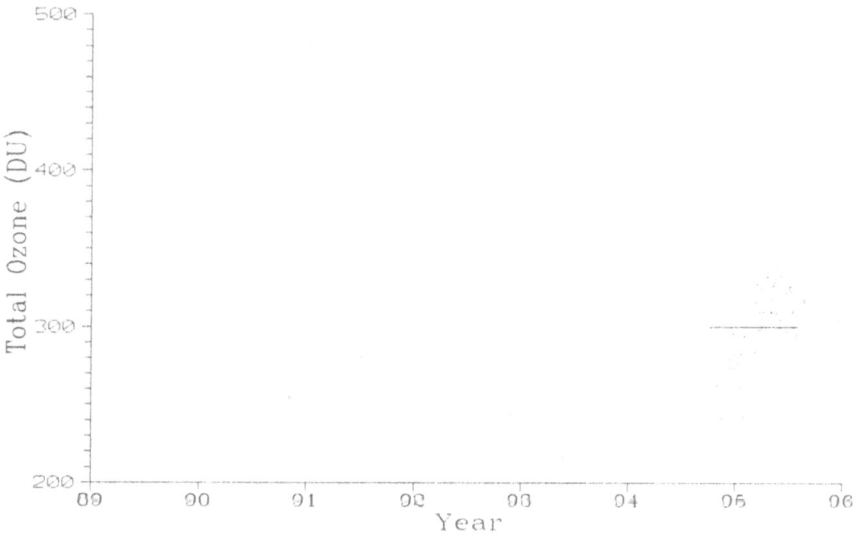

Figure 4.1.4. Total ozone at Penhas Douradas station, measured with Brewer #102. Dots: daily averages of DS measurements; solid line: 365 days moving averages.

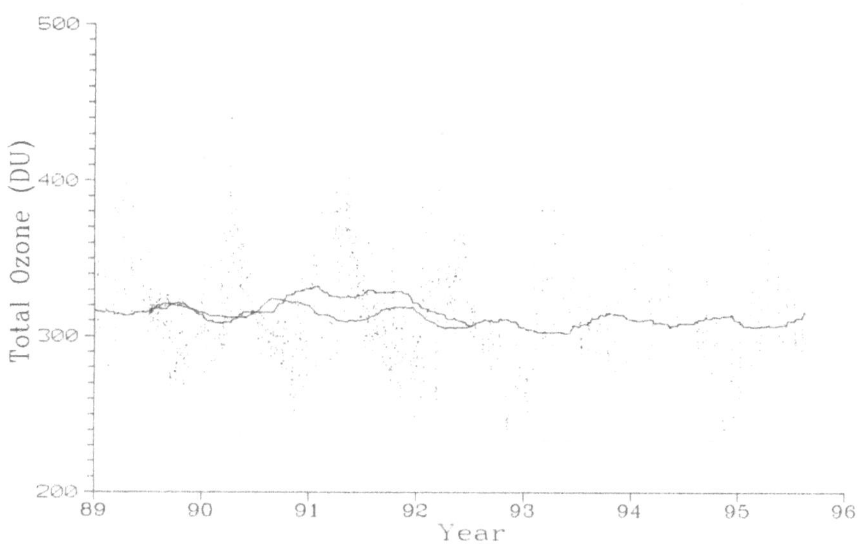

Figure 4.1.5. Total ozone at Lisbon station, measured with Brewer #047 and Dobson #13. Dots: daily averages of DS measurements; solid line: 365 days moving averages.

Figure 4.1.5 shows the daily results for Lisbon during 1989 to 1995 obtained with both Dobson and Brewer instruments and it can be seen a relative good consistency between them, except for 1991. The relative high values observed in 1993/1994 fall season can be also noted in all the stations.

4.2 Total SO_2

In a similar way, the Figures 4.2.1-4 shows the DS total SO_2 results obtained with Brewer instruments at Lisbon, Funchal, Angra do Heroismo and Penhas Douradas stations respectively. Negative values means SO_2 amounts below the zero calibration level. As can be noted, Penhas Douradas station shows very low values as compared with the low altitude stations. In addition, the short term record of Lisbon shows a light increase of the total SO_2.

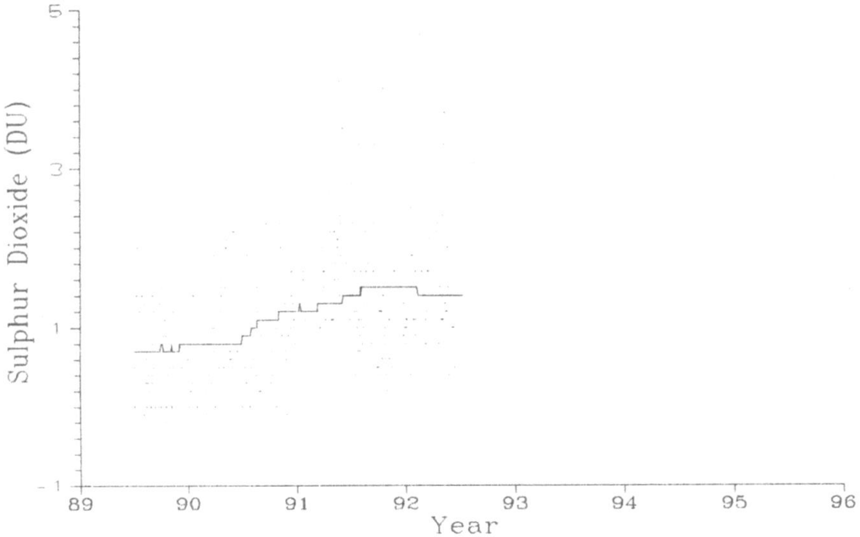

Figure 4.2.1. Total SO_2 at Lisbon station, measured with Brewer #047. Dots: daily averages of DS measurements; solid line: 365 days moving averages.

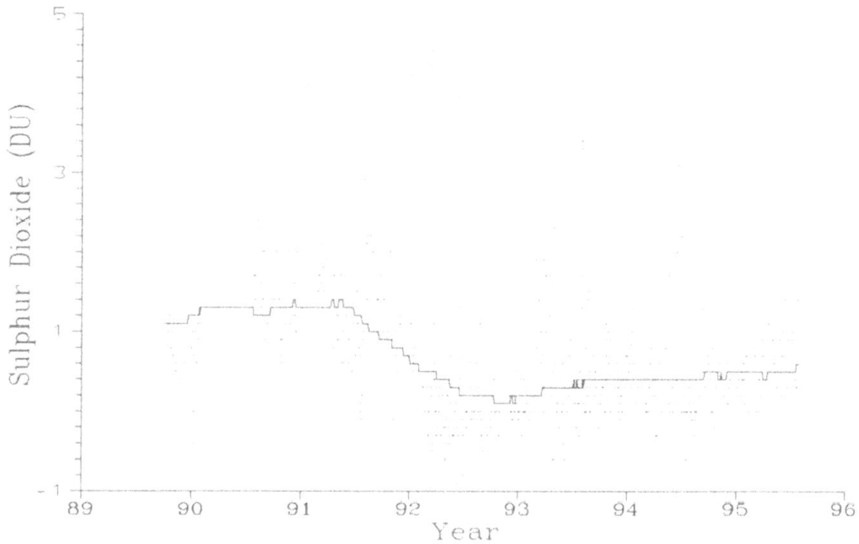

Figure 4.2.2. Total SO_2 at Funchal station, measured with Brewer #048. Dots: daily averages of DS measurements; solid line: 365 days moving averages.

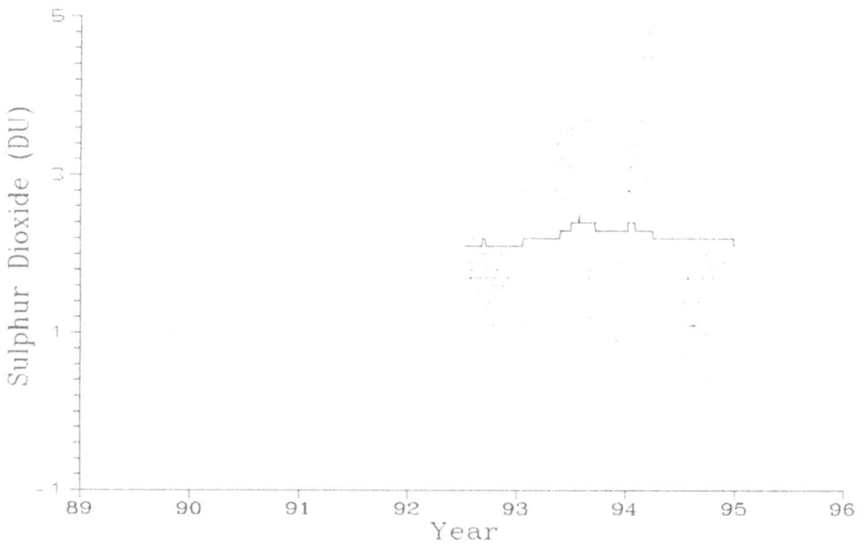

Figure 4.2.3. Total SO_2 at Angra do Heroismo station, measured with Brewer #047. Dots: daily averages of DS measurements; solid line: 365 days moving averages.

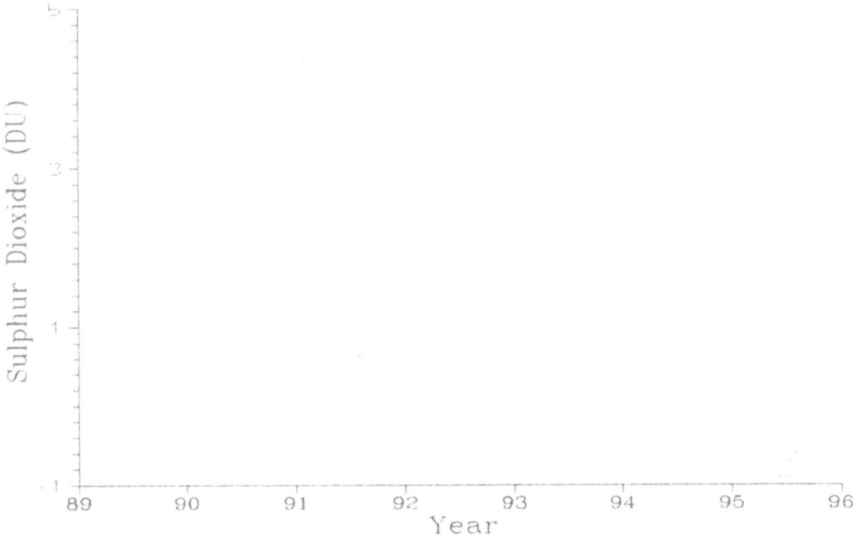

Figure 4.2.4. Total SO_2 at Penhas Douradas station, measured with Brewer #102. Dots: daily averages of DS measurements.

5 Future developments

The negative trends observed in global total ozone suggests that the UV-B levels at surface could be changing. However, the available UV-B data is some times inconsistent, probably due to different instrumentation and methods used, non calibrated instruments, etc. In other hand, there is a strong public interest to know how are and how will be the UV-B levels in the future. In this way, the IM is planning to install an extended UV-B network over the mainland territory, using broad-band or multi-filter instruments. This network will consist in 4 to 6 stations covering mainly the coastal and south regions with interest for the tourism in summer time and other activities. This monitoring could be complemented also with UV-B daily forecastings, using the available ozone and meteorological information. However, the development of this network will depend partially on an international definition of standard instrumentation and methods that are currently matter of discussion.

Photochemical Air Pollution: A regional scale problem

George Kallos, Vassiliki Kotroni and Kostas Lagouvardos

University of Athens, Department of Applied Physics,
Meteorology Lab., Ippocratous 33, Athens 10680, Greece

Abstract. The air pollutant transport and transformation processes and especially the photochemically driven ones consist a very complicated problem due to the numerous related factors such as the spatial and temporal distribution of the emissions, the air pollutant transportation and removal mechanisms. Photochemical air pollution is a regional scale problem as this will be argued. In the frame of this presentation the role of the meteorological conditions, the landscape variations and the distribution of emissions in controlling the photochemical pollution are emphasised.

Keywords: photochemical pollution, transport of air pollutants

1. Introduction

During the last century, several areas have been plagued by air pollution problems. Initially, industrial activities and central heating were mainly responsible for creating the "industrial-type" air pollution. During the last few decades, photochemical air pollution, which is mainly associated to the traffic, tends to create serious problems. The principal primary pollutants entering in the photochemical cycles are the nitrogen compounds (mainly NO and NO_2), usually called NO_x, and the organic compounds (the VOCs). The most important by-products of these primary pollutants are the oxidants (e.g. O_3 or PAN). The air pollutants can enter in complicated chemical reactions, some of which are induced by insolation. The various chemical cycles (e.g. nitrogen or carbon cycles) compete and the products can be completely different.

NATO ASI Series, Vol. I 53
Atmospheric Ozone Dynamics
Observations in the Mediterranean Region
Edited by Costas Varotsos

As it is well known, the transformation and especially the photochemical processes of the air pollutants, are highly non-linear due to the sensitivity of the primary cycles of reactions (mainly the nitrogen and organic compounds). The distribution of the various air pollutants entering into the photochemical cycles exhibit significant spatial and temporal variations because of the different rate of production/destruction of oxidants from the primary cycles of reactions but also because of the spatial and temporal distribution of the emissions. Moreover, among the important factors controlling the distribution of air pollutants and as a matter of fact the air quality in urban or rural areas, are the prevailing meteorological conditions and the spatial variation of the landscape (orography, vegetation cover, soil properties). The role of each one of the aforementioned factors will be discussed in the following section.

2. Factors affecting photochemical air pollution

One of the primary factors which is directly influencing the pollutant observed concentrations and therefore the transformation processes, is the spatial and temporal distribution of the sources. The effects on air quality due to the spatial and/or temporal variations of emissions are more pronounced in urban areas where new technologies in automobile industry (e.g. the use of catalytic converters) and changes in traffic regulations or modifications in city planning have been introduced. The construction of an emission inventory is a very complicated task and the uncertainties are still very important. The spatial/temporal variation of emissions might deviate significantly from the average conditions which are represented in the existing emission inventories.

The rates of production/destruction of oxidants within the primary cycles of reactions are controlled by various factors such as the ratios of the concentrations of the primary pollutants. An uncertainty on the estimated ratio NOx/VOCs might lead through the photochemical cycle to a large uncertainty on the estimation of the resulting oxidants. It is obvious that the uncertainties in the emission estimates might introduce large uncertainties in the photochemical model results.

A very important factor which controls the transport, transformation and removal processes and therefore the air quality over an area, is the prevailing meteorological conditions. The detailed description of the meteorological fields prevailing at local but also at regional scale are necessary in order to achieve a correct description of the photochemical processes. As the local weather conditions (at urban scale) and the associated circulation result from the synoptic, regional, meso and microscale interaction, an inaccurate consideration of any of these scales in air pollution modelling might affect the local circulations and therefore the local conditions (e.g. Kallos and Kassomenos, 1994; Kallos et al., 1993, Rao et al., 1995).

The landscape characteristics in the vicinity of an urban area affect significantly the local meteorological conditions. The shape and the orientation of the coastline and/or the topographic ridges (e.g. hills, mountains) control the development of local circulations. Large-scale topographic ridges might create channelling of the synoptic and regional-scale flow, modifying significantly the local atmospheric conditions and the urban and regional air quality. Moreover the vegetation cover, the land water distribution, the land use, soil type and especially their variation are very important in defining the boundary layer growth (e.g. Avissar and Pielke, 1989; Pielke et al., 1991; Pielke and Uliasz, 1993; Segal et al., 1988, 1989 ant references therein). The role of the landscape variability on the dispersion characteristics over an area has been studied during the last years (Pielke and Uliasz, 1993). Kallos (1995) discussed the relation between the land-use changes, city-planning and air quality. It was demonstrated that land-use changes might cause climatic changes at local or regional scale according to the scale of intervention. In an urban environment, such landscape variations might control the local atmospheric conditions and therefore the air quality.

Air quality at local and regional scale has been the subject of many research programmes (e.g. MECAPIP, SECAP, Transport and Transformation of Air Pollutants from Europe to the Mediterranean/T-TRAPEM). According to the findings of these research projects (Millan et al., 1992; Kallos et al., 1995) photochemical pollution is not only an urban-scale problem but also a regional-scale problem. A quick description of the results obtained in the frame of these research programs in the Eastern Mediterranean Region is given in the following section.

3. Characteristics of urban plumes in Eastern Mediterranean

Air quality in the Greater Athens Area has been the subject of several studies during the last two decades (e.g. Lalas et al., 1982; Kallos et al., 1993; Pilinis et al., 1993; Moussiopoulos et al., 1994; Kassomenos et al., 1995). Air quality standards are significantly violated during 5-10 times per year (Kassomenos et al., 1995) creating an air pollution episode. These air pollution episodes can be attributed to the prevailing meteorological conditions in association with the topography and the landscape characteristics, as well as to the emissions, mainly from the traffic. The insolation helps significantly the photochemical activity resulting in the development of a photochemical cloud called "nephos" over the Athens Basin. The photochemical pollution over the Athens Basin presents spatial and temporal variations as this has been discussed in Pilinis et al. (1993) or Moussiopoulos et al. (1994) and in the references therein. Kallos et al. (1993) showed that the worst air pollution episodes in Athens are associated with almost stagnant meteorological conditions, mainly due to the advection of warm air masses which prevent the regular development of the typical thermal circulations.

In the frame of the T-TRAPEM (DG-XII, EU) program, airborne measurements were performed during summer 1993 offshore of Israel, during summer 1994 over the Aegean Sea and Athens, and during summer 1995 over the Southern Aegean and the Libyan Sea, in order to study the behaviour of the urban plumes over the Eastern Mediterranean. Airborne measurements revealed that the plume of Athens is not confined over the Saronic Gulf as it was believed during the last 15 years but it can influence the air quality of the Eastern Peloponnese, Crete and some other islands. For example on 7 July 1994 an O_3 episode occurred over Athens and Saronic Gulf (Figs 1a,b). The following day, the urban plume of Athens was found advected all the way down at the coastal area East of Peloponnese and Southwest of Crete (Fig. 2).

The behaviour of the urban plumes over Eastern Mediterranean has been also studied through the modelling of selected case studies. The models used are the Regional Atmospheric Modelling System (RAMS) and the Hybrid Particle and Concentration Transport Package (HYPACT), (Pielke et al., 1992).

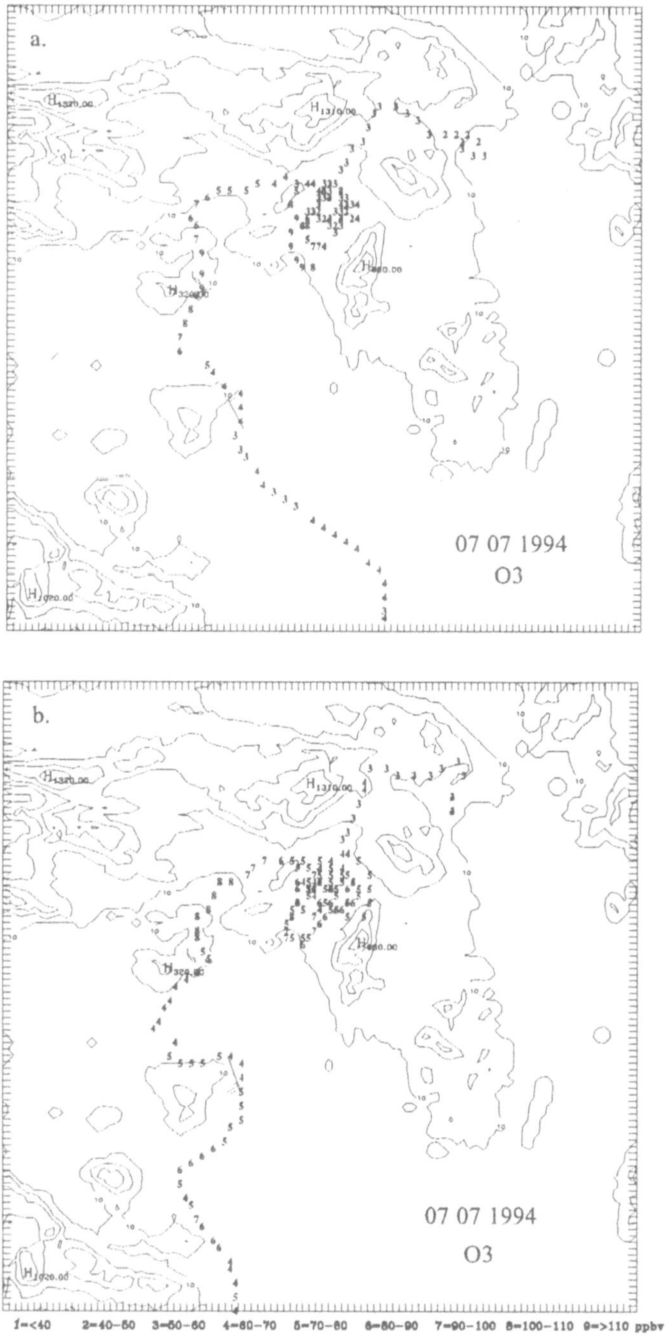

Fig. 1: Ozone concentrations measured by the research aircraft on July 7 1994 for the time period (a) 11:20 to 12:20, (b) 14:40 to 15:50. One full wind-barb equals 5 ms^{-1} and one half barb 2.5 ms^{-1}.

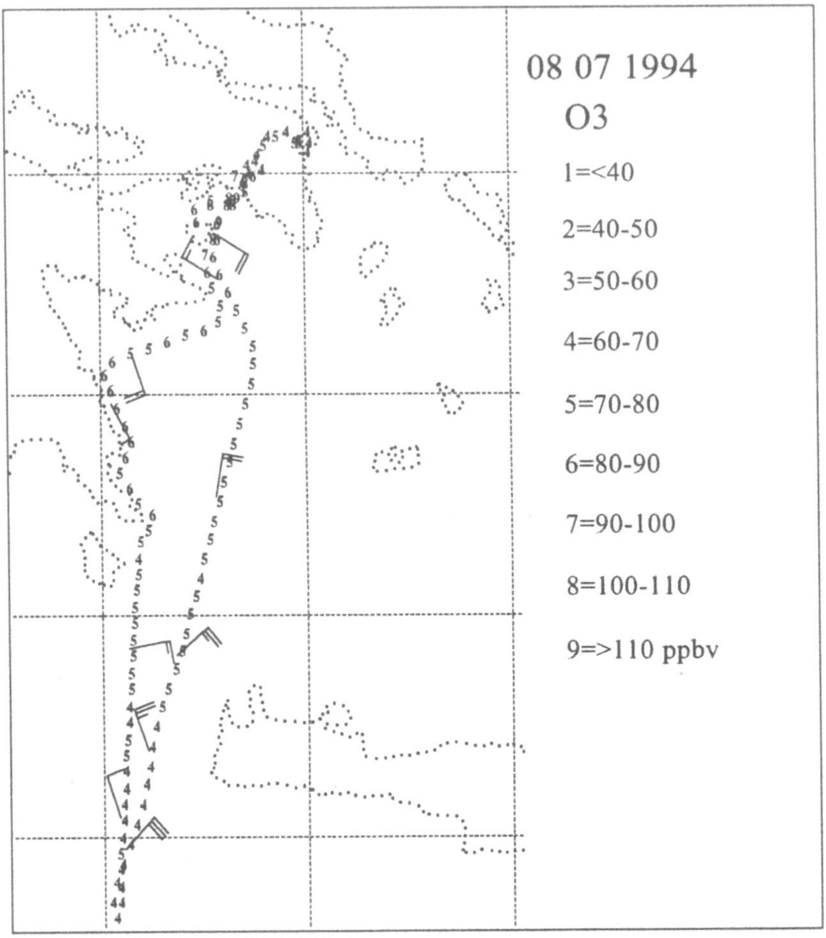

Fig. 2: Ozone concentrations measured by the research aircraft on July 8 1994 for the time period 10:05 to 13:55. One full wind-barb equals 5 ms^{-1} and one half barb 2.5 ms^{-1}.

Model results showed that air quality in the area East of Peloponnese and North of Crete can be influenced by plumes travelling from longer distances. Using the Lagrangian Dispersion mode of HYPACT, and defining particle releases over urban and industrial areas in southern Europe for selected meteorological cases it was found that plumes from Southern Italy (from the area of Messina and Catania) can travel towards Greece and reach the straights between Peloponnese and Crete within approximately 40 hours (Fig. 3).

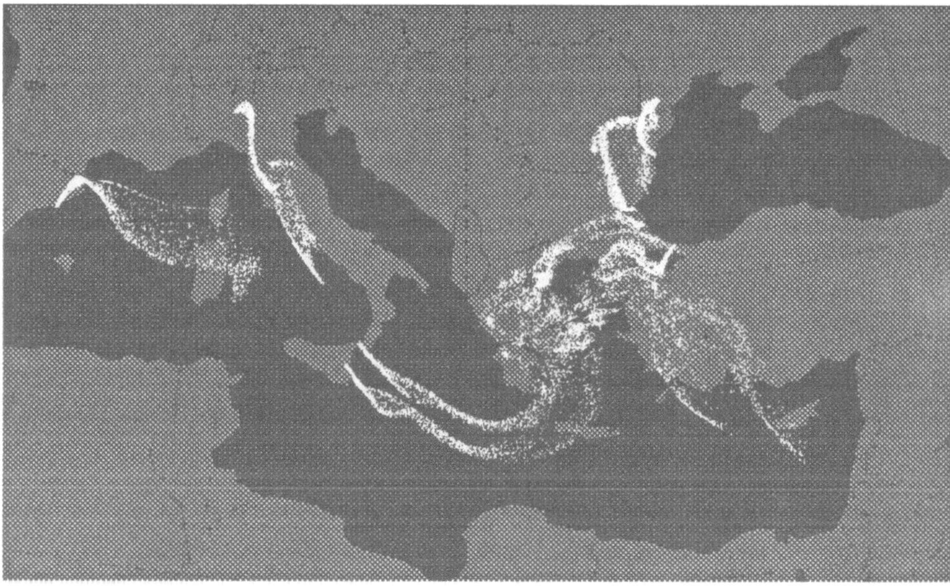

Fig. 3: Particle projection at the first model level from the HYPACT dispersion model at 1500 UTC, 16 June, 1990 (after 39 hours of particle release).

For another meteorological case, the plumes of Athens, Thessaloniki, Messina and Catania were studied (Fig.4). Within 48 hours, the plumes from Italy affect the area South of Peloponnese. Over the Athens Area, the particles released the second day are found at the northern part of the city and some of them out of the Athens Basin. The particles released during the first day have been transported over the sea East of Peloponnese in the morning of the second day, while in the afternoon of the second day these particles are advected back over the Saronic Gulf. From these results it is obvious that within a time period of two days the plume of Athens can recirculate over a large area (more than 100 km far away). Moreover, it has been shown that plumes from remote locations influence the air quality of urban areas.

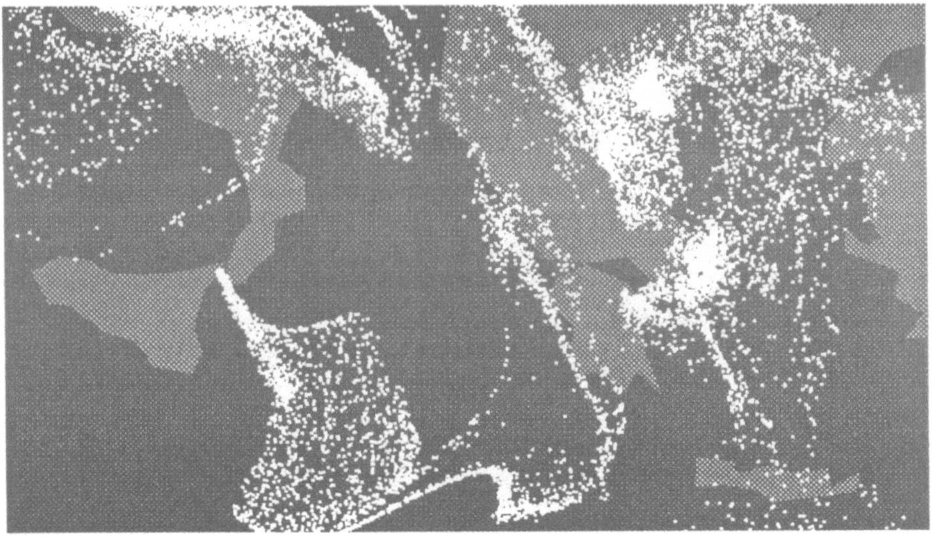

Fig. 4: Particle projection from the HYPACT dispersion model at 1500 UTC, 7 July, 1994, (after 39 hours of particle release).

3. Conclusions

In the frame of this presentation, the behaviour and characteristics of the urban and industrial plumes, at regional scale, have been discussed, as well as their relation to photochemical pollution and air quality. Model results and airborne observations analysis showed that these plumes can travel large distances keeping their characteristics for some days, affecting, thus, the air quality in remote locations. The contribution of remote sources in the air quality conditions of an area can be quite significant, especially when the remote sources are over coastal areas and the meteorological conditions favour the transport of plumes towards/and over the sea.

More specifically the transport of urban and industrial plumes from South Europe towards Eastern Mediterranean is favoured mainly during summer because of the stable conditions prevailing in the region (large-scale subsidence), the absence of removal processes, the appearance of trade winds and the insolation. South Europe urban and industrial plumes can affect not only the Northern part of the Mediterranean but also the Eastern and Southern part of it.

Acknowledgements. This research was funded by the DG-XII of the EU through the research contracts AVI-CT92-0005 and EV-5VCT91-0050. The members of the Israeli team participating in the T-TRAPEM project Drs M. Luria, M. Peleg, G. Sharf and Mr A. Vanger are kindly acknowledged, especially for the airborne measurement part.

References

Avissar R, and R.A. Pielke, 1989: A parameterization of heterogeneous land surfaces for atmospheric numerical models and its impact on regional meteorology. Mon. Wea. Rev., 117, 2113-2136.

Kallos, G., P. Kassomenos, and R.A. Pielke, 1993: Synoptic and mesoscale weather conditions during air pollution episodes in Athens, Greece. Boundary-Layer Meteorol. 62, 163-184.

Kallos, G., and P. Kassomenos, 1994: Effects of the selected domain in mesoscale atmospheric simulations and dispersion calculations. Proc. of the 20th ITM of NATO/CCMS on Air Pollution Modeling and its Application. Editors: S. Gryning and M. Millan, Plenum Press, X, 35-44.

Kallos, G., Kotroni V., Lagouvardos K., and Varinou M., M. Luria, M. Peleg, G. Sharf, V. Matveev, D. Alper-SimanTov, A. Vanger, G Tuncel, S. Tuncel, N. Aras, G. Gullu, M. Idrees, F. Al-Momani, 1995: Transport and Transformation of Air Pollutants from Europe to the East Mediterranean Region. Environmental Research Program AVICENNE. Intermediate Technical Report prepared for the DGXII of the EU, pp 97.

Kallos G., 1995: Urban air pollution and city planning. Presented at the COST 616 Workshop, Brussels, March 1995.

Kassomenos P., V. Kotroni and G. Kallos, 1995: Analysis of climatological and air quality observations from Greater Athens Area. Atmos. Envir., 29, 3671-3688.

Lalas, D.P., V.R. Veirs, G. Karras, and G. Kallos, 1982: An analysis of the SO_2 concentration levels in Athens, Greece. Atmos. Envir., **16**, 531-544.

Millan, M., B. Artinano, L. Alonso, M. Castro, R. Fernandez, and J. Goberna, 1992: Mesometeorological cycles of air pollution in the Iberian peninsula. Final Report prepared for the CEC, DG-XII, pp.219.

Moussiopoulos N., P. Sahm, C. Kessler, and R. Kunz, 1994: Simulation of mesoscale wind flow and photosmog formation in the Greater Athens area. In the EUMAC Zooming Model (EZM) (edited by Moussiopoulos N.) pp. 67-86. EUROTRAC Special Publication.

Pielke, R.A., G. Dalu, M. Uliasz, T.J. Lee and R.A. Stocker, 1991: Impacts of land surface characteristics on atmospheric dispersion. Preprints, Seventh Joint Conference on Applications of Air Pollution Meteorology with AWMA, New Orleans, Louisiana, AMS, January 14-18, 1991, 302-307.

Pielke, R.A., W.R. Cotton, R.L. Walko, C.J. Tremback, W.A. Lyons, L.D. Grasso, M.E. Nicholls, M.D. Moran, D.A.Wesley, T.J. Lee and J.H. Copeland, 1992: A comprehensive meteorological modelling system - RAMS. Meteorol. Atmos. Phys. 49, 69-91.

Pielke, R.A., and M. Uliasz, 1993: Influence of landscape variability on atmospheric dispersion. J. Air and Waste Manage. Assoc., 43, 989-994.

Pilinis C., P. Kassomenos and G. Kallos, 1993: Modeling of photochemical pollution in Athens, Greece. Application of the RAMS-CALGRID modeling system. Atmos. Envir., 27B, 353-370.

Rao, S.T., E. Zalewsky and I.G. Zurbenco, 1995: Determining temporal and spatial variations in ozone air quality. J. Air and Waste Manage. Assoc., 45, 57-61.

Segal M., R. Avissar, M.C. McCumber and R.A. Pielke, 1988: Evaluation of vegetation effects on the generation and modification of mesoscale circulations. J. Atmos. Sci., 45, 2268-2292.

Segal M., J.R. Garratt, G. Kallos and R.A. Pielke, 1989: The impact of wet soil and wet canopy temperatures on daytime ABL growth. J. Atmos. Sci., 46, 3673-3684.

Re-evaluation of Hohenpeissenberg Ozone Records

Ulf Köhler

Deutscher Wetterdienst, Meteorological Observatory Hohenpeissenberg
Albin-Schwaiger-Weg 10, 82383 Hohenpeissenberg, Germany

Abstract

The Hohenpeissenberg measurements of total ozone (with a Dobson spectrophotometer since 1967) and of ozone vertical profile (with Brewer/Mast ozonesondes since 1967) were examined and recalculated according the WMO-recommendations and afterwards re-analyzed. The applied methods of examination and recalculation are briefly described. New trend analyses of both total ozone and vertical ozone profile resulted in an enlarged trend of total ozone by +50%, which is now in better agreement with other data records (near-by stations, satellite), whereas the trends in the vertical ozone are nearly constant. This contradiction compared with the large change in the total ozone trend can be explained by a completely new sonde evaluation, which comprises not only the application of the recalculated total ozone, but also a quality check inclusively exclusion of bad data points or even complete soundings.

Keywords. Total ozone, vertical ozone profile, homogenization, trend analyses.

1 Introduction

Ozone measurements were started at the Meteorological Observatory Hohenpeissenberg (MOHp) more than 25 years ago. A Dobson spectrophotometer (D104) and Brewer/Mast (B/M) ozonesondes have been in use for total ozone and vertical ozone profile measurements since 1967. All observations have been carefully conducted in agreement with the officially WMO-regulations. In the eighties scientific and public interest concerning the environmental ozone problems increased, especially the provision with homogeneous long term data series for reliable trend analyses became more and more important. This strong interest caused WMO and concerned scientists to start new and to intensify already existing programmes (*NOAA* 1993, *WMO* 1993), in order to improve and to assure the quality of the ozone data in the ***G***lobal ***O_3*** ***O***bserving ***S***ystem (GO_3OS). The foundation of the ***G***lobal ***A***tmosphere ***W***atch (GAW) programme and the inclusion of the GO_3OS intensified this ***Q***uality ***A***ssurance and ***Q***uality ***C***ontrol (QA/QC) activities. The MOHp was one of the first stations, which examined and recalculated the long term series of its Dobson and B/M-soundings (s. *Köhler* 1995).

NATO ASI Series, Vol. I 53
Atmospheric Ozone Dynamics
Observations in the Mediterranean Region
Edited by Costas Varotsos

2 Methods

- *Dobson Spectrophotometer*:

External sources (other instruments like Brewer 10 at MOHp, TOMS satellite), near-by Dobson stations like Hradec Kralove, Potsdam or Arosa, and other data sets (100 hPa-temperature) were used to detect suspicious periods in the long term total ozone series of Hohenpeissenberg. Figure 1 (D104-TOMS Version 6 comparison) reveals an obvious break in the relative difference in 1986.

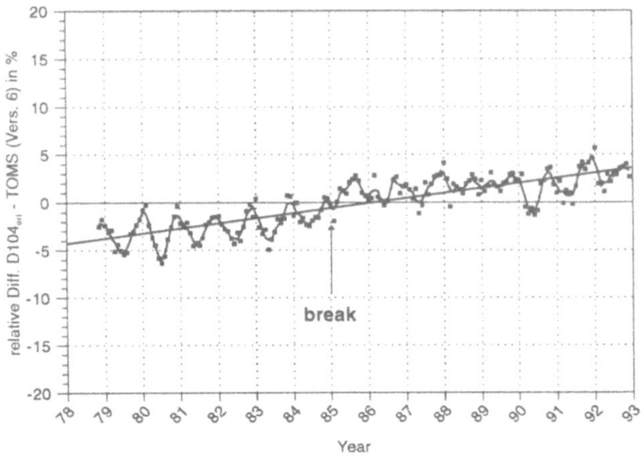

Figure 1: Comp. D104 (original, Vigroux) with TOMS (Vers.6, B/P), monthly means (squares), annual course (curve), trend (straight line)

Internal data sources like the lamp test records (mercury lamp HG, standard lamp SL) (s. Figure 2) were used to determine corrections to the instrumental constants (extraterrestrial constants ETC). The success or improvement of the recalculated data set was again checked with the above mentioned external data.

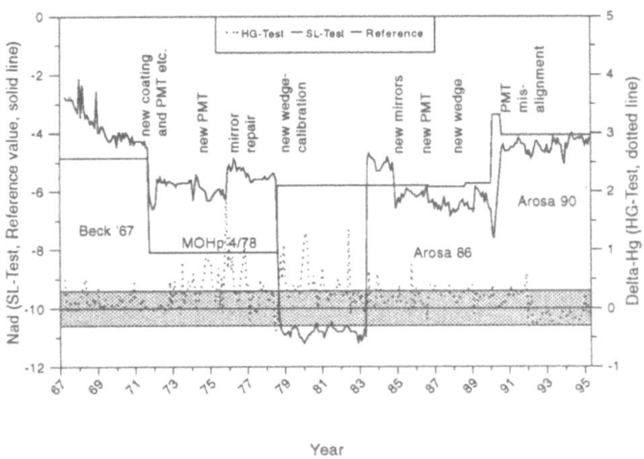

Figure 2: D104 SL-(solid), HG-(dotted) testvalues and reference values (straight lines)

- *Brewer/Mast Ozonesondes*:

The ozonesonde record is strongly depending on the total ozone record of the Dobson 104 via the normalization procedure. Thus, it was necessary to replace the originally used total ozone values in each sounding by the homogenized data in order to get consistent data sets.

A further step was to examine all soundings, whether their correction factor (ratio of the total ozone measured with spectrometers and sonde total ozone as sum of integrated and residual ozone) lies inside the limits of 0.9 and 1.25. This normalization procedure is applied to overcome the sonde short-comings or possible erroneous assumptions for the calculation algorithm like pump efficiency, residual ozone, sensitivity to other trace gases, response time, temperature dependence, wall losses and collection efficiency. The last data points of "bad soundings" just before the balloon burst level were checked, whether the basic condition for the calculation of the residual ozone above the balloon burst level is fulfilled: Constant ozone mixing ratio above 17 hPa. The examined soundings were truncated, when the last significant points differed too much from this constant mixing ratio-condition.

Thus, by omitting few dubious data points a number of soundings fulfilled the WMO-criteria and could additionally be used for publication. On the other hand a lot of soundings had to be excluded after this procedure.

3 Results

- *Dobson Spectrophotometer:*

Figure 3 shows the effect of the homogenization inclusively change from Vigroux to Bass/Paur (B/P) absorption coefficients:

The original, linear trend for the period 1968-94 increased from -1.86 % per decade up to -2.92 %, related to the start values of the linear regression. Both records were rather similar in the first ten years between 1968 and 1978, the homogenized total ozone is higher for the next five years until 1983 and remarkably lower after 1984. This behaviour reflects the combination of homogenization, which mainly causes an enhancement in the period 1967 until 1983 (up to + 5% in 1980), and the Vigroux → Bass/Paur-change (-2.6 %).

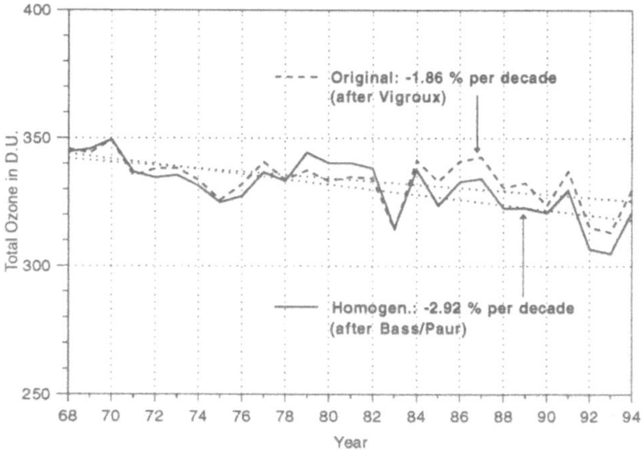

Figure 3: Annual means of D104 total ozone, trends of original and homogenized data are dotted

This trend of the homogenized data fits much better with the results of trend analyses of the Dobson network and of TOMS version 6 for the northern mid-latitudes. This statement is confirmed by Figure 4: The annual course of the total ozone trend (maximum in winter/early spring, minimum in summer/fall) between 1979 and 1990 of D104 coincides with the corresponding TOMS-trends in the midlatitudes (45^0 - 55^0 N, data from *Herman et al.* 1993).

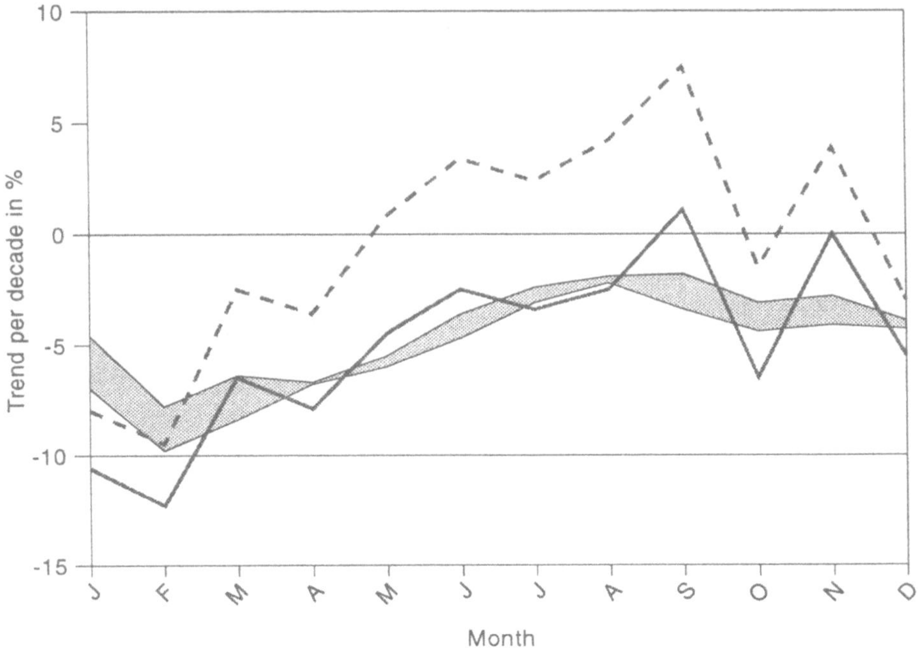

Figure 4: Annual course of D104 (ori=dashed, homo=solid) and TOMS (Vers. 6) trends (shaded area) at 45^0-55^0 N between 1/79-12/90

- *Brewer/Mast Ozonsondes*:

Trend analyses of the original record were already published by *Logan* (1994), *Claude* 1994, *Claude et al.* (1994a), *Wege and Vandersee* (1991) and *Köhler et al.* (1988). In most of these publications a stratospheric trend of approximately -5% per decade is frequently mentioned and well known as the ozone layer trend above Hohenpeissenberg. Here in this paper, however, a different altitude range and a longer period was used for the trend analysis of both original and homogenized data. Thus, the original stratospheric trend is now a little lower with -3.63% per decade and is changed to -3.61% in the homogenized data set. The tropospheric trend is slightly reduced from -20.3% (original) down to -17.8% (recalculated) per decade (s. Fig. 5).

These original and recalculated trends are very similar, although the total ozone trend has been changed significantly (s. Figure 3). This seeming contradiction can be explained easily: The recalculation comprises not only the new Dobson values but also a complete examination of the last significant points of the soundings above 30 km with respect to the assumption of constant mixing ratio. The number of soundings being used for the trend analysis changed especially in the early years.

The overall ozone trends of the homogenized ozonesonde records of the entire period 1967 - 1994 amount to -10% in the stratosphere and +50% in the troposphere.

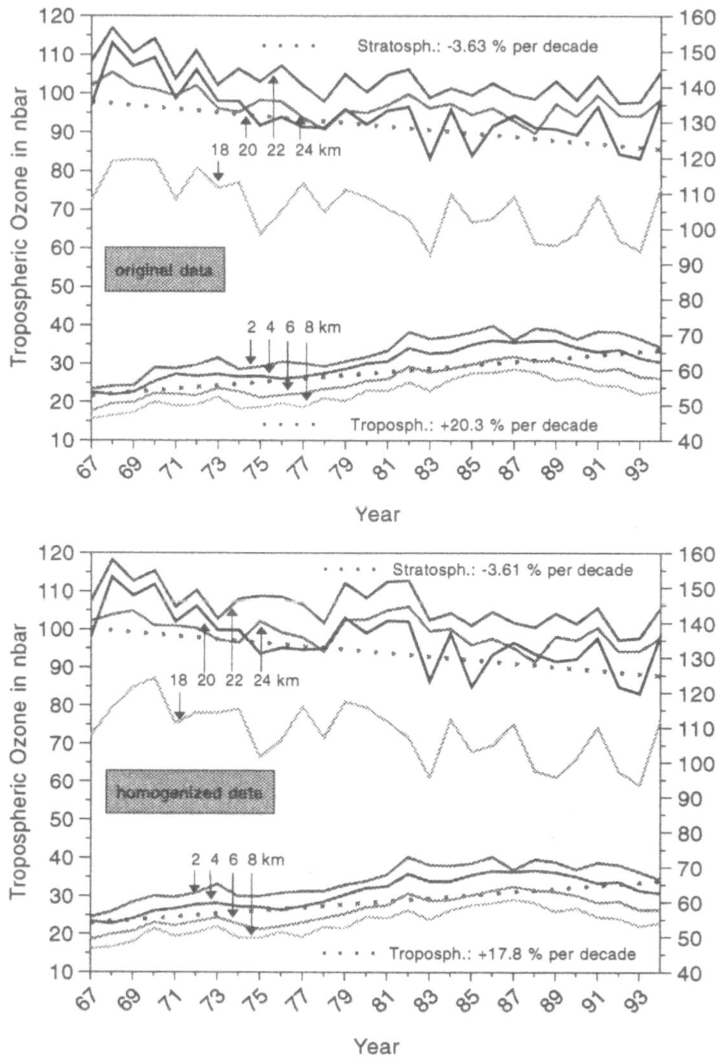

Figure 5: Original and re-evaluated B/M ozone trends

4 Conclusion and Outlook

It can be stated as résumé, that the homogenized data sets of Dobson 104 total ozone and of the Brewer/Mast-soundings have been clearly improved. It turned out as very important and useful to follow the WMO-recommendation, that all re-evaluations should be done completely independent from external sources. They should be used only for detecting possible breaks and examination of the recalculation.

The success and advantage of these independent correction procedures can be seen in Figure 6. The comparison with TOMS version 6 data shows an obvious improvement after the homogenization (upper and central picture). A mean bias and even a trend change, however, were remaining. As no internal, plausible correction could be found to achieve a better agreement, this homogenized D104 total ozone record was accepted as the best and most reliable one. No attempts were made to fit these data to TOMS Version 6.

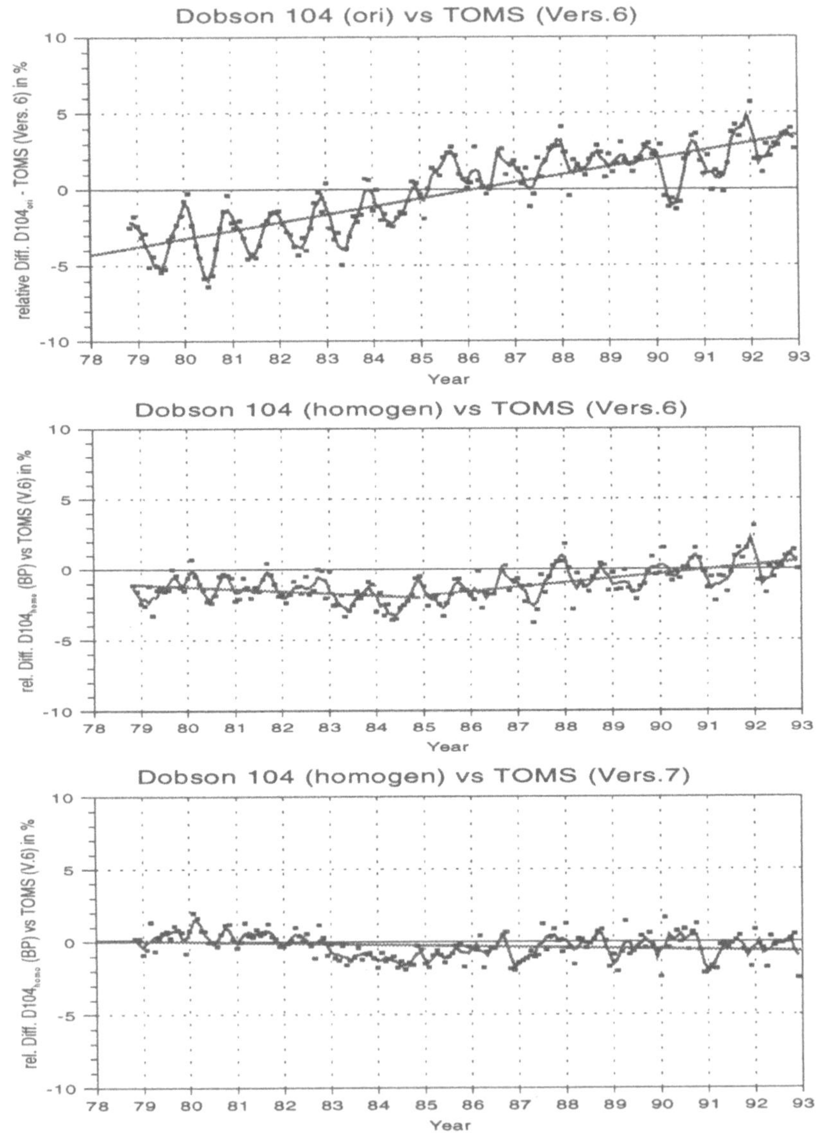

Figure 6: Comp. of D104 (original and homogenized) with TOMS (Vers. 6 and 7), relative difference in %, monthly means (squares), annual course (curve), trend (solid line)

Now the preliminary Version 7 has been released and the comparison (lower picture of Figure 6) confirms, that TOMS was responsible for most of the differences and their trends. A hasty fit of D104 to Version 6 would have had fatal consequences.

The present state of the B/M-record can be assumed only as preliminary. The above described correction procedures are not more than a first step to achieve an optimal data quality, as it is demanded in the QA/QC (Quality Assurance and Quality Control) activities within the GAW-programme of WMO. Intensive investigations to quantify the above mentioned shortcomings of the sonde and to find improved correction algorithms will be done in the next years.

References

Claude, H., 25 Jahre Ozonforschung auf dem Hohenpeissenberg, Entwicklung und Highlights. In "Observation und Simulation der Atmosphäre", Annalen der Meteorologie Nr. 29, pp. 313-320, 1994.

Claude, H., W. Vandersee, K. Wege, On Long-Term Ozone Trends At Hohenpeissenberg. Ozone in the Troposphere and Stratosphere Part 1, pp. 190-194, NASA Conference Publication 3266, Quadrennial Ozone Symposium 1992 in Charlottesville, USA, 1994a.

Herman, J. R., R. McPeters, D. Larko, Ozone Depletion at Northern and Southern Latitudes Derived From January 1979 to December 1991 Total Ozone Mapping Spectrometer Data. J. Geophys. Res., Vol. 98, No. D7, pp. 12783-12793, 1993.

Köhler, U., K. Wege, R. Hartmannsgruber, H. Claude, Vergleich und Bewertung von verschiedenen Geräten zur Messung des atmosphärischen Ozons zur Absicherung von Trendaussagen. BPT-Bericht 1/88, 1988.

Köhler, U., Homogenization and Re-evaluation of the Long-Term Ozone Series at the Met. Obs. Hohenpeissenberg. Final Report of the DWD-Project K/U 31. Arbeitsergebnisse der Abteilung Forschung Nr. 31, Deutscher Wetterdienst, 1995.

Logan, J. A., Trends in the vertical distribution of ozone: An analysis of ozonesonde data. J. Geophys. Res., Vol. 99, No. D12, pp. 25553-25585, 1994.

NOAA, Dobson Data Re-Evaluation Handbook. NOAA Technical Report NESDIS 74, 1993.

Wege, K., W. Vandersee, Ozonbeobachtungen am Nordrand der Alpen. Meteorol. Rdsch. 44, pp. 138-146, 1991.

WMO, Handbook For Dobson Ozone Data Re-Evaluation. GAW, WMO Global Ozone Research and Monitoring Project Report No. 29, 1993.

A review on Greenhouse Effect and Ozone Dynamics over Greece

K. Ya. Kondratyev[1] and C. Varotsos[2]

[1] Russian Academy of Science, Research Centre of Ecological Safety, 18 Korpusnaya St., 197110, St. Petersburg, Russia
[2] University of Athens, Department of Physics, Division of Applied Physics, Laboratory of Meteorology, Panepistimioupolis, Build. PHYS-5, GR 157 84, Greece

Abstract. Great interest to the problem of the atmospheric greenhouse effect (not only in scientific publications, but also in mass media), on one hand, and undoubtfully overemphasised contribution of the greenhouse effect to the global climate change, on the other hand, motivate a necessity to analyse the role which the greenhouse effect plays as a factor of climate change. Significant progress in the analysis of existing observational data as well as succesfull development of numerical climate modeling which have been achieved during the recent few years create a basis for new survey of the atmospheric greenhouse effect in the context of global climate change. Such a survey is the principal purpose of this paper. After discussing a notion of the greenhouse effect the detailed analysis of the present-day and paleoclimatic observational data has been conducted with subsequent consideration of numerical modeling results. A special attention has been paid to assessments of the greenhouse warming versus aerosol cooling. Then possibilities of the early detection of a greenhouse climate signal have been analysed and a few comments on the global climate observing system have been made with the general conclusion that more observations and further numerical modeling efforts are necessary to more reliably assess contributions of various mechanisms to observed global climate changes. It is only in the context of a coupled totality of significant climate forming factors and processes that the contribution of the greenhouse effect may be estimated.

A comprehensive analysis of the records of surface ozone available for Athens, Greece, (38°N, 24°E) for the periods 1901-1940 and 1987-1990 is presented. Both records are analysed to explore the intraseasonal fluctuations and the harmonic components of surface ozone and are also compared to other historical surface ozone records. The variation in surface ozone concentration during rainfall is also investigated, using the hourly measurements of the surface ozone concentration obtained by a network of four stations within the Greater Athens area. The results indicate that, during rainfall events which are associated with the passing of a cold front, an important decrease of the surface ozone concentration is observed. Daily measurements of surface ozone and NO_x from five stations in the Greater Athens Basin over the period 1986-1990 are also used in order to examine the main features of basin-wide O_3-HC-NO_x relations. A simple regression model between the surface ozone concentration and the temperature at the 850 hPa level, which was first tested

NATO ASI Series, Vol. I 53
Atmospheric Ozone Dynamics
Observations in the Mediterranean Region
Edited by Costas Varotsos

in Los Angeles, gave satisfactory results in reproducing the mean monthly ozone variation in Athens, when coefficients extracted from local data were used in the regression equation. A series of vertical ozone soundings over Athens has been also performed in order to explore the tropospheric ozone variations and to examine further the transport that occurs at the 700 hPa level with advection from the north-western sector. The relevant results are discussed. The existing uncertainties concerning the stratosphere-troposphere exchange of ozone which mainly occurs during mid-latitude tropopause folding as well as during cut-off low events are also discussed. The examination of the role of the atmospheric circulation in the lower stratosphere in relation to the laminated structure of ozone is also attempted. The data collected during the balloon ascents have been compared with those during the balloon descents. Both profiles are compared with the total ozone measurements derived from the TOMS on the Nimbus-7 satellite and the Dobson spectrophotometer. The data collected for the vertical distribution of ozone and temperature have been compared with the satellite - derived reference models which provide the monthly latitudinal variations of vertical structure of both ozone and temperature. We have also used total ozone measurements obtained with a Dobson spectrophotometer (No. 118) which have been instituted in Athens from 1989 in order to examine the consistency of data from TOMS with the corresponding Dobson data on a daily basis. Furthermore monthly mean total ozone data were first estimated for the entire period and were then Fourier analysed to obtain the amplitude, phase and percentage contribution to the total variance of the first, second and third harmonics.

Keywords. Greenhouse effect, surface ozone, tropospheric ozone, stratospheric ozone, total ozone, ozone dynamics

1 Introduction

It has now become clear that the only potential way to understand the nature of climate changes is to consider the atmosphere-hydrosphere-lithosphere-cryosphere-biosphere climatic system in all its interactive complexity under conditions of strong anthropogenic impacts. To realise such an approach it is equally important to further develop both observations and numerical modeling.

As far as observations is concerned the basic difficulty is that a combined system of satellite and conventional (in situ) observations is necessary to get global-scale representative datasets for long enough time-periods (not less than 30 years). The first operational meteorological satellite TIROS-1 was launched, however, only on April 1, 1960 and it took a long time since to improve the quality and to broaden the range of parameters retrieved on the basis of satellite measurements with the use of various remote sensing techniques (the only parameters which have been measured directly are the Earth's radiation budget components). This process still continues and only recently it has been decided to develop the Global Climate Observing System (GCOS) including such components as Global Ocean Observing System

(GOOS) which faith is not quite clear as yet, however, and Global Terrestrial Observing System (GTOS). Thus, it is still a long way to go before complete enough long term series of important climatic parameters will be accumulated on the basis of the combined use of conventional and satellite climate observing systems. It should be pointed out in this context that the first phase of the climate-oriented Earth Observing System (EOS) programme will only begin before the end of this century.

The fundamental way of understanding climate changes and development of climate prediction methods is the use of simulation climate modeling which was progressing very rapidly but still survives a preliminary stage of development. Even the most sofisticated and complex three-dimensional global climate models are still able to simulate only certain basic climate features, because, in many respects, they are not complete enough with regard to consideration (parametrisation) of numerous climatically significant processes, although a very important progress has been achieved in assessments of nature of climate changes (including anthropogenic impacts) in the course of preparation of the Intergovernmental Panel on Climate Change (IPCC) reports, published in 1990 and 1992 (Intergovernmental Panel on Climate Change, 1992). Numerous productive efforts within the World Climate Research Programme (WCRP) and the International Geosphere-Biosphere Programme (IGBP) were most instrumental. A very important contribution was made by the Second U. N. Conference on Environment and Development which was held in Rio-de-Janeiro, June 1992 (Kondratyev et al., 1994).

It has so happened that the atmospheric greenhouse effect (AGE) has become the focal point of the outcome of the Second U. N. Conference on Environment and Development (UNCED). The most important practical recommendation of the UNCED in general and of the Framework Convention on Climate Change (FCCC), approved by the UNCED, in particular has been formulated as a necessity to reduce emissions of greenhouse gases (GHG's) to the atmosphere (United Nations Framework Convention on Climate Change, 1993). Such a recommendation has been based on the assumption that the global climate warming during the recent century was due to GHG's concentrations increases and, therefore, to avoid undesirable climate change (potential sea level rise requires special attention in case of such low-land countries like Bangladesh and small island states) it is necessary to reduce emissions of GHG's (first of all, - carbon dioxide) to the atmosphere. This recommendation is most serious from the economic point of view. To follow the recommendation, the U. S. A. alone will have to spend approximately 3.6 trillion US dollars (others mentioned another figure: up to $150 billion per year). It is obvious that many developing countries will not be able to satisfy similar requirements.

It is quite clear, however, that the hypothesis of a catastrophic greenhouse climate warming is based on information subject to serious uncertainties and inconsistensies. This has been persuasively shown in the documents of the Second World Climate Conference (1990) and, in particular, in the Intergovernmental Panel on Climate Change Report prepared for the consideration by the Conference and supplemented in 1992 by relevant updated documents (Intergovernmental Panel on Climate Change, 1992). These important international documents as well as the most recent scientific publications form a basis for an updated consideration of the role and place of the AGE in the context of global climate change on the basis of the analysis of both observations and numerical modeling results. Such a consideration, which has not been accomplished before, is a principal aim of this survey paper.

The paper is based on the consideration of basic priorities of the global climate change problem which may be formulated in the following way:

1. What do observations say concerning paleoclimate and contemporary climate: proven results, uncertainties and controversies, global means versus regional variability. Conventional versus satellite observations. In this context an urgent necessity arises to substantiate requirements to the Global Climate Observing System (GCOS), which general aims have been approved by the World Meteorological Organization (WMO) Congress in 1991. The following specific aspects have to be taken into account:

1.1 Priorities of climate parameters (since the number of parameters is great, this task is very important);

1.2 Requirements to space and time resolution as well as substantiation of acceptable measurement errors;

1.3 A necessity to optimize a combined system of satellite and conventional observations, including problem oriented (focused) field experiments. The core of this problem - optimization of surface observational networks of stations, including not only standard meteorological observations, but also a number of specific observations (dust - and ozonesondes, aircraft measurements e.t.c.).

2. Consideration of those climatically significant processes which are the most important from the viewpoint of overcoming existing uncertainties of numerical climate modeling:

2.1 Interaction between the atmosphere and ocean: energy - and mass - exchange (with special emphasis on such problems as energy - active zones, the role of polar regions, carbon cycle, DMS transformation e. t. c.).

2.2 Parameterization of cloud cover dynamics. Interaction between cloudiness and radiation, focused field experiments, including the ARM (Atmospheric Radiation Measurement programme), and numerical climate modeling (including nested models to simulate mesoscale processes).

2.3 Colloidal nature of the atmosphere in the context of "aerosol and climate" problem. Global and regional scale models of natural and anthropogenic aerosols. Interaction between aerosols and clouds. "Aerosol cooling" versus "greenhouse warming" under conditions of variable cloudiness.

2.4 Biogeochemical cycles (carbon, sulphur and nitrogen, in particular) and their interaction with climatically significant processes. Fundamental significance of "biosphere and climate" problem.

3. Principal causes of numerical climate modelling uncertainties and the problem of predictability.

3.1 Parameterization of subgrid processes (including application of nested approach) and interaction between cloudiness and radiation.

3.2 Verification of numerical modeling results through intercomparisons with satellite observations (with special emphasis on the earth's radiation budget data) and results of focused field experiments.

3.3 Development of coupled models for the whole climatic system "atmosphere - hydrosphere - lithosphere - cryosphere - biosphere".

3.4 Computational aspects of numerical climate modeling (application of splitting and conjugated equations technique, removal of flux corrections).

4. A fundamental revision of the Framework convention on climate change which is supposed to become interactive with regard to future Conventions on biodiversity, forestry, and desertification.

Of course, the authors do not intend to exaust the subjet mentioned. Fortunately, a rather detailed annotaded bibliography has been recently published. A great number of important new publications have been out (Bernard, 1993; Pitt D. and S. Nilsson, 1993; Sakellariou et al., 1993; Varotsos et al., 1992c; Varotsos and Deligiorgi, 1991; Hughes, 1993). A special place is occupied by the IPCC Report (Report Intergovernmental Panel on Climate Change, 1992). The purpose of the survey is to briefly discuss some conceptual aspects of the problem.

Stratospheric and tropospheric ozone trends which have been observed during the recent years are of great concern from the viewpoint of their impact on climate and human health as well as on ecosystems.

Studies of tropospheric ozone variability are motivated by its significance as a greenhouse gas influencing climate formation as well as its serious impact on human health under conditions of increasing surface ozone concentration due to smog photochemical reactions in the presence of growing atmospheric pollution (nitrogen oxides, hydrocarbon compounds etc.).

As far as the stratospheric ozone dynamics are concerned the main problem is anthropogenically induced depletion of the total ozone content which has been the most dramatically expressed in the formation of the spring total ozone content minimum in Antarctica (the so-called "ozone hole") as well as in miniholes in high latitudes of the northern hemisphere. Even in low and mid-latitudes the decreasing trend of total ozone content is of serious concern because of the impact on human health (skin cancer etc.). That is why the Montreal protocol and further similar documents have been approved and the use of the ozone destroying chlorofluorocarbons CFC-11 and CFC-12 prohibited. There are also substantial climatic impacts of ozone variability in the stratosphere which deserve serious attention.

The complex nature of environmental ozone impacts requires one to apply a system approach to studying ozone changes and the relevant impacts. The purpose of this paper is to discuss as a case study an experience of complex ozone studies (both stratospheric and tropospheric ozone dynamics) in the Athens region.

2 Definition of the atmospheric greenhouse effect

As has been suggested by Kondratyev and Moskalenko (1984) and later on by others the greenhouse effect (G) may be defined by the difference between thermal emission from the surface (E) and outgoing longwave radiation (F):

$$G = E - F \tag{1}$$

Assessments of global/annual average values on the basis of satellite observations led to the following figures: $E = 390\ Wm^{-2}$; $F = 235\ Wm^{-2}$, so that $G = 155\ Wm^{-2}$.

The clear-sky greenhouse effect (G_A) may then be defined as

$$G_A = E - F_{clear} \tag{2}$$

with global/ annual average value $F_{clear} = 265\ Wm^{-2}$ so that $G_A = 125\ Wm^{-2}$. Although it may seem a paradox, but the cloud longwave radiative forcing is thus equal to only $30\ Wm^{-2}$.

Let us introduce normalised quantities to remove the strong temperature dependence:

$$G' = E/F_{clear} \tag{3}$$

There is a very simple possibility to qualitatively assess the influence of atmospheric temperature and humidities on the clear-sky greenhouse effect. Obviously,

$$F_{clear} = \varepsilon_A \sigma T_A^4 + (1 - \varepsilon_A)\, \sigma T_S^4 \tag{4}$$

where ε_A is atmospheric emissivity (σ-Stefan-Boltzman constant), T_A - is the mean temperature, T_S - the mean surface temperature.

Now, from equation (3), we have

$$G' = T_S^4/(\varepsilon_A \sigma T_A^4 + (1 - \varepsilon_A)\, \sigma T_S^4) \tag{5}$$

Or it is possible to introduce parameter g [57]:

$$g = \varepsilon_A\, [1 - (T_A/T_S)^4] \tag{6}$$

Hence:

$$G' = 1/(1 - g) \tag{7}$$

$$G' = 1/[1 - \varepsilon_A(1 - (T_A/T_S)^4)] \tag{8}$$

Here it is assumed that the emissivity ε_A is proportional to the total column moisture content w. Although it is a clear simplification, equations (6) and (8) illustrate the essential result that there is a separation between the effects of the atmospheric humidities, as measured by w and represented by ε_A in these equations, and the temperatures, as represented by T_A. As is seen, the increase of w and, consequently, ε_A leads to the enhancement of G or g, while the T_A increase results in the decrease of temperature contrast between the atmosphere and the surface and, thus, to the G' decrease. It is quite obvious that the roles of w and T_A have to be geographically specific.

On the basis of the qualitative analysis it has been pointed out that, for instance, since the temperature field in the tropics is comparatively homogeneous, dominating role in the greenhouse effect variability belongs to w, which is controlled by sea surface temperature (SST). The opposite situation takes place in middle and high latitudes where air temperatures are highly variable and therefore exert much more powerful control over G' than at low latitudes.

The analysis of satellite data on the earth radiation budget (Earth Radiation Budget Experiment) and the total column moisture (Special Sensor Microwave/Imager), as well as the results of radiative transfer simulations, has confirmed the conclusions of the simplified qualitative assessment. At low latitudes the clear-sky greenhouse effect varies mainly due to w changes (which, on the other hand, are controlled by the SST) and seasonal variations are small. In contrast, at middle and high latitudes both G' and w exhibit strong seasonal variations. The clear-sky greenhouse effect variation are controlled by the seasonal changes in atmospheric temperatures, which are strong enough to overcome the opposing effect of the moisture variations. There are strong seasonal variations of the greenhouse effect which has the maximum in winter (when surface-atmosphere temperature difference increases which leads to the enhancement of the greenhouse effect, inspite of small moisture content) and the minimum in summer. The combined impact of both temperature and column moisture results in the formation of the meridional profile of the greenhouse effect which is characterised by the decrease of G' towards the winter pole at a much slower rate than it does towards the summer pole.

It has been justly pointed out a special role of water vapour as a greenhouse gas (see also Kondratyev, 1949, 1956, 1969, 1988; Kondratyev and Moskalenko, 1984). In this context they discussed the so called "super greenhouse effect" (SGE) defined, in accordance with the earlier definition, as such a situation in which atmospheric greenhouse trapping increases more rapidly with spatially increasing SST than do infrared surface emission ("spatially increasing" means that if one were to move from a region of lower SST to a region higher SST, one would generally find a large AGE at the new location).

Under clear sky conditions the observed values of AGE rise abruptly with spatially increasing SST at SSTs above roughly 298K. The derivative $dG_A / dT_S \approx 8\ Wm^{-2}\ K^{-1}$ (T_S - SST) becomes even larger than the change in surface emission $dE/dT_S = 6.12\ Wm^{-2}\ K^{-1}$ at 300 K, i.e. the loss of infrared energy to space (in the absence of clouds) is reduced in regions of high SST. This is the phenomenon which is called super greenhouse effect. Its existence has been discovered on the basis of satellite observations in various parts of the World Ocean.

On the basis of the analysis of numerical modeling results and satellite observations it has been shown that four processes contribute to the formation of SGE, but the principal contribution belongs to water vapour continuum absorption and thermodynamically controlled increases in water vapour concentration at constant relative humidity with increasing atmospheric temperature. Besides, such processes are significant as the increase of atmospheric moisture content (in the upper and middle troposphere in particular) over the warmest SST, while the atmospheric vertical temperature profile increasignly unstable. Regions with these high SSTs are also increasignly subject to deep convection, which suggests that convection moistures the upper and middle troposphere in regions of convective relative to nonconvective regions, resulting in the super greenhouse effect. Hailberg and Inamdar (1993) note that for the explanation of the observed SGE an impact of dynamic processes have to be also taken into account.

Uncertainties in climatological tropical humidity profiles have been analysed to demonstrate that they are very important for assessements of the greenhouse effect. An important potential consequence of the enhancing greenhouse effect is global climate warming. In connection with the substantiation of potential mitigation measures against global warming some scientists have calculated Global Warming Potentials (GWP's) for 10 alternative hydrohalocarbons, which may be used as substitues for chlorofluorocarbons CFC II and CFC 12, on the basis of new measurements results for infrared cross sections of the substitutes. Table 1 illustrates relevant results (GWP's were calculated relative to CFC-II as a reference gas). As it is seen from the table, for three compounds (HCFC 22, HCFC 142b, HFC 125) the GWP's are higher than unity for a 5- to 10-year period. Moreover, these compounds maintain a relatively strong GWP for a very long time. Thus, it may be concluded that the problem of mitigation measures has not been solved as yet.

Table 1. Lifetimes (years), radiative forcings, and GWP's relative to CFC II.

Halocarbon	Lifetime	Radiative forcing		Global Warming Potential (years)						
		per molecule	per kilogram	5	10	20	50	100	200	500
CFC II	57.0	1.00	1.00	1.00	1.00	1.00	1.00	1.00	1.00	1.00
HCFC 22	14.3	0.85	1.35	1.19	1.06	0.86	0.56	0.41	0.35	0.34
HCFC 123	1.5	0.90	0.81	0.25	0.13	0.07	0.04	0.03	0.02	0.02
HCFC 124	6.0	0.94	0.95	0.67	0.50	0.32	0.17	0.12	0.10	0.10
HCFC 141b	9.7	0.66	0.77	0.63	0.53	0.39	0.22	0.16	0.14	0.13
HCFC 142b	21.1	0.82	1.12	1.04	0.97	0.86	0.64	0.50	0.43	0.11
HCFC 225ca	2.4	1.08	0.73	0.32	0.19	0.10	0.05	0.04	0.03	0.03
HCFC 225cb	6.8	1.31	0.88	0.65	0.50	0.34	0.18	0.13	0.11	0.11
HCF 125	33.9	0.91	1.04	1.01	0.98	0.93	0.82	0.71	0.64	0.62
HCF 134a	13.1	0.78	1.06	0.92	0.81	0.64	0.41	0.29	0.25	0.24
HCF 152a	1.5	0.50	1.03	0.31	0.17	0.09	0.05	0.03	0.03	0.03

3 Observed contemporary climate changes

Recent years have been marked by a number of attempts to assess the reliability of existing datasets which were used for empirical analyses of climate changes. For instance, for this purpose a time series of 43 years of observed monthly mean surface air temperature at 109 sites in the 48 contiguous United States has been compared to monthly mean air temperature specified from hemispheric gridded 700-hPa heights. Such a check of mutual data consistency showed that in both datasets cooling (of about 0.5 °C) from 1951 to about 1970 and subsequent warming (also by 0.5 °C) that continues through the present took place. This allows to make a conclusion that interdecadal temperature changes considered are probably real. Another important conclusion is that a comparison of the full set of 109 stations to a "clean" subset of 24 has indicated that the influence of common problems in surface data (station relocation, urbanisation, etc.) is quite small. This conclusion is in the contradiction, however, with the results of the detailed analysis of urban bias in air surface temperature of China's northern plains, which indicates that despite past efforts to remove the effects of the urban heat islands from land-surface datasets, large urban biases may still remain. Thus, the problem of observation errors still deserves serious attention.

After the analysis of the bias in the global mean temperature estimated from sampling a greenhouse warming pattern with the current surface observing network the principal conclusion is that the observations prove adequate to estimate the globally averaged temperature change associated with the pattern of CO_2 warming from a general circulation model with a bias whose absolute value is generally less than 2%. The calculated pattern of climate change was obtained as a result of 60-year run of the global coupled atmosphere-ocean NCAR R15 model (4.5^{o} latitude by 7.5^{o} longitude resolution) with linear increase in atmospheric CO_2 of 3.3 ppm yr^{-1}. The CO_2 signal is positive all over the globe with largest values exceeding 1°C north of 60°N. The global average (last 30 years of the 60-year run) CO_2 signal is 0.808 ^{o}C.

Some scientists have concluded recently that for global temperature maps generated by a long (120 ensemble members), perpetual January GCM run, empirically estimated mean-square deviation (MSD) between perfectly and imperfectly spatially sampled temperatures varies from 0.050 $^{o}C^2$ before the turn of the century to 0.002 $^{o}C^2$ after 1950. Corresponding RMSD are 0.224^{o}C and 0.045^{o}C. Since 1950 imperfect spatial sampling is less of a problem than imperfect temporal sampling. Some initial calculations to assess sampling errors for real observational data have shown that in case of 1000-hPa temperatures for the time period 1982-1987 spatial variance is near 2 $^{o}C^2$ for every month of the year. It is believed that such a small spatial variance in combination with averaging the twelve months data will result in real erroes in annual means which are smaller than obtained on the basis of numerical modeling results.

Although existing observational networks produce reliable enough information and historic data series for the previous century which are more or less representative from the climatic viewpoint, an important fact is that over the past decades observational networks have been designed for a variaty of purposes but rarely have been designed to detect and monitor climate change and variations. Therefore the problem of data quality is still urgent. In this connection the following principal aspects of managing climate data from weather observing systems have been pointed out:

1) long-term homogeneous databases;
2) resolution of datasets for various time and space scales;
3) information about the observing systems, data collection systems and data reduction algorithms, broadly defined as "metadata" ("important information about the data"); and
4) enhancing weather observing systems to reduce the uncertainties about how climate has (or is) changed and varied.

A number of recent global climate diagnostics studies have arrived at several unexpected conclusions. For instance, although most global climate models expect the strongest impact of the CO_2 rise in the high latitudes (see details in Kondratyev, 1992; Kondratyev and Grassl, 1993) neither systematic change of minimum and maximum temperatures nor an overall warming has been observed in the Arctic over the last 50 years or so. The analysis of tropospheric temperature trends in the Arctic during the time period 1958-1986 has shown that absolute trends of 3^oC / 30 yr or higher were found, with both cooling and warming tendencies observed in the four layers considered: 850-700, 700-500, 500-400, 400-300 hPa. The majority of the trends, however, are not statistically significant at the 90% confidence level. Therefore it has been concluded that "greenhouse-induced warming is not detectable in the Arctic troposphere for the 1958-1986 period". This is, of course, a rather surprising conclusion, since northern high latitudes have always been considered as the zone of maximum chances to detect a greenhouse climatic signal. Such a conclusion is especially important because it was made on the basis of the most reliable observational data for the recent three decades.

Another climatic surprise has been a discovery of asymmetric trends of daily maximum and minimum temperature. Some scientists have pointed out that monthly mean maximum and minimum temperatures for over 50% (10%) of the Northern (Southern) Hemisphere landmass, accounting for 37% of the global landmass, indicate that the rise of the minimum temperature has occured at a rate three times that of the maximum temperature during the period 1951-1990 (0.84^oC versus 0.28^oC). The decrease of the diurnal temperature range (DTR) is approximately equal to the increase of mean temperature. The asymmetry is detectable in all seasons and in most of the regions studied. Table 2 data illustrate observed trends of DTR. For the part of the globe (37%) the rate of the decrease in the DTR (-1.4^oC / 100 years) is comparable to the increase of the mean temperature ($1.3\,^\circ C$ / 100 years).

A very long record of maximum and minimum temperatures for the Klementinum-Observatory in Prague (Czech Republic) shows an increase of the DTR from the early to the mid-twentieth century, with a substantial decrease since about 1950. The increase coincides with the increase of mean temperature since the turn of the century and the decrease occurs when the mean temperature reflects little overall change.

Although the variability of DTR is affected by many factors it has been shown that two variables are most important: cloud amount and cloud height. The decrease in the daily temperature range is partially related to increases of cloud cover. More complete indentification of various causes requires further studies. It is clear, however, that "the direct radiative effect of increasing CO_2 alone is unlikely to explain the current trends".

The results of "Minimax Workshop" have been summarised, which confirmed conclusions concerning the decrease of DTR made earlier as well as the conclusion that this decrease is due to the rise of the night temperatures. Although, in accordance with earlier findings, it was confirmed that the decrease in daily temperature range is related closely with the increase in cloud cover, this is not always the case. The undiniable fact is however that the cloud cover is indeed increasing in most places where the decrease of DTR has been reported and that the preferential increase of night temperatures is in line with such development. Recent results show that the cloud cover increased between 1952 to 1981 by 2.3% in the Northern Hemisphere and 1.2% in the Southern Hemisphere. The daytime As and Ac between 30° and 50°N latitude increased by 27%.

Since global climate models do not predict the increase in cloud amount over land, it may be assumed that the increase of cloud cover (frequency and density of clouds) could be due to an impact of atmospheric pollution.

Of course, surface air temperature (SAT) should be not the only parameter to be used for climate diagnostics. The problem, however, is that time series for other parameters are too short for climatological analysis. In fact, only sea surface temperature (SST) data merged with SAT data have been extensively used for the analysis of global climate changes. The results of the analysis of four components of the climatic system: SST, SAT, surface wind and sea level pressure applying the empirical orthogonal functions (EOF) technique have been discussed. The results show that the patterns of SST and SAT change between the time periods of 1900-1929 and 1939-1968 and demonstrate that the climate warming was concentrated along the Gulf Stream east of Cape Hatteras and may have been a result of altered ocean current. Warming also occured over the Greenland Sea and the eastern subtropical Antlantic.

A number of efforts have been undertaken to analyse available data on precipitation. Combined datasets on rainfall, total water and water vapour over sea from polarised microwave simulations and Special Sensor Microwave/Imager (SSM/I) data have been discussed. Recently, some scientists considered global oceanic precipitation from the Microwave Sounding Unit (Tiros-N satellites) during 1979-1991 on a 2.5^{o} grid and made comparisons to other similar climatologies. The analysis of the global maps shows that peak annual rainfall (5600 nm) occurs in a quasistationary portion of the Intertropical Convergence Zone (ITCZ) over the eastern Pacific, while peak monthly raifall (over 900 mm) occurs in the northeastern Bay of Bengal in June. Comparisons with other data (both conventional and satellite) revealed several important differences with existing ocean rainfall climatologies which interpretation allows to judge about higher reliability of the MSU data. It is important, however, that at the 2.5^{0} gridpoint level, the correlation between two satellites monthly anomalies (MSU and GOES data for 81 months) is generally above 0.8 in the tropics, reaching 0.99 in the central Pacific.

Table 2. Trends of temperature (^{0}C/100yr) for annual and three-month mean maximum (MAX), minimum (MIN), and diurnal temperature range (DTR). Percent of the land area covered for the Northern and Southern hemisphere and the globe is denoted within parenthesis.

Northern Hemisphere (50%) 1951-1900			
Seasons:	MAX	MIN	DTR
D-J-F	1.3	2.9	-1.5
M-A-M	2.0	3.2	-1.3
J-J-A	-0.3	0.8	-1.1
S-O-N	-0.4	1.3	-1.7
Annual	0.5	2.0	-1.4

Southern Hemisphere (10%) 1951-1900			
Seasons:	MAX	MIN	DTR
D-J-F	1.6	2.2	-0.6
M-A-M	1.7	2.5	-0.8
J-J-A	1.0	1.3	-0.4
S-O-N	0.8	2.1	-1.3
Annual	1.3	2.0	-0.8

Globe (37%) 1951-1900			
Seasons:	MAX	MIN	DTR
D-J-F	1.3	2.9	-1.6
M-A-M	1.9	3.1	-1.2
J-J-A	-0.2	0.8	-1.1
S-O-N	-0.3	1.4	-1.7
Annual	0.7	2.1	-1.4

In connection with the problem of the Antarctic polar ice sheets dynamics, changes of the precipitation rate over Antarctic have been studied which appears to have increased by about 5% over a time period spanning the accumulation means for the 1955-1965 to 1965-1975 periods. During the same time period over Greenland the precipitation rate decreased by about 15% since 1963 with secondary increase over the southern part of the ice sheet starting in 1977. At the end of the 10-year overlapping period, the global sea level impact of the precipitation changes over Antarctica dominates that for Greenland and yields a net ice-sheet precipitation contribution of roughly -0.2 mm yr^{-1}. These conclusions should be considered as tentative, however.

Important studies of climate changes have been accomplished by many scientists with the use of satellite information on sea ice and snow cover dynamics. In the context of global warming contemporary large-scale changes in solid and total precipitation and satellite-derived snow over the North American continent have been examined. It is shown that snow cover extent over the last 19 years decreased up to 6×10^5 km^2 relative a 0.93°C (0.33°C) increase in North American (Northern Hemisphere) temperature. Over the last two decades the decrease in snow cover during winter (December-March) has largely occured through reduced frequency of snow cover in areas that typically have a high probability of snow on the ground with little change in the frequency of snow cover in other areas. Similar characteristics were observed during spring (April-May) in areas with high snow cover probability except for an expansion of the snow-free regions. Anomalies in these two seasons dominate the interannual variability (nearly three-fourth of the variance) of snow cover.

It has been emphasised that the apparent unprecedented global warmth of the 1980s was accompanied by a retreat of the mean annual North American snow cover, a 10% increase in annual Alaskan precipitation, a significant decrease (-7%) in annual snowfall over southern Canada (while the total precipitation remained above normal) and a more than twofold increase in the variance of tha ratio of frozen to total precipitation over the contiguous United States, where also precipitation has significantly (2%-3% per decade) increased during the last four decades, but on a century time scale the increasing trend is not yet statistically significant. Karl et al. (1993) have cautioned against oversimplification of relationship between precipitation and snow cover extent. An increase of precipitation with enhanced greenhouse gases may result both in the retreat of snow cover (in case of rainfall) and in the snow cover expansion (in case of snowfall). Therefore, a priori, the early evolution of future changes of snow cover is uncertain in high latitudes as the climate warms. Thus a careful investigation of the climatology of snowfall and snow cover is necessary. This is equally true for ice sheets.

A very important source of information are satellite measurements of the earth's radiation budget, especially in the context of the atmospheric greenhouse effect dynamics. This problem has been discussed in detail in Kondratyev et al. (1988) and others.

4 Paleoclimate dynamics

Various information on paleoclimate dynamics serves a very important source of information not only on how climate changed in the geological past but also gives numerous possibility to interpret this dynamics with the purpose of better understanding of the present-day climate changes (although paleoclimatic changes can not be correctly used as analogues to predict future climate). This is why scientific literature on past climates is so abundant. We shall mention here just a few recent publications.

An active direction of research are studies based on the use of such a proxy climate indicator as tree rings. Secular trends in high Northern latitudes temperature variations using boreal-tree ring records from high latitude North America, Scandinavia and Russia for 16/82-1968 have been studied. The principal finding is that the overall evaluation confirms that the high northern latitudes are now in anomalously warm state relative to the past three centuries. A 2000-year tree ring record of annual temperatures in the Sierra Nevada (California) mountains has been utilized to reconstruct the mean late seasons (June through January) temperature variations. Retrieved long -term trends are indicative of a 125-year periodicity that may be linkked to solar activity as reflected in radiocarbon and aerosol records (we are not going to do well upon a very important problem of solar activity climatic impact but would like to emphasise with satisfaction a new wave of interest to the problem; an indication of this wave is an application of general circulation models for relevant assessment: see, for instance, a detailed survey in the monograph by Kondratyev and Cracknell (in press).

An important subject of paleoclimatic studies is the faith of ice-sheets. GCM experiments have been accomplished which suggest that the sources of Greenland precipitation varied with different climate states, allowing dymamic atmospheric mechanisms for influencing the ice core isotope shifts which compliinterpretations of ice-core data.

In a conceptual study of geological assessment of the greenhouse effect has been emphasised that geological studies provide a valuable perspective on the importance of greenhouse forcing for climat change. On both Pleistocene and tectonic time scales in climate are positively correlated with greenhouse gas variations. During the last glacial maximum (14000-22000 BP), when global sea surface temperature decreased by about 2°C, ice core data indicate significant decreases in carbon dioxide, methane, and nitrous oxide concentrations. Longer records also indicate good correlation, but there are lags between temperature and CO_2 concentration over some time intervals. The total change in radiative forcing from ice-age trace gas variations is on the order of 2.5-3.0 Wm^{-2}. Some scientists believe that the sensitivity of the climatic system to greenhouse gas changes can not yet be constrained by paleoclimatic data below its present last range (1.5-4.5 global temperature increase in the case of CO_2 doubling). Geologic records do not support one of the major predictions of greenhouse models-namely, that tropical sea surface temperature will increase. Geological data also suggest that winter warming in high latitude land areas is stronger than predicted by models. The conclusion is obvious: further improvement of climate models as well as more complete analysis of paleoclimatic data are necessary.

The key problem of paleoclimatic reconstructions is to assess reliability of such reconstructions. This problem has been discussed in detail by some scientists, who pointed out that the 19th century measurements of CO_2 in the atmosphere were carried out with an error of up to 100%. Even observations at the Mauna Loa (Hawaii) observatory since 1957 which are considered as globally representative may be subject to substantial biases, because the observatory is located near the top of an active volcano, which has, on average, one eruption every three and a half years. There are permanent CO_2 emmissions from a rift zone situated only 4 km from the observatory, and the largest active volcanic crater in the world is only 27 km from the observatory, and the largest active volcanic crater in the world is only 27 km from the observatory.

Scientists have also criticized data on CO_2 concentrations in air bubbles trapped in glacier ice which are often interpreted as previous atmospheric concentrations, assuming that the composition of the air in the bubbles remained unchanged. Numerous studies indicate that, due to comparison with the original atmospheric level in the air inclusions from pre-industrial ice the CO_2 concentrations were found to range between 135 and 500 ppmv. These and other circumstances lead to the conclusion that glacier ice cannot be regarded as a steady state matrix suitable for observation of long- term atmospheric trends. Thus, the results of CO_2 determination in air inclusions in ice cannot be accepted as representative the original atmospheric composition. The similar conclusion can be made concerning retrieval of the paleoatmospheric CO_2 content from 13C / 12C carbon stable-isotope ratios in tree rings.

Scientists have emphasised large uncertainties of paleo-temperature calculations based on light stable isotope ratios (D/H and $^{18}O/^{16}O$) in ice mentioning that after the discovery of liquids between ice crystals in the deeply frozen Antarctic ice, considerable isotopic exchange and fractionation should be expected in the ice, making calculated paleotemperatures meaningless if phase changes occured in the presence of a mobile fluid phase.

The results mentioned allowed to conclude that "the so-called increasing greenhouse effect signal, i.e. the anthropogenic increase of the global air temperature, which was claimed to have been observed during the last decades, is not confirmed by recent studies of long temperature series". As we have seen earlier such a conclusion is in accordance with the analysis of climate diagnostic data. This is why it was written that: "There is no conclusive evidence that global warming, associated with the enhanced greenhouse effect, is occuring. We have a long way to go in our understanding of and ability to model the complex climate system before confident predictions of the aspects of climate change that need to be known".

5 A few comments on numerical climate modeling results

Although a number of numerical modeling experiments to assess climatic consequences of doubling CO_2 concentrations has been excessive (too many unnecessary repetitions took place: (see Kondratyev and Cracknell (in press)) the problem is far from being solved. The results obtained were very useful to analyse climate sensitivity to CO_2 increase but hardly realistic because even the most sofisticated three-dimensional global climate models can not reliably simulate real climate changes. The principal reasons of such a situation are that parametresations of atmosphere-ocean interaction, cloud dynamics and cloud-radiation interaction can not be considered as adequate enough as yet, although, undoubtedly, important progress has been recently achieved (Viner and Hulme, 1993a; Viner and Hulme, 1993b).

As has been pointed out in the Report (1993), only recently first global coupled (for the atmosphere-ocean system) climate modeling has been accomplished in the four scientific centres: the Max Planck Institute fur Meteorologie, Germany), the Geophysical Fluid Dynamics laboratory-GFDL (USA), National Center for Atmospheric Research-NCAR (USA), the United Kingdom Meteorological Office - UKMO. There are only preliminary but very important results of these modeling efforts. It has been discovered, for instance, on the basis of four 50-year climate change simulations, starting at different times during a 100-year control integration but otherwise with identical initial conditions including 1% per year rise of CO_2 concentration (the Max Planck Institut fur Meteorologie) that the period before appreciable warming begins to occur is different in each case, ranging between 6 and 31 years. The geographical patterns of the responses may be significantly different, especially at middle and high latitudes where natural variability is larger. This indicates that climate changes computed with a coupled-atmosphere model may depend significantly on the initial conditions up to a few decades ahead and different experiments may therefore give notably different results. It has been emphasised in the Report (1993) that the actual variation of the Earth's climate observed over the next several decades might be as different from any computed response as the differences between various simulations.

The numerical experiments conducted in GFDL with the model intergration which has been extended to 1000 years (with a low resolution coupled ocean-atmosphere model) suggested a physical mechanism for low frequency climate changes in the north Antlantic region (a specific feature of previous model has been the failure to maintain cold water in the north Antlantic and north Pacific). The results obtained with a coupled NCAR model allow to simulate warm and cold events in the tropical Pacific which resemble the observed characteristics of the Southern Oscillation, although their amplitude is generally smaller. Similar results are found with the UKMO model.

As an illustration of the most recent efforts we shall mention a recent study on the intra-annual and inter-annual variability of the greenhouse climate change using a low resolution four layers (900, 650, 350 and 100 hPa) spectral (R21) global climatic model. The model included the diurnal and seasonal cycles, smoothed orography, computed sea ice amount and cloud cover and used implied oceanic heat fluxes to represent transport processes in the oceans. A highly responsive 2-layer soil moisture formulation was also incorporated.

Greenhouse induced climate change has been analysed as a difference between twenty year equilibrated simulations for greenhouse (2 x CO_2) and control (1 x CO_2) conditions. Such a comparison revealed considerable differences from the time-averaged results with occasions of marked positive and negative temperature deviations. Of particular interest were the periods of negative temperature departures compared to the control run which were identified, especially over the Northern Hemisphere continents. This is clear demonstration of the complexity of the possible climatic responses under greenhouse conditions.

Table 3 data illustrate annual means and standard deviations for time series which characterize sea surface or land surface temperatures at certain points. Average figures here show greenhouse warming in all the points, but the analyses of temperature variability for individual points reveals large differences between calculated variability for various points as well as between calculations and observations. For instance, the central Pacific gridpoints characterized an enormous range of simulated sea surface temperature variability. In case of the northern point rather regular behaviour with interannual variability of about 2°C in the winter temperature of the control run (1 x CO_2) and the greenhouse simulation represented a simple warming of about 3°C superimposed on the control. But for the low latitude point, a completely different situation prevailed with very marked interannual variability consisting of warm and cold periods of few year duration. Cosiderable variability also took place in the low latitudes of the Indian Ocean. Under conditions of the land points an additional source of variability exists in that soil moisture changes can have a noticable impact on surface temperature, particularly for regions where drought conditions tend to prevail.

The presence of only small differences between the standard deviations of the control and greenhouse (2 x CO_2) runs implies that the same underlying physical processes were operative in the two runs despite the considerable changes in their respective climatologies. The existence (in a number of situations) of very similar variations for both the greenhouse and control runs in the spatial characteristics and amplitude of these deviations suggests that, under greenhouse conditions, occasions will arise when natural variability will override the greenhouse warming. It is important to emphasize in this connection that any clear greenhouse "fingerprint" except changes in the mean values could not been identified. Further careful analysis is necessary, however, with the application of elaborate statistical methods.

Since one of the most dangerous global warming consequences may be sea level changes (Warrick et al. (1993)) a transient (1% CO_2 concentration increase per year) coupled ocean-atmosphere GCM developed at the U. K. Meteorological Office has been applied to conduct 75-yr numerical experiment for sea level rise computations. The results show that over the final decade global sea level rise of 90 mm due to thermal expansion of the ocean takes place. An important conclusion is also that there most be significant geographical variability of the sea level rises. The largest rises are found in the northwest Antlantic.

Table 3. Annual means and standard deviation (SD) for time temperature series at various points.

Point	$1 \times CO_2$		$2 \times CO_2$	
	Mean	SD	Mean	SD
North Pacific	285.29	0.29	288.36	0.35
Tropical Pacific	300.98	0.61	301.71	0.41
Indian Ocean	300.71	0.29	303.34	0.41
Canada	273.67	0.86	277.92	0.89
China	299.46	0.41	301.64	0.38
South Africa	290.06	0.31	293.79	0.44

A significant progress has been achieved through the analysis of satellite earth's radiation budget and cloud observations as well as similar computational results to assess the role of cloud-radiation interaction.

Some scientists have compiled Table 4 data which characterise hemispheric and global averages of cloud radiation forcing (i.e. differences of radiation budget components for cloudy conditions and clear sky) at the surface CRF(S) for shortwave (sw) and longwave (lw) components as well as for the atmosphere: CRF(A). Seasonal change has been defined as the difference between July 1985 and January 1986 values. $CRF(S)_{sw}$ data show that the SW effect at the surface is a cooling for all seasons in both hemispheres. The strongest cooling (-120 to -180 Wm^{-2}) occurs over midlatitude storm tracks of the summer hemisphere. In contrast, $CRF(S)_{lw}$ is always positive in both hemispheres, being maximum (60 to 75 Wm^{-2}) over storm tracks of the winter hemisphere. Zonal averages (not shown here) indicate that the entire summer hemisphere is dominated by shortwave cooling, the middle and high latitudes of the winter hemisphere are dominated by longwave warming, with a broad transitional zone in between.

The globally averaged total CRF is cooling throught the year, ranging from a low of about -12 Wm^{-2} for July 1985 to a high of about -25 Wm^{-2} for January 1986. Table 4 data show that SW cooling exceeds LW warming for all seasons in both hemispheres with the exceptions of the SH in July and the NH in January.

On the basis of the International Satellite Cloud Climatology Project (ISCCP C1 data). Shortwave cloud radiative forcing has been calculated at the top of the atmosphere (TOA) and at the surface and of the atmospheric column to arrive at the following estimates for area weighted global averages of the monthly mean values for three consecutive July's (Wm^{-2}):

	July 1983	July 1984	July 1985
TOA:	-43.6	-43.1	-39.1
Surface:	-44.6	-44.3	-40.9
Atmosphere:	1.0	1.2	1.8

Table 4. Seasonal variation of hemispheric and global averages of surface and atmospheric cloud radiative forcing, Wm^{-2}.

	April 1985	July 1985	October 1985	January 1986	Annual Average	Seasonal Change
			CRF(S)sw			
NH	-54.8	-63.8	-41.5	-30.1	-47.6	-33.7
SH	-41.8	-27.4	-70.2	-89.8	-57.3	62.4
Global	-48.3	-45.6	-55.9	-60.0	-52.5	14.4
			CRF(S)lw			
NH	34.1	28.9	30.5	33.5	31.8	-4.6
SH	36.9	38.3	38.2	36.2	37.4	2.1
Global	35.5	33.6	34.3	34.9	34.6	-1.3
			CRF(S)			
NH	-20.7	-34.9	-11.0	3.4	-15.8	-38.3
SH	-4.9	10.9	-32	-53.6	-19.9	64.5
Global	-12.8	-12.0	-21.5	-25.1	-17.9	13.1
			CRF(A)lw			
NH	-2.6	4.7	3.6	-7.3	-0.4	12
SH	-5.1	-10.9	-7.4	-1.6	-6.3	-9.3
Global	-3.9	-3.1	-1.9	-4.5	-3.4	1.4

An important finding is that relative to global, annual mean (GAM) value of the solar radiative forcing (a negative quantity for an increase of liquid water path of clouds) the forcing is more negative in low latitudes throught the year and during summer at the high latitudes. In contrast, the forcing is more positive that the GAM value during the winter season at the higher latitudes (poleward of 40%). A significant aspect of the problem is possible cirrus cloud feedback. Some scientists have accomplished an intercomparison of observed cloud radiative forcing in the context of zonal and global perspectives.

Further improvement of cloud dynamics parametresation techniques is most urgent. An important step towards the solution of this problem is the development of three-dimensional numerical models of cloud dynamics, microphysics and chemistry.

It has been undertaken by some scientists a new analysis of the cloud optical feedback in dependence on the cloud optical thickness, τ. The feedback is characterised by its intensity S, expressed in percent: $S = (\Delta T_\tau / \Delta T - 1)\ 100$, where ΔT_τ and ΔT are changes of surface air temperature with and without taking account of the feedback. Assessments which had been made earlier in various studies showed that CO_2 concentration growth results in variations of S within broad range of values from small positive to large negative. Such controversial results may be explained by differences in parametresation techniques for simulation of cloud dynamics and cloud phase changes as well as differences in control climates.

The value of S may depend on a number of specific parameters which determine climate changes. For example, in cases of assessments of inbfluences of CO_2 concentration or solar constant increases an important role belongs to variations of vertical profiles of radiative forcings. Since vertical radiative forcings distributions for CO_2 and other greenhouse gases (CH_4, N_2O, CFC's) are not similar, new assessments have been obtained by scientists of the contribution of the cloud-optical feedback in the climate change due to GHGs increase under considerations of one- and multilayer cloudiness.

For some cloud types the parametresation of the feedback is acceptable which is based on the consideration of the dependence of τ on cloud liquid water content (LWC) through their temperature (CT scheme). Another possibility is the CP scheme with the consideration of the dependence of τ on LWC through latent heat due to precipitation formation. Calculations have been made of the basis of one-dimensional radiative-convective climate model which show that in case of CP-scheme stronger cloud-optical feedback forms if CH_4, N_2O and CFC's concentration increase takes place than under considerations of CO_2 concentration growth (Table 5).

As it is seen from the table this difference is especially strong between CO_2 and CFC's: in the first case the cloud-optical feedback is weakly negative, while in the second case it is strongly negative. The use of CT scheme shows that the transition from strongly negative to much weaker negative feedback takes place if one-layer cloudiness is substituted by multilayer cloudiness.

Table 6 data illustrate changes of cloud optical thickness, $\Delta\tau$, radiative flux divergence at the tropopause level, ΔF (Wm^{-2}) and cloud temperature, Tc, for various cloud conditions in case of CO_2 doubling. In the case considered S = -33%. As it could be expected, the enchancement of upper level (cirrus) clouds results in weakening the feedback. In case of the upper level clouds $\Delta F > 0$ that is their contribution to the feedback is positive. However, negative radiative forcing due to low and middle level clouds results in the significant negative S value.

It is, of course, necessary to take into account latitude dependence of the contributions considered. In the tropics the role of cirrus clouds is, naturally, the strongest which results in weakening the negative feedback, while the opposite situation is typical for high latitudes.

Table 5. Impact of cloud-optical feedback (COF) on the surface air temperature change (^oC). Figures in brackets show relative S values (%).

Case	Without COF	With COF	
		CP Scheme	CT Scheme
CO_2: 330 ⟶ 660 ppm	2.88	2.49 (-13)	1.68 (-42)
CFC-11: 0 ⟶ 2ppb	0.57	0.39 (-31)	0.34 (-41)
CFC-12: 0 ⟶ 2ppb	0.57	0.39 (-31)	0.34 (-41)
N_2O: 0.3 ⟶ 0.6 ppm	0.72	0.55 (-22)	0.44 (-39)
CH_4: 1.7 ⟶ 3.4 ppm	0.48	0.58 (-20)	0.33 (-30)

Table 6. Changes of cloud optical thickness, $\Delta\tau$,radiative forcing, ΔF, and cloud temperature, ΔT_c, in case of CO_2 doubling (CT scheme).

Parameter	Cloud levels		
	Upper	Middle	Low
τ	1.78	5.18	11.84
$\Delta\tau$	0.22	0.58	0.87
ΔF	0.25	-0.7	-0.84
ΔT_c	1.93	1.93	1.93

Satellite microwave soundings results confirm the existence of correlation between τ and precipitation which has been used in the CP scheme. On the other hand, retrieval of cloud characteristics indicates that in low latitude ocean regions a decrease of τ with the increase of temperature of low and middle levels clouds is observed which shows a deviation from adiabatic conditions. Strong cloud-optical feedback sensitivity to vertical distributions of clouds and τ demonstrates a necessity to further develop the reliability of cloud cover climatology data, especially from the viewpoint of cloud liquid water content. It is, probably, also important to get observational data on cloud microphysical parameters and three-dimensional structure of cloud cover.

Some scictists have calculated the cloud diabatic heating under average and in clear-sky conditions, from time series of simultaneous observations of diabatic heating (ECMWF initialised analyses data) and cloud amount (ISCCP data). The results of calculations indicate that the CDF is dominated by release of latent heat in the storm track regions and intertropical convergence zone (ITCZ). Over summer continents the CDF is weakly negative, due to the decrease of sensible heating with increasing cloud amount.

Since a basic feature of climate change is its inhomogenety in space and time an important task is to consider subgrid-scale inhomogeneties in the climatic system and to develop relevant parametresation techniques. Various kinds of approaches have been suggested to solve such problems. For example, it has been realised an idea to interrelate the characteristic patterns of observed simultaneous variations of regional climate parameters and of large-scale atmospheric flow using the canonical correlation technique. The possibilities of scaling water and energy fluxes in climate systems have been analysed and it was concluded that the surface fluxes and land characteristics can be scaled and microscale models based on effective parameters are sufficient to account for the small-scale heterogeneties investigated. The scale at which a microscale model becomes valid, the representative elementary scale (REA), is on the order 1.5-3 correlation lengths, which for land processes investigated appears to be about 1000-1500 km.

It has been considered the scaling problem for the purposes of land biosphere-atmosphere interactions parametresation. The principal aim has been to study the effects of subgrid variability in leaf area index, minimum and maximum stomatal resistance and soil moisture on grid-scale fluxes. Comparisons between energy fluxes obtained using parameter values for the average of 100 subgrid points and the average fluxes for the 100 subgrid points showed minor differences for emitted IR radiation and reflected solar radiation but large differences for sensible heat and evapotranspiration. Leaf area index was the most important parameter (stomatol resistances were only important on wet soils). Interactions among parameters considered increased the non-linearity of land-atmosphere energy exchange. It should be emphasised that further studies are most necessary to improve land surface processes parametrisation techniques.

Predictions of greenhouse-induced future climate changes depend on the reliability of global carbon cycle simulations. For the time being global carbon cycle models are still not adequate and an important indication of this situation is the so called "missing carbon" (Bolin, 1993) i.e. the fact that the global carbon budget has not been closed in the most of studies. An urgent necessity in this respect is the development of climate models with coupled consideration of the biosphere dynamics, which influences climate dynamics very substantially. In this connection an important discovery has been an increasing turnover through time in tropical forests due to (probably) CO_2 increase and subsequent fertilisation effect.

The simulation of the biosphere as a coupled component of the climate system has become a very important task. This has also stimulated the interest to assessments of an impact of climate changes on the biosphere dynamics. It is in this context that some scientists used a biome model to predict global patterns of potential plant formations, or biomes, from climatologies simulated by ECHCAM, a model used for climate simulations at the Max - Planck - Institute fur Meteorologie. The purpose of the modelling is twofold: a qualitative test of simulated climatologies and an assessment of the effects of climate change.

The first important result is that a good overall agreement between simulation and observation of global patterns of biomes has been found. There are also discrepancies for Australia, Kalahari Desert and in the Middle West of North America, which can be traced back to failures in simulated rainfall as well as summer or winter temperatures.

As an example of potential vegetation changes in response to simulated changes in climate global patterns of biomes have been computed from an ice age simulation 18.000 years BP. These calculations indicate that North America, Europe, and Siberia should have been covered largely by tundra and taiga, whereas only small differences are seen for the tropical rain forests. Since a new improved ECHAM-3-T42 model has become available it will be used for continuation of similar modelling.

In accordance with the IPCC scenario A of the CO_2 concentration increase simulation has been accomplished to predict relevant biome dynamics. The results show that the largest changes occur for boreal biomes, whereas little change is seen for the Sahara and the tropical rain forests. It was noted, however, that since the biome model is not capable of predicting changes in vegetation patterns due to a rapid climate change, this simulation has to be taken as a prediction of changes in conditions favourable for the existence of certain biomes, not as a prediction of a future distribution of biomes. Because the biome model does not take into account changes in vegetation due to changes in CO_2 and soil feltilization, subsequent studies are necessary to couple a dynamic model of vegetation succession to a climate model. A next step of the study will be to use nested approach to simulate smaller scales dynamics. The sensitivity of a climate model to changes in global patterns of biomes will also be investigated.

6 Aerosol cooling versus greenhouse warming

An old problem of aerosol impact on climate (Kondratyev, 1988, 1992) has recently survived a new wave of interest with the purpose to assess natural and anthropogenic climatic effects (see also Fong ., 1993; Schwartz, 1993).

In connection with the problem "aerosol cooling versus greenhouse warming" a great attention has been attracted by the hypothesis of dimethylsulphide (DMS) climatic impact which is based on the assumption that phytoplankton could affect cloud albedo by producing DMS, that the latter is a precursor to aerosols and cloud condensation nuclei (CCN) and that cloud albedo could in turn effect the productivity of the phytoplankton thus creating a climatically significant feedback cycle (see Kondratyev and Cracknell (in press)).

It has been undertaken an empirical analysis of the strength of this phytoplankton-dimethylsulphide-cloud-climate cycle by considering available data on the relationships between individual components of the feedback and developing an empirical model of the cycle as a whole which allows an assessment of the strength of the cycle to thermostatically regulate climate.

The feedback considered includes three components:

1) the coupling between phytoplankton, DMS and CCN;

2) the effect of changes in CCN levels on albedo, incident irradiance at the surface and surface temperature;

3) the response of phytoplankton to changes in incident irradiance and seawater temperature.

The first and third components of the cycle have been derived primarily from observations of their respective chemical, physical and biological constituents. The second component was based on the estimate that a 30% increase of CCN in the area covered by marine stratiform clouds would cause a decrease in the global average surface temperature of 1.3°C, as well as enhanced sulphur emissions should cause roughly the same magnitude of cooling via the direct aerosol backscatter effect as via the indirect cloud albedo effect.

The first step of the consideration has been the substantiation of the relationship between phytoplankton population and seawater dimethylsulphide concentration $[DMS]_{sw}$. The following relationship for the DMS flux has been chosen:

$$F_{DMS} = K_L \left([DMS]_{sw} - \frac{[DMS]BL}{H}\right) \tag{9}$$

where K_L is the piston velocity, an empirical constant; $[DMS]_{BL}$ is the boundary layer DMS concentration: and H is the Henry's law constant for DMS.

Since for the range of atmospheric and oceanic DMS concentrations which were observed the second term of (9) is negligible, one can simply use:

$$F_{DMS} = K_L\, [DMS]_{sw} \tag{10}$$

where area-weighted global annual mean K_L value (which, in general, depend on meteorological conditions) may be prescribed as equal to 2.5 ± 0.2 md^{-1}.

The first step of the empirical parametrisation is the choice of the following relationship between F_{DMS} and aerosol number density (CN concentration):

$$[CN] = A_1 \ln (F_{DMS} + 1) + B_1 \tag{11}$$

If F_{DMS} is expressed in $\mu molm^{-2}d^{-1}$ and [CN] in cm^{-3}, then the constants are $A_1 = 184 \pm 34$ in cm^{-3} $(\mu molm^{-2}d^{-1})^{-1}$, $B_1 = 43 \pm 61$ in cm^{-3}. The final step is the parametrisation of the growth of CN to CCN:

$$[CCN] = A_3\, [CN] \tag{12}$$

where $A_3 = 0.46 \pm 0.24$ (number densities are expressed in cm^{-3}).

For sea water DMS concentration the following formula has been substantiated:

$$[DMS]_{sw} = \frac{\mu\beta P_p}{K_L \frac{AB}{VB} + L_{wc}} \tag{13}$$

where P_p - the phytoplankton population density; μ is the phytoplankton specific growth rate; β - is the mean DMS content per phytoplankton cell, AB/VB is the ratio of the area of the bloom at the ocean surface to the volume of the bloom; L_{wc} $[DMS]_{SW}$ represents various losses of DMS in the water column (microbial degradation etc.) apart from zooplankton grazing and to a lesser extent, senescence which are considered as basic factors responsible for the release of DMS into surrounding waters.

Since L_{WC} is significantly less light and temperature dependant than β in (13), the simplified expression may be used for assessments of changes due to incident irradiance and temperature variations:

$$[DMS]_{SW} = \eta\mu \qquad (14)$$

where η = 2.3 nmol DMSl^{-1} d (the potential range of η values is from 1.7 to 21 nmol DMSl^{-1}).

The phytoplankton specific growth rate (μ) in dependence of incident irradiance and temperature has been parametrised on the basis of existing empirical models. The first order feedback strength, FS, is computed as the ratio of the change in [CCN] predicted by the model to the change in [CCN] necessary to counteract the initial perturbation:

$$FS = 100 \times a_d \frac{[CCN]p - [CCN]o}{[CCN]c - [CCN]o} \qquad (15)$$

where FS is in per cent; a_d is the parameter which has been introduced to take into account not only the feedback through indirect (cloud albedo) cooling, but also direct cooling due to enhanced sulphur emissions (enhanced backscattering) which has approximately the same value (therefore $a_d = 2$).

Calculations of FS values were made for several different cases gave the "expected" FS (for an initial temperature of 15 ^{o}C) equal to 23% which indicates that the marine biogeochemical cycle could be playing a significant role in the global climate change (for example, in aerosol cooling versus greenhouse warming). It is important to note that an FS of 20% represents a change in [CCN] of only 3% for a temperature perturbation of 1.3 ^{o}C. Thus, the climate is extremely sensitive to the changes in aerosol and CCN levels. The potential range of FS values is from 10% to 50%. It is obvious that a 50% value implicates significant revisions of global climate change assessments, but, of course, a degree of uncertainty is very high which defines a necessity of further studies. It has been pointed out that two areas in particular are in need to reduce uncertainties: 1) phytoplankton ecology and the concomitant DMS production and 2) the relationship between DMS emissions and [CN].

Important studies of biogenic sulphur aerosol in the Arctic troposphere have been made on the basis of long-term observations from 1980 to 1990 of aerosol methanesulphonate (MSA), sulphate, sodium and other related chemical species as well as a shorter time series of aerosol sulphur isotopic composition ($\delta^{34}S$) at Alert, Northwest Terratories, Canada. The principal aim of the studies was to determine the contributions of sea salt, biogenic sources and pollution to aerosol SO_4^{2-}. The utilisation of two different processing techniques gave similar values of MSA/biogenic SO_4^{2-} mass ratio within 0.2 to 0.9 for the summer months and much lower in October to March (<0.08). The analysis of the isotopic composition data show that aerosol SO_4^{2-} in summer is 25 to 30% biogenic, 1 to 8% sea salt and the rest (62 to 74%) anthropogenic in origin. At other times of year it is <14% biogenic, 1 to 8% sea salt and the rest anthropogenic in origin.

Various aspects of anthropogenic aerosol - climate problem has been recently studied. An important finding has been that the satellite data (low-level cloud albedo) demonstrate enhanced cloud albedo near the costal boundaries where sulphate concentration are large. Similar trends are absent over ocean regions of the Southern hemisphere that are removed from anthropogenic sulphate sources.

Some scientists have assessed an impact of fossil fuel and biomass burning on climate. They have pointed out that although coal and oil emit 120 times as many CO_2 molecules as SO_2 molecules, each SO_2 molecule is 50-1100 times more effective in cooling the atmosphere (through the effect of aerosol particles on cloud albedo) than a CO_2 molecule is in heating it. They concluded that the cooling effect from coal and oil burning presently range from 0.4 to 8 times the heating effect. Within this large uncertainty, it is presently more likely that fossil fuel burning causes cooling of the atmosphere rather than heating. Biomass burning associated with deforestation, on the other hand, is more likely to cause heating of the atmosphere than cooling since it aerosol cooling effect is only half that of fossil fuel burning and its heating effect is twice as large. For a doubling in the CO_2 concentration due to fossil fuel burning, the cooling effect is expected to be 0.1 to 0.3 of the heating effect.

In a later publication it has been concluded (using a conservative approach) that the sulphate aerosol induced cooling can presently counteract 50% of the CO_2 greenhouse warming. Their assessment also showed that a complete cessation of SO_2 emissions will result in a warming surge of 0.4°C in the first few years after the elimination of the SO_2 emission. Table 7 data illustrate results of numerical modeling for two extreme IPCC scenarios for the CO_2 growth: BAU (business as usual: the strongest increase of CO_2 concentration) and D (minimum increase of CO_2 concentration).

Table 7. The CO_2 - and SO_2 - induced changes in the surface air temperature (^0C) for years 1980 and 2060 for Southern and Northern hemispheres.

Emission Scenario	Change from 1900 to 1980			from 1900 to 2060		
	NH	SH	Global	NH	SH	Global
			BAU scenario			
CO_2 only	0.31	0.29	0.30	2.04	1.73	1.89
$CO_2 + SO_2$	0.01	0.21	-0.20	1.31	1.54	1.43
SO_2-induced cooling	0.30	0.08	0.19	0.73	0.19	0.46
			Scenario D			
CO_2 only	0.31	0.29	0.30	1.13	0.98	1.06
$CO_2 + SO_2$	0.01	0.21	0.11	0.89	0.89	0.89
SO_2-induced cooling	0.30	0.08	0.19	0.24	0.09	0.17

Some scientists believe that the uncertainties in the SO_2-induced radiative forcing and climate cooling is a factor of 3-4. According to the BAU scenario a potential of offsetting the CO_2-induced warming is equal to 60% at the present time and 25% by the year 2060.

Some scientists have studied seasonal, latitudinal and secular variations in temperature trend with the purpose to identify influence of anthropogenic sulphate. These data indicate that prounounced minima in the rate of temperature increase in summer months in Northern Hemisphere midlatitudes are consistent with the latitudinal distribution of anthropogenic sulphate and changes in the rate of SO_2 emissions over the industrial area.

To summarise, one can say that all existing empirical and theoretical assessments of aerosol climatic impact are still highly speculative. As far as numerical modeling results is concerned it is highly surprising that they have ignored aerosol absorption. The observations made within the CAENEX programme long ago (Kondratyev, 1972) showed that under various conditions studied aerosol absorption of shortwave radiation in the troposphere was, on the average, equal to water vapour absorption. Therefore it is very important to undertake new field measurement of the chemical and physical properties of atmospheric aerosol particles.

Undoubtedly, aerosol versus greenhouse gases climate change problem is very important. There are many reasons to believe that aerosol contribution to climate formation is comparable with the impact of greenhouse gases. Some scientists assessments gave the interval for current climate forcing due to anthropogenic sulphate from -1 to -2 Wm^{-2}; estimates by Rodhe (1994) has led to the interval from -0.2 to -1 Wm^{-2} versus 2-2.5 Wm^{-2} greenhouse warming.

However, the conclusion made by some scientists, sounds most persuasive "because of the strong role of cloud dynamics, the independent sulphate aerosol concentration effects on cloud microphysics, cloud amount and rainfall are difficult to assess".

It is also wortwhile to emphasise that more attention in the context of aerosol and climate problem should be spent to the analysis of the results of an impact of volcanic eruptions on climate (Kondratyev, 1991; Kondratyev, 1993) as well as to the climatic consequences of the Arctic haze.

7 Greenhouse warming climate signal

A controversial problem of detecting climatic change signals is a very important aim of the climate theory to identify climate changes due to various causes. This problem may be defined as controversial because, as has been mentioned earlier, the most important specific feature of the climatic atmosphere-hydrosphere-lithosphere-cryosphere-biosphere system is that it is based on strong coupling of its components. Therefore an isolation of a contribution of certain component in the formation of climate change is a formidable task. Nevertheless, numerous attempts have been recently undertaken to analyse possibilities of early detection of CO_2 climatic signal (see Kondratyev, 1992; Kondratyev and Grassl, 1993). The principal purpose has been to filter out a greenhouse warming signal due to the increase of greenhouse gases concentrations in the atmosphere. Here we shall limit the discussion by just a few comments on the most recent results.

The subjet of global warming and the problem of testing for trend in time series data which solution was an aim of a number of earlier studies have been considered. The standard approach of testing for the presence of a linear trend in global temperature datasets (to identify greenhouse warming) is to assume the model

$$Y_i = a + bt + E_t \qquad (16)$$

where Y_i represents the data at time t and E_t is the deviation of the data from a straight line. It has been made the assumption that E_t is a stationary, zero-mean process. If $b \neq 0$ (at an appropriate significance level) then it is generally accepted that a linear trend is present which defined as deterministic or long term. Although in a number of studies a significant deterministic trend in the temperature data discovered, it has been shown that certain autoregressive-moving average (ARMA) model may also be very reasonable models for these data due to the random trends present in their realisations. It has been demonstrated, thus, that trend based on models such as (9) for the purpose of prediction or inference concerning future behaviour should be used with caution. Based solely on the available temperature anomaly series, it is difficult to conclude that the trend will continue over any extended length of time.

Summarising the present-day understanding of climate change causes some scientists have emphasised that "century-long global warming (or cooling) trends of

0.5 °C appear to have occured infrequently over the past several years - perhaps only once or twice a millenium, as proxy records suggest. This implies an 80 to 90 percent heuristic likelihoud that the 20th century 0.5 ± 0.2 °C warming trend is not a wholly natural climatic fluctuation". These scientists believe that "a goal of 99% statistically significant signal detection over the next decade or two is unrealistic, unless the global net anthropogenic forcing is known to better that 0.5 Wm^{-2} and is thus large enough that "noise" from solar, volcanic and other "minor" natural perturbations can be overcome".

As far as a methodological approach to identifying fingerprints is concerned it should be mentioned that it has been derived perspective optimal linear filter for the detection of a given time-dependent, multivariate climate change signal in the presence of natural climate variability noise.

8 Surface ozone observations at Athens, Greece

There are few long-term surface ozone observations. The historical ozone series considered consists of measurements of surface ozone mixing ratios taken at the National Observatory in Athens for the period 1901-1940, using De James colometric papers. Paper strips were exposed to atmospheric air for varying periods and were shielded from sun and rain. Depending on the ozone mixing ratio, there was a corresponding colour change which was then correlated to a chromatic scale from 0 to 21.

Measurements were influenced by relative humidity, an effect which was found to be significant in several studies (Varotsos and Cartalis, 1991). This was taken into account using a conversion chart which provides the correlation between the De James scale and the surface ozone mixing ratio and involves the relative humidity (Varotsos and Cartalis, 1991). It should be mentioned that the construction of the conversion chart was based on test runs for Schoenbein papers at varying relative humidity levels.

The 1987-1990 ozone record consists of daily averages of measurements taken at 30 minute intervals at several monitoring stations of the National Service for Air Pollution Monitoring located at the perimeter of the National Observatory of Athens. For the purposes of this study, only highest measurements were used.

Daily mean ozone mixing ratios were analysed to yield:

- the annual cycle of surface ozone mixing ratio for the period from 1901 to 1940 and from 1987-1990 and
- the amplitude, phase and the percentage contribution to the variance of the first three harmonics for both periods.

It should be mentioned that the harmonic analysis for the period 1901-1940 was applied only for the corrected ozone mixing ratios.

The annual cycle of the corrected mean surface ozone for the period 1901-1940 shows that maximum and minimum corrected surface ozone values are observed during the periods from April to June and from October to December respectively. The uncorrected surface ozone for the time period demonstrates that there exist large differences which result in the event that the effect of relative humidity and exposure time on measurements is neglected. It should be noted here that for several subperiods (1901-1903, 1907-1908, 1934-1935 and 1938-1940) the 4-month cycle is stronger in magnitude and has a larger percentage contribution to the total variance than the 6-month cycle. In terms of the period 1987-1990 the 6-month cycle is comparable to the 4-month cycle.

In the following the annual cycles of the daily surface ozone mixing ratios at Montsouris (1877-1907), Hohenpeisenberg (1971-1988) and Arosa (1954-1958) are compared to the annual cycle of the daily surface ozone mixing ratios in Athens for the periods 1901-1940 and 1987-1990. One interesting feature is the relatively small differences of the ozone levels between the 1901-1940 and 1987-1990 Athens records. This is because daily (24 h) averaged ozone values are used for the interpretation. If the average morning ozone values are used instead, differences increase and the 1987-1990 ozone values are almost double the 1901-1940 values. The large differences are directly attributed to the increase of the concentrations of the precursors of ozone, namely NO_X and hydrocarbons (Varotsos and Kondratyev, 1994).

8.1 The surface ozone variations due to changing meteorological conditions

Many efforts have been made to interpret the daily and seasonal variations in the amount of surface ozone, due to the influence of various meteorological phenomena. In India, for example, surface ozone concentration variations have been correlated with rainfall in a non-polluted area and concluded that the day-time rainfall is accompanied by a decrease in the surface ozone concentration, while nocturnal rainfall is accompanied by an increase of surface ozone concentration.

Galbally (1968) conducted efforts to explain the correlation between surface ozone concentration and variations of surface air temperature during the day and attributed this correlation to the thermal turbulence and mixing in the lower troposphere. Later Decker et al. (1976), Shreffler and Evans (1982) and Evans et al. (1983) proved that a direct relation exists between the daily variations in air temperature and surface ozone concentration. We believe, however, that since temperature variations are usually very well correlated with solar radiation changes the latter produce variations of the photochemical activity which leads to ozone variability.

Recently Gusten et al. (1988) observed variations on surface ozone concentration during vertical motions of air masses. Short-term fluctuations which often appear in the surface ozone concentration daily variations have also been studied in recent years. Their investigation in relation to various meteorological phenomena is particularly difficult since the simultaneous interaction with more than one meteorological parameter often leads to inconsistent conclusions. For instance, the association between the surface ozone concentration and the occurence of rain has led to the formulation of contradictory conclusions among researchers (Sanhuenza et al., 1985).

We considered the hourly surface ozone concentration measurements obtained by the existing network of ground-based stations in the Greater Athens area during

1987. These measurements have been performed at four points within the Greater Athens area. The hourly values of rainfall, air surface temperature and wind speed have been obtained by the National Observatory of Athens, which is located on the top of a small hill, 107 m above sea level, in the centre of the urban area.

From the set of rainy days in 1987 we selected the cases with total rainfall greater than 5 mm. For these cases we studied the variations of hourly surface ozone concentration values in combination with the hourly rainfall amounts, surface air temperature and wind speed variations. We chose finally three rainfall events. Two of them occurred on 13 October, 1987 and one on 11 March, 1987. Two of these events were associated with the passing of a cold front and they were diurnal events while the third one was a nocturnal case without a cold front passage.

During all rainfall events a decrease of the surface ozone concentration was observed. In the third case (11 March, 1987) rain occurred during the night and the early morning hours. Simultaneously a significant decrease in air temperature was observed. As a result of this overnight cooling, a temperature inversion was formed which lasted throughtout the day. The very low air temperatures were due to the existence of a low pressure with an associated cold front in the Aegean Sea area. The observed frontal rainfall occurred just before the passing of the cold front.

What we observed during this rainfall event was a significant decrease of the social ozone concentration in all stations. This station in the centre of the city (P) showed unusually high ozone concentrations at the beginning of the event while the social ozone concentration minima appeared after the passage of the cold front and not during the rain event.

In the first and second cases (13 October, 1987) rain occurred during the morning hours 06.00-08.00 LT and a heavy rain at 16.00 LT. The decrease of the air temperature during the morning rain cannot be considered as an important one: a substantial decrease of the surface ozone concentration was observed only at one station. This decrease, however, cannot easily be attributed to the rain because it coincided with the maximum values of the normally observed daily course of the surface ozone concentration in Athens (Gusten et al., 1988). Moreover during the maximum intensity of the rain at 08.00 LT a small increase of the surface ozone concentration was observed. On the contrary, during the heavy evening rain (17.3 mm per hour) which was accompanied by very significant decrease in air temperature (4.3^{o}C per hour), and which was due to the passing of a cold front over Athens, we observed a substantial decrease in the surface ozone concentration.

The increased ozone values during the midday hours which are observed before the passing of the cold front are due mainly to the diurnal photochemical activity cycle which is also enhanced by the southern air flow coming from the Saronicos Gulf, which is considered to be a zone of ozone formation as a result of the sea-breeze circulation. Downward ozone transport from the upper troposphere before the passing of the cold front could also contribute to this increase.

The surface ozone concentration during the rain event is influenced by two physical mechanisms: vertical mixing and advection of ozone from the higher altitudes accompanied by its partial destruction at the ground. The first mechanism reveals itself during the day-time rain, as a result of the air temperature decreases near the ground causing an attenuation of the vertical mixing and therefore a decrease of ozone concentration. The influence of the mixing mechanism usually exceeds that of rain.

It seems as though the variations in ozone concentration during a rain event vary significantly, depending on the type of rain. A usual rain event in the Athens region is associated with the passing of a cold front. It is known that before the passing of a cold front downward motions of air masses prevail. These air masses are usually richer in ozone at higher altitudes because in general over urban areas we observe an increase of ozone concentration with altitude. This may be partially due to the ozone destruction at the ground especially during the night but the main reason are NO emissions due to human activities which rapidly destroy ozone.

In both frontal rain events mentioned we observed a strong ozone decrease. It should be mentioned, however, that unusually high values were observed at the Patission station (in the city centre) which were about three times higher than the average value for that period of the year and for that hour. In this connection, of special interest is the heavy evening rain of short duration on 13 October, 1987 when very little ozone increase was observed as it could be expected from the decrease of photochemical activity. This is also in agreement with the literature data which reports a very low solubility of ozone in water.

The rain event which occurred during the morning hours of 13 October, 1987 was not associated with a cold front. It connot give us significant information because ozone concentrations for all stations were already very low. Nevertheless one can see a slight ozone increase for all stations when the maximum rain rate was observed. In this case the change in the horizontal wind direction may be responsible for the ozone increase because, as we mentioned earlier, southern winds in this area are usually transport air rich in ozone.

The phenomenon of strong surface ozone concentration which decreases during the passages of a cold front has been observed very frequently in Central Europe. The suggested explanation is that gas-phase reactions of ozone with nitrogen oxides, in combination with the fast heterogeneous ozone removal under the cloudy conditions which are characteristic for the air mass travelling behind a cold front, may lead to a sharp decrease of the surface ozone concentration. This decrease is enhanced during the night hours. Under such conditions the ozone concentration may be reduced to 1/4 of its initial value. These observations have been made in a rural place of Central Europe (Hohenpeissenberg) which is often influenced by polluted air coming from the city of Munich. It is remarkable though to observe a similar strong surface ozone concentration decrease in the middle of a dense urban area in the Athens region which can be even more pronounced (reduction to 1/7 of the surface ozone concentration initial value). If we examine the absolute surface ozone concentration values in Athens they can be even lower than in Central Europe because the fresh nitrogen oxide emissions destroy the remaining ozone in the cold front zone of the air mass.

One has to emphasise, however, that the cloudy conditions decelerate the photochemical ozone formation which makes it impossible to observe any increasing ozone signal within 2-3 hours after the cold front passage.

As far as the overnight rains impact is concerned the surface ozone concentration in a polluted area does not increase as it does in a non-polluted area, although such an increase was mentioned in the recent literature. This may, possibly, be due to the fact that the mechanism of vertical ozone advection by the rain during the night is of a smaller order of magnitude than the other mechanisms or, alternatively, to the fact that the vertical gradient of the ozone concentration in the area is not important. The second alternative appears to be supported by the measurements obtained by Gusten et al. (1988).

8.2 Surface ozone and its precursors

Photochemical pollution is a very complex process involving meteorological, topographic emission and chemical parameters. It is characterised by the atmospheric formation of photochemical oxidants (mainly ozone) which are produced by the irradiation of mixtures of hydrocarbons and nitrogen oxides (precursors). The most important chemical mechanisms involved in these atmospheric processes have already been identified and studied. But many unknown parameters still exist because of the large number of participating chemical reactions with a great variety of atmospheric hydrocarbons (Logan et al., 1981; Finlayson-Pitts and Pitts, 1986).

For the investigation of this complex atmospheric phenomenon various mathematical models called Air Pollution Simulation models have been developed. These models come into two main categories: prognostic models which are based on the physicochemical principles determining photochemical pollution and diagnostic models which are statistical descriptions of the observed air pollution data.

Significant progress has recently been made in the development of the prognostic Air Quality Simulation Models which represent the most fundamental approach relating the concentrations of photochemical oxidants to those of their precursors (Kuntasal and Chang, 1987). Nevertheless, many uncertainties remain regarding such models because of the difficulties encountered in representing real atmospheric processes and because of the extensive field data requirement for their verification.

The substantiation of statistical models from data already available, which is the second modelling approach (diagnostic models), contains one important weakness. Since photochemical oxidant formation involves complex spatial and temporal characteristics of emissions, meteorological variables and photochemical reactions, it is difficult to choose realistic databases for the statistical substantiation of the O_3-HC-NO_x relation (Lonneman et al., 1986). The meteorological variables which were usually used for the development of statistical models were the inversion base height, the wind speed and direction and the temperature at the 850 hPa level. It is difficult to deal however with inversion base height and wind speed and direction because their fluctuations appear complex and uncertain. But the temperature at 850 hPa, which is considered to be a good indicator of the air mass stability, has been shown to correlate well with ozone levels in California. For this reason it is used as a meteorological index to characterise the ozone-forming potential (Chock et al., 1982).

It is of great importance to investigate whether similar statistical models could be applied in Athens, which is one site in the Mediterranean region where very frequent photochemical pollution episodes occur. Therefore, five-station composite monthly averages of daily concentrations in the Greater Athens area are used for regression analysis, in order to examine the broad features of basin-wide O_3-HC-NO_x relations based on the monthly variation of ozone and the concentrations of its precursors.

Daily maximum hour measurements of surface O_3 and NO_x concentration over the period 1986-1990 at five stations of the operating network in the Greater Athens Basin have been considered. These measurements have been performed by using the conventional methods of UV-absorption for O_3 and chemiluminescence for NO_x, while the background detection levels of the instruments were 0.01 ppm and 0.04 ppm for O_3 and NO_x respectively. Use has also been made of radiosonde-derived temperature data at the 850 hPa pressure level, these data being held in archive by the Hellenic National Meteorological Service over the period 1986-1990.

Kuntasal and Chang (1987) found that a linear correlation exists between the surface ozone concentration and the temperature at the pressure level of 850 hPa. The correlation coefficients are shown in Table 8. Kuntasal and Chang also proposed a linear relation between the surface concentrations of O_3, NO_x, nonmethane hydrocarbons (NMHC) and T_{850}. The coefficients of this relation are also shown in Table 8. The annual variation in surface ozone concentration derived by averaging the mean monthly values at five stations at various locations within the Athens Basin and the expected variation in surface ozone concentration applying Equation (17), proposed by Kuntasal and Chang, indicate that during the April-November period the observed surface ozone concentration is lower than that which could be predicted on the basis of the Kuntasal and Chang model, whereas the observed values exceed those predicted for the winter months. The discrepancy may be attributed to the different initial levels of the chemical precursors NO_x and HC as well as to the different NO_x/HC ratios due to the different profile of human activities in the two urban areas. The seasonal variation of the solar radiation may be another factor due to the fact that solar ultraviolet radiation exerts a direct influence on surface ozone concentration. We should also keep in mind that the Kuntasal and Chang model is derived from measurements in an urban region (Los Angeles), which experiences much higher levels of air pollution than Athens (the ozone concentrations are three times higher). The topography and the meteorological conditions in both areas may also be other factors causing differences.

Table 8. Regression results for O_3-NO_x-NMHC relations. The correlation coefficients (r) are significant at the 5% level.

	a (pphm)	b (pphm/°C)	c	d	r
(1) $[O_3] = a + b\,T_{850}$	-0.25	0.62			0.82
(2) $[O_3] = a + b\,T_{850} + c\,[NO_x] + d\,[NMHC]$	-5.87	0.68	0.14	0.012	0.90

The annual variation of $[O_3]$-d[NMHC] which equally represents the annual variation of surface ozone concentration as a function of $[NO_x]$ and of T_{850} was also calculated taking into account the mean monthly values of T_{850} and corresponding mean monthly values of $[NO_x]$. Comparison of these data with that given by Equation (17) indicates that the latter is systematically lower. This relation shows that the presence of NMHC exerts a positive influence on the ozone concentration in an urban region, which is in agreement with the theoretical expectations (Finlayson-Pitts and Pitts, 1986).

In order to define the relation between T_{850} and surface ozone concentration for the Athens Basin, linear correlations have been studied between:

(i) Mean monthly values of T_{850} and corresponding mean monthly values of daily maximum hour surface ozone concentration. (These measurements were averaged over the five stations for the period 1986-1990).

(ii) Mean monthly values of T_{850} and corresponding mean monthly values of (daily maximum hour) surface ozone concentration at the Nea Smirni Station and at the Liossion Station.

The Nea Smirni and Liossion Stations were selected for further analysis because in these stations the highest and lowest values of surface ozone concentration have been observed.

If we examine Table 9, we observe that the strongest correlation appears to be for the set of five stations and for 1988, whilst the weaker is that for the Liossion station.

Table 9. Values of constants a, b of Equation: $[O_3] = a + b\, T_{850}$ and the correlation coefficients between O_3 and T_{850} for various stations.

Station	Year	a (pphm)	b (pphm/^{o}C)	r
Athens Basin	1987	0.11	0.20	0.92
Athens Basin	1988	1.62	0.18	0.94
Nea Smirni	1987	0.99	0.26	0.81
Nea Smirni	1988	1.07	0.25	0.81
Liossion	1987	2.56	0.11	0.50

By applying in Eq. (17) values for T_{850}, a and b obtained from the five stations for the year 1987 (see Table 9) the expected value of surface ozone concentration was calculated.

Finally, mean monthly values of NO_x concentrations for the five stations were used in order to permit linear correlation of T_{850}, surface ozone concentration and NO_x concentration:

$$[O_3] = a + b\, T_{850} + c\, [NO_x] \qquad (17)$$

where $a = -1.72$ pphm, $b = 0.2$ pphm/^oC and $c = 0.6$. The result is that the constant b has a comparable value as in the equation $[O_3] = a + b\,T_{850}$ while the constant a is lower than the average corresponding value in the previous equation.

In conclusion we can say that a significant linear correlation exists between the surface ozone concentration in the Athens basin and the temperature at the 850 hPa pressure level. A linear relation, whose coefficients are estimated by processing local data permits an improved forecasting of the surface ozone concentrations in the Athens Basin in comparison with the coefficients chosen for the Kuntasal and Chang (1987) model. This is probably due to the differences in precursor concentrations and meteorological conditions between the two areas. Detailed prognostic models should be developed for the Athens Basin in order to be able to assess the reasons of this discrepancy.

9. Tropospheric ozone concentration at Athens, Greece

9.1 Vertical ozone distribution in the troposphere

Long term vertical soundings of ozone in Central Europe have shown an increase in tropospheric ozone over the last twenty five years (Bruhl and Crutzen, 1989). This effect could be attributed to either tropospheric photochemical production and transport, or to variations in the troposphere-stratosphere exchange pattern (Reid and Vaughan, 1991).

In order to investigate the origin of the free tropospheric ozone, systematic long-term measurements should be carried out in the widest possible latitudinal and longitudinal ranges. These measurements should be taken at the surface and in the free atmosphere.

The ozone sensors which have been used for the vertical ozone soundings were of the ECC type. The Electrochemical Concentration Cell (ECC) ozone sonde is an ozone measuring device designed to be flown with and on the same balloon as a standard National Weather Service meteorological radiosonde. The sensing unit for the ozone sonde is an iodine/iodide redox concentration cell composed of two platinum electrodes immersed in neutral buffered iodine solutions of different concentrations in the anode and cathode chambers. The sensor operates as a galvanic cell that causes a current to flow in an external circuit while it drives to equilibrium. In operation a continuous stream of ambient air is bubbled through the cathode cell of the sensor by a constant volume pump. Ozone oxidizes iodide in the cathode to iodine unbalancing the equilibrium of the cell.

As the cell returns to equilibrium, the resulting current in the external circuit gives a measure of the amount of ozone in the air sample (Komhyr, 1969). The ozone sonde is connected to the meteorogical radiosonde (RS-80, Vaisala). The data were taken during the ascent of the balloon (Totex, 1200 gr). The ascent velocity was about 5 mxs^{-1}. The soundings were performed at about 1000 UTC. The observational results showed that the average vertical profile of tropospheric ozone up to about the seventh kilometre was around 50 ppt.

We also observe no vertical gradient of ozone mixing ratios in the troposphere, which is in accordance with other observations in the northern hemisphere.

If one compares mean values with the average winter values which have been measured at Julich, Germany, (50.6^{o}N, 6.2^{o}E) two years before (Volz-Thomas et al., 1991) one can observe that between the first and the fourth kilometre our values are about 20% higher than those obtained at Julich, while between the fourth and the eight kilometre the difference becomes less than 10% and is considered insignificant since it is below the relative standard error (10-15%). The above remarks should be handled with caution due to the fact that our data set does not cover the winter season completely and because there are also year to year variations of the seasonal average of ozone mixing ratios.

We also tried to examine whether, from our limited data, we could observe any signal of pollution transport in the free troposphere coming from western Europe to our site. In order to minimise the influence of local pollution we examine the transport that occurred at the 700 hPa level. Since we did not dispose yet of the corresponding back-trajectories of air masses we made a first approach by considering the wind direction at this level. Advection from the north-western sector was dominant as it was expected for our region covering 60% of the cases. The ozone mean value is then (51 ± 4) ppb. Advection from the south-southwestern sector was also relatively frequent covering 25% of the cases. The average ozone value is (47 ± 2) ppb.

The difference becomes much more significant if we examine the same behaviour within the same month in order to eliminate the month to month variation. March serves as a good example because during three out of eight soundings that we performed the wind at the 700 hPa level was coming from the south-southeastern sector while during the other days the wind was blowing from the north-northwestern sector. The average value in the first case is (46 ± 2) ppb. The decrease observed in the ozone mixing ratio at 3 km in the monthly curve of March could be attributed to this influence.

9.2 Cross-tropopause ozone exchange during a cut-off low event and the polar vortex

Many recent studies show that there is an increasing trend of the background values of ozone in the troposphere (Bojkov, 1986). This creates serious concerns because of the damage caused by high ozone concentrations to human beings, animals and plants and also because of possible climatic effects. Stratospheric ozone leaks into the troposphere forming a natural background. In addition man-made and natural emissions of nitrogen oxides and hydrocarbons lead to the production of ozone in the sunlit troposphere.

In order to assess the anthropogenic component of free tropospheric ozone, the nature of stratosphere-troposphere exchange must be well understood. It was believed until recently that the main transfer of stratospheric ozone occurs in tropopause folding events associated with rapid cyclogenesis (Reiter, 1975).

Major tropopause folding events occur at the flanks of large-scale northern flow. They also proposed a three-fold tropopause structure model. According to this model our sounding station in Athens is situated at such a latitudinal band (36°-38^oN) where stratospheric air intrusions should occur by means of tropopause foldings between the polar and the mid-latitude troposphere. This fact makes the Athens region subject to the influence of the polar front and sometimes to the impact from the subtropical jet stream. Influence of the meteorological situation have been revealed by observations as the laminated structure of the vertical profile of ozone over Athens.

Observations made near a cut-off low in the upper troposphere suggest that a cut-off low may also contribute to the removal of ozone from the stratosphere. But a limited number of studies have been conducted so far and there still remain many uncertainties about this type of stratosphere-troposphere exchange.

In the following we examine the stratosphere-troposphere ozone exchange over Athens, Greece, during the winter 1991-1992 EASOE campaign, in relation to the presence of a cut-off low in the upper troposphere or a possible tropopause folding, when the station is close to the flank of the upper level trough. We should mention that such an intensive sounding campaign has been performed for the first time in Athens. We examined four ozone soundings which correspond to the above meteorological situations. It is also important to emphasise that the nearest ozone sounding station to Athens (Hohenpeissenberg, Germany: 48^oN, 11°E) is far enough in order to obtain information about the situation over our region, which seems to be quite interesting, from the atmospheric dynamics point of view, as mentioned above.

9.2.1 Cross-tropopause exchange during a cut-off low event

A cut-off low system developed over Athens region on 18 and 20 March. On the first day (18 March) the lowest height for the pressure level of 200 hPa during our campaign has been observed. This resulted in a very low tropopause (approaching the 400 hPa level).

We also observed very high ozone values with a maximum of 190 nb and a strong secondary ozone maximum (about 120 nb) in the lower stratosphere (about 5 km deep). The synoptic map shows that our site is influenced by the northern circulation which brings ozone-rich air to our region. From another point of view, we may say that a cut-off low event is combined with high ozone partial pressure at low levels. Because the stratospheric air is then at higher pressure than normal, the ozone partial pressure will be similarly increased. So the remarkable fact is the minimum in partial pressure near 15 km, which our ozone data clearly shows (Vaughan, personal communication). The ozone profile on that day does not show any characteristics of detectable cross tropopause exchange.

On 20 March (the day of the next ozone sounding) we observed that the cut-off low system had been moving eastwards and that our sounding site was influenced by the polar air. The intensity of the low had been weakened and consequently the height of the tropopause had been increased.

On 18 March, the already observed laminated ozone structure in the region of 100-300 hPa becomes much more pronounced. This is in agreement with the laminar structure in ozone profile over Athens is strongly associated with the establishment of the northwestern circulation over the region. If we examine the tropospheric temperature profile of both days we find that the tropospheric temperature on 20 March is higher by 3-5°C. This agrees fairly well with the findings of similar studies of cut-off low systems. This might be associated with a cooling of the lower stratosphere and a lifting of the tropopause which is in fact observed on that day. The most important feature observed on 20 March is a significant tropopause peak of ozone (about 10 nbs) at 600 mbs associated with a sharp decrease in relative humidity suggesting that this air might originate from the stratosphere. One can also see on the same day that the ozone partial pressure increases sharply before arriving to the tropopause which demonstrates another example of stratospheric ozone intrusion into the tropopause.

9.2.2 Troposphere-stratosphere exchange at the flank of the polar vortex

On 27 March and 8 April our sounding site was situated on the eastern edge of the upper level trough. The strong laminae observed during those days is also an indication that we are close to the vortex boundaries. As mentioned above, in this type of situation tropopause folding occurs which creates a potential for troposphere-stratosphere exchange. In fact, we observe troposphreric ozone peaks of 5-10 nbs in the region of 400-600 hPa level. These peaks are less pronounced than the broad peak observed on 20 March under a cut-off low regime. Also these peaks are not associated with a sharp decrease of relative humidity. In addition, on 8 April we observed a sharp tropospheric ozone increase before the beginning of the cross-tropopause exchange. A slight positive temperature gradient right above the tropopause as observed on the same day is another indication of the ongoing troposphere-stratosphere exchange (Dobson, 1973).

Further analysis of the synoptic situation and the history of the relevant air parcels is essential and it will be the subject of future publications. For instance, plots of potential vorticity and cross-sections of potential temperature should indicate strongly the exchange that occurred.

10 Stratospheric ozone over Athens

10.1 Lamination in ozone profiles

During the last thirty years many researchers reported the existence of layers with enhanced and depleted ozone amounts (laminar structure) in the vertical ozone profiles (Reid and Vaughan, 1991). In particular, Dobson (1973) made a systematic analysis of the laminated ozone structure in the stratosphere over a wide latitudinal and longitudinal range. In Dobson's analysis the criterion for the detection of laminae, was that the change of ozone partial pressure in a certain height interval should be greater than 3 mPa. Also, in the same study, the ozone profiles were separated into three groups: (i) with profiles containing almost no laminae, (ii) with profiles exhibiting moderate lamination, and (iii) with profiles extensively laminated. Among the findings of Dobson's work is the fact that the features of the laminated structure vary with latitude and season. In particular, between January and April the lamination phenomenon is present most frequently. Also, at latitudes below about 20°N during spring and below about 30°N during autumn a laminated structure of the ozone is very seldom found. In more detail, about 35% of all profiles were classified as group (iii) at high latitudes in spring, while group (ii) profiles reached a peak frequency of 70% in June. Group (iii) profiles were rarely observed at any latitude during the summer, while group (ii) profiles were found in polar regions throughout the year. In Dobson's paper no explanation was given for both the incidence of the minima at a preferred height region of 14-17 km and its constancy with latitude. It has been mentioned though, that there is a strong correlation between the existence of the characteristic ozone minimum at 14-17 km and the occurrence of the double tropopause, especially at latitudes around 40°N. Finally, it should be emphasised that Dobson made the assumption that the laminated structure and the characteristic ozone minimum are of the same origin, since the variations in the frequency of appearence with both season and latitude are very similar.

Reid and Vaughan (1991) examined the ozone laminar structure in the altitude range of 9.5-21.5 km. Laminae were classified according to their vertical extent (depth), as well as the difference in ozone partial pressure (magnitude). They also used some criteria to separate genuine laminar features from instrumental noise or large-scale features in the ozone profile. The acceptable depth for laminae was chosen to be between 200 m and 2.5 km while the minimum magnitude was set to 2 mPa. They also showed that laminae are most frequently found below 18 km at high latitudes during winter and spring, which is in agreement with the findings of Dobson's study. Reid and Vaughan (1991) disagree with Dobson about the origin of laminae. Based on the observation that their magnitude is greatest and their depth least during winter and spring, they ruled out the suggestion of Dobson that they all originate near the subtropical jet-stream, although the possibility remains that a small number of laminae may be generated by stratosphere-troposphere exchange (Begum, 1989). Reid and Vaughan (1991) suggested that laminae may represent evidence for a process which can cause exchange of air in and out of the polar vortex. This argument is supported by recent polar ozone campaigns, where laminae particularly sharp and deep near the winter polar vortex have been detected (Mc Kenna et al., 1989).

Given the ozone destruction which is known to take place within the vortex, the possibility that such exchange can spread ozone depletion to lower latitudes is examined in the following.

A number of vertical ozone profiles measured at Athens station were classified in four groups according to their specific features. The first group is consisting of profiles in cyclonic flow at the 200 hPa level, generally on the western or southern side of an elongated trough. The main characteristics of the profiles of the first group are: a) high tropopause, b) small secondary maximum in ozone (<7 mPa) and c) ozone minimum at 100 hPa.

The second group is similar in meteorological respect to the first group, but it consists of profiles with rather larger secondary layer in the lower stratosphere and again an ozone minimum at 100 hPa.

The third group is mainly characterised by many thin layers in the lower stratosphere (laminated structure). The profiles of the third group were observed during a fairly cyclonic flow of trough or on the eastern side of the trough.

Finally the fourth group contains profiles with one thick layer in the lower stratosphere and an ozone minimum at 14-17 km. The profiles of the fourth group were usually observed during cut-off low events.

For the identification of laminar structures the same criteria (laminar depths less than 2.5 km and magnitudes higher than 2 mPa), as in the study of Reid and Vaughan (1991), have been used. Features having depths higher than 2.5 km are considered as ozone maxima and minima.

The first important finding is that the lamination phenomenon has been very frequently observed over the Athens region, which is in contradiction with previous measurements for those latitudes (Reid and Vaughan, 1991). This difference becomes much more pronounced in February and March, when the lamination frequency over Athens is about twice that observed over Cagliari, Italy, (39.1°N, 9.1°E). It should be noted that the data for Cagliari were collected between 1972 and 1975, while the Athens data were obtained later (1991-92). However, it is reasonable to assume that a similar regime in the lower stratosphere took place over Cagliari and Athens during the period 1972-75. This fact, in combination with the observation that the laminar events are much more pronounced at the edge of the polar vortex, can be explained by the observation that the vortex expanded in 1992 due to the Mount Pinatubo eruption (Angell 1992, Angell and Korshover 1985). It should be stressed here that the nearest ozone sounding station to Athens (Hohenpeissenberg, Germany: 48°N, 11°E) is too distant to get more information about the situation over Athens.

Inspection of the weather maps, provided by the German Weather Service, shows that there is a significant correlation between the appearance of laminar events and the establishment of a north-northwest circulation in the lower stratosphere over the site. In this case, it is clearly seen that the site is under the influence of the polar air. The previous observation is in agreement with the findings mentioned above concerning the enhancement of laminar phenomena at the vortex edge. A characteristic example of such a day is the 20 March, 1992. Eight days with similar characteristics have been observed in the data set considered.

By examining the weather maps of the days with the characteristic ozone minima it is evident that during such days Athens was influenced by the subtropical jet-stream. These days are in agreement with the observations made by Dobson (1973) that the 15 km minimum is associated with air coming from the subtropical region. Finally, there are cases where Athens was influenced by both types of circulation of air masses in the lower stratosphere.

10.2 Stratospheric ozone profiles as deduced from ozonesondes and satellites

Various types of wet chemical ozonesonde are currently used for in situ measurement of the vertical ozone profile at various sites around the world. The most widespread types are the Brewer-Mast type (Brewer and Milford, 1960) and the Electrochemical Concentration Cell (ECC) type (Komhyr, 1969).

One would expect a decrease in the temperature of the chemical solution to occur during a flight of a Brewer-Mast type ozonesonde and a decrease from 23°C to -2°C. This may raise questions about the influence of chemical solution temperature on the output current and subsequently on the calculated ozone partial pressure.

The aim of this section is to examine the differences between the vertical ozone profiles obtained from measurements made during the ascent and the descent of the balloon. Furthermore, these differences have been examined in relation to the box temperature (containing the ozonesonde) and the corresponding total ozone data obtained by remote sensing ozone measurements from both the TOMS instrument on the Nimbus-7 satellite and the Dobson Spectrophotometer No. 118 installed at Athens University.

The comparison of a typical vertical ozone profile obtained during the balloon ascent with that measured during the balloon descent shows that at the middle stratosphere, where the atmospheric pressure ranges between 40 hPa and 10 hPa, the ozone values obtained during the balloon descent are somewhat lower (5 nbars) than for the ascent, although they follow the general trend.

In the region between 100 hPa and 40 hPa the ozone values obtained during balloon descent still follow the same general trend as for the ascent, but they are higher by about 10 nbars. In the lower stratosphere (100-200 hPa) the ozone values during the balloon descent are almost always higher than the corresponding ozone values obtained during the balloon ascent. In almost all the ozone profiles collected at Athens the same behaviour was observed.

As has been mentioned in the section above, a decrease in the chemical solution temperature during the descent of a Brewer-Mast ozonesonde was observed. The commercial ECC ozonesondes do not provide a direct reading of the chemical solution temperature but they can measure the box temperature; this may give an idea of possible variations occurring in the chemical solution temperature. It should be noted that the box temperature during the balloon descent was almost constant (between 7-9°C). This was the temperature observed for most cases. Possible explanations of the discrepancies between ascent and descent will now be presented.

In most cases the difference between the box temperature during ascent and descent was about 7-8°C. However, in two cases when quite unusual extreme temperatures (15°C and 0°C) were observed, the ascent-descent pattern remained the same. This indicates that the box temperature does not play a significant role in the differences in the retrieved ozone amounts.

One possible reason for the difference is that the ozonesonde during the balloon descent has a velocity about ten times higher than during the balloon ascent (ascent velocities: 5-6 mxs^{-1}, descent velocities: 25-60 mxs^{-1}). Given that the ECC ozonesonde gets 90 per cent of a standard value in 45 sec, then it should be expected that during the balloon descent a shift in the measurements due to the above described factor should be reflected.

The observation that the lower part of the descent profile usually shows the maximum discrepancy, should be combined with a possible increase of the background current of the ozonesonde with time, which leads to an increase of the measured ozone partial pressure.

We shall now relate the ozonesonde results to the total ozone data over Athens obtained by the Nimbus-7 TOMS instrument and the Dobson No. 118 spectrophotometer installed in Athens.

During the winter campaign the weather was favourable for Dobson observations on only 13 days. The linear correlation between total ozone obtained from the Dobson instrument and through integration of the ozonesoundings during the balloon's ascent is given by the equation $Y = 1.04\ X - 11.86$, where Y and X denote the total ozone amount obtained from the ozonesonde and Dobson spectrophotometer, respectively.

It should be mentioned here that because the descent profile is incomplete one cannot calculate the total ozone amount from the descent data.

The linear relation between the TOMS and Dobson total ozone values obtained during the WMO campaign in May 1993 is given by the equation $Y = 0.83\ X + 60.53$, where Y and X denote the total ozone amount derived from the TOMS and the Dobson spectrophotometer, respectively. For instance, for a typical value of 350 D. U. obtained by the Dobson instrument the corresponding values from the TOMS is 351.03 D. U. while from the integration of an ozonesonde profile during the ascent it is 352.14 D. U.

Based on the above analysis we can also investigate the accuracy of the ozone concentration values obtained during the descent of the ozonesonde. We have already mentioned that in the region between 10 and 40 hPa the values collected during the descent of the ozonesonde are somewhat lower than those obtained during the ascent, while in the region between 100 and 40 hPa the situation is reversed. This observation is repeated in all our ozonesonde data. A more detailed examination of all these results shows that the total positive difference at the lower altitudes is greater than the total negative difference at the higher altitudes. If we make the reasonable assumption that the variations of the ozone values for different profiles in the troposphere are very low we could say that the corresponding total ozone values from the descent values would be higher than the ascent ones. The difference from the total ozone obtained by the spectrophotometric techniques would be even greater for the descent than for the ascent of the balloon. This leads to the suggestion that the descent values are less accurate than the ascent ones.

In conclusion, the intercomparison of the total ozone measurements made by the Dobson spectrophotometer, TOMS and ozonesondes during ascent shows that there is a close agreement among them, while the descent ozonesonde values seem to be less accurate than the ascent ones.

10.3 Comparison of the local ozone and temperature profiles at Athens with ozone and temperature models for middle latitudes

During the last five decades a series of ozone measurements in the atmosphere has been obtained by using in situ and remote-sensing instrumentation. The in situ measurements were mainly performed by using balloon soundings (i.e. ozonesondes) and rocketsondes while the remotely-sensed observations were made by using both ground-based and space-borne instrumentation (i.e. spectrophotometers).

A number of models have been developed to characterise ozone measurements results in view of the importance of data on the spatial and temporal variability of ozone for studies of atmospheric circulation, climate, radiation balance and the photochemistry of the atmosphere. In addition, models based on balloon and rocket soundings and relating the vertical ozone distribution to the total ozone were generated as a first guess for satellite data retrievals (Bhartia et al., 1984). Also, models based not just on one satellite measurement, but on multiple data sets from satellites have been generated, allowing the high precision determination of ozone variation with pressure, latitude and time covering the height range from 25 to 90 km. For instance, the most recent model of monthly latitudinal variations of vertical ozone structure is based on ozone data from five satellite experiments and it is used nowadays as the Satellite Ozone Reference Model (SORM).

Remote sensing techniques also contribute a great deal as far as the temperature observations in the atmosphere are concerned, with the reduction of the radiance data to equivalent temperature measurements (Varotsos and Helmis, 1992). This reduction is usually performed by sampling the radiation emitted at a number of wavelengths which correspond to different atmospheric absorptivities. The most recent model of monthly latitudinal variations of vertical temperature distribution is based on radiance data from the satellite experiments and it is used nowadays as the Satellite Temperature Reference Model (STRM).

The reliability of the above-mentioned ozone and temperature reference models is very important for various atmospheric applications. The comparison between these satellite-derived models with in situ measurements at a given latitude is essential and it is the principal aim of this paragraph. In particular, a number of vertical ozone and temperature profiles obtained over Athens, Greece, (38°N, 24°E) are compared with the corresponding reference models SORM and STRM in order to investigate the plausible explanations for any inconsistencies between them.

The SORM, which provides the monthly latitudinal variations of vertical ozone structure, has been constructed using the results from the following satellite experiments: the Nimbus-7 Solar Backscatter Ultraviolet (SBUV), the Nimbus-7 Limb Infrared Monitor of the Stratosphere (LIMS), the Applications Explorer Mission-2 Stratospheric Aerosol and Gas Experiment (SAGE), the Solar Mesosphere Explorer UV Spectrometer (SME UVS) and the Solar Mesosphere Explorer 1.27 ìm Airglow (SME-IR). The 4-year mean from November 1978 to September 1982 of the SBUV data was given a weight of 2 due to the combination of extensive spatial and tempolar coverage.

The STRM, which provides the mean monthly latitudinal variations of vertical temperature profile, has been substantiated using the results of radiance measurements made by the Nimbus-5 Selective Chopper Radiometer (SCR), the Nimbus-6 Pressure Modulator Radiometer (PRM), the Nimbus-7 Stratospheric and Mesospheric Sounder (SAMS) and the Limb Interferometer Monitor of the Stratosphere (LIMS) radiometers.

10.3.1 Comparison between SORM and in situ ozone measurements

The vertical ozone profiles at the beginning and at the end of the months of February, March and April 1992, as derived from the ozone soundings performed over Athens, were compared with the ozone mixing ratio at constant pressure levels of the upper stratosphere, as given from the SORM. This comparison showed a fairly good agreement.

It should be stressed here that although the SORM has been compiled from radiance data from the period 1978-1982, ten years later the ozone of the upper stratosphere (above the ozone maximum) does not show signs of depletion following the general trend of the last decade. This result is in agreement with the theory of the stratospheric ozone depletion, according to which the main ozone reduction happens just below the ozone maximum (Stolarski et al., 1991; Reinsel et al., 1988).

10.3.2 Comparison of STRM and in situ temperature measurements

The vertical temperature profiles at the beginning and the end of the months of February, March and April 1992, as they are derived from the radiosoundings performed over Athens, have been compared with the temperature at constant pressure levels in the troposphere and stratosphere as given from the STRM. This comparison shows that the in situ and the model values are in a very good agreement. It is worth noting at this point that this conclusion is consistent with the recently published results (Varotsos and Helmis, 1992) for the deviation of the temperature models based on remote-sensing and in situ data for the global middle atmosphere. According to the latter publication the main deviations between the two models were observed (a) at high latitudes at altitudes higher than 35 km, and (b) at low and middle latitudes of the southern hemisphere at the upper mesosphere. Regarding the temperature deviations between these models at the latitude of Athens, it is evident that they did not exceed the value of 2°C for all months.

11 Total ozone content (TOC)

11.1 Comparison of TOC obtained using Dobson spectrophotometer and satellite observations

The Dobson ozone spectrophotometer serves as a standard for atmospheric total ozone measurements. It has been an important instrument for measuring global ozone since its development by G.M.B. Dobson around 1927 (Dobson, 1931).

It is well known that this instrument is based on making ozone measurements using direct or scattered sunlight of the Huggins bands. It is a double monochromator, with one monochromator used to disperse the radiation and the second used to reject interfering scattered radiation. A pair of wavelength-isolating slits within the instrument is coupled to an optical wedge that is adjusted at one slit to give a null difference reading. Total ozone observations using direct sunlight are usually made on AD double-pair wavelengths where the A pair is at 305.5 and 325.4 nm, while the D pair is at 317.6 and 339.8 nm. Also, with the Dobson spectrophotometer, measurements of ozone vertical distribution can be made. The technique is called Umkehr and observations are made on the clear zenith sky as the Sun is rising or setting. Ozone vertical profile data are obtained through an inversion technique that takes into account absorption and molecular and aerosol scattering of light that passes through the atmosphere from the Sun to the instrument (Mateer and De Luisi, 1981).

Because of the reasonable accuracy of the Dobson spectrophotometer approach (Basher, 1982), a widespread network comprising approximately 100 Dobson-equipped stations has allowed many important observations to be made over many parts of the globe. The Dobson network also serves a valuable function in providing ground truth for satellite-based measurements such as for the Total Ozone Mapping Spectrometer (TOMS) (Bojkov et al., 1988).

The Dobson spectrophotometer No. 118 has been in use for measuring total ozone at the Athens University since 1989. Athens University participated in the recent International Dobson Intercomparison during the period 22 July-10 August 1990, at Arosa, Switzerland; this was intended to provide for the maintenance of a well-operating Dobson network in Europe. It was the seventh intercomparison of Dobson ozone spectrophotometers organised by the World Meteorological Organization (WMO) in which individual instruments providing continuous total ozone data from all parts of Europe were involved.

Changes in total ozone have been measured by several different techniques from the ground and from satellites. These datasets give a clear enough picture of the ozone changes which have taken place. For instance, a significant depletion in the springtime total ozone content over the Antarctic was first measured by the British Antarctic Survey (Farman et al., 1985), using the Dobson station data at Halley Bay (76°S). The careful re-examination of TOMS and Solar Backscatter Ultraviolet data (Stolarski et al., 1991) not only confirmed the decrease reported from the ground station data but demonstrated that the phenomenon is of large regional scale, at least of the size of the Antarctic continent.

In this paragraph, the comparison of the ground-based total ozone observations, as they are derived from Dobson spectrophotometer No. 118, and the satellite total ozone measurements, as they are deduced from TOMS instrument flown on the Nimbus-7 satellite during its overpasses above Athens, is shown. This intercomparison took place during May 1991 within the framework of the trial period of the WMO Global Ozone Observing System (GO_3OS).

11.1.1 Short description of the theory of Dobson measurement

Total ozone observations are made with the Dobson spectrophotometer by measuring the relative intensities of ultraviolet radiation from the Sun, moon or zenith sky at selected wavelength pairs. By using one pair of wavelengths a differential measurement is made that includes the differential absorption of sunlight by ozone but also includes the effects of differential scattering by atmospheric molecules and aerosols.

If measurements at two wavelength pairs are combined, the scattering effects can essentially be excluded, because they are similar for different wavelength pairs (e.g. AD pair). In this case the general data reduction equation is:

$$x_{AD} = \frac{N_A - N_D}{1.388\,\mu} - 0.009\,\frac{mP}{\mu P_0} \qquad (18)$$

where

x_{AD} : total amount of ozone in Dobson Units (1 DU = 10^{-5} mat STP = 1 atmo-cm)

m: ratio of the actual and vertical paths of solar radiation for refraction and earth's curvature

P: station pressure

P_0: mean sea level pressure

μ: ratio of actual and vertical path of solar radiation through the ozone layer

N_A, N_D: functions of the intensities of solar radiation at the top of atmosphere at the short (A) and long (D) wavelengths, respectively.

11.1.2 TOMS instrument

The TOMS instrument on the Nimbus-7 satellite makes observations in six wavelength regions from 312.5 to 380 nm using a single monochromator and scanning a cross track over 105° providing total ozone values. In order to avoid problems from clouds in the measurement it uses an IR cloud-cover photometer.

The precision of the TOMS instrument is 2% or better and there is a small underestimation on the data owing to the drift in the reflectance of the diffuser plate. This underestimate is being corrected in the data processing by using an algorithm which assumes that the drift is proportional to wavelength separation (Stolarski et al., 1991).

11.1.3 Temporal and spatial distribution of total ozone during trial GO_3OS campaign - Results

During May 1991 under the WMO GO_3OS programme a trial period for near-real-time total ozone reporting system in Europe took place, which also supported the European Arctic Stratospheric Ozone Experiment (EASOE) by providing a timely source of maps of the total ozone field over Europe and the Arctic.

It should be noted that the satellite overpass always occurs near the local noon. Also a new method in the TOMS data processing has been used which is based on a requirement of internal consistency in ozone measured with different wavelength pairs. These reprocessed data (version 6), were obtained from the NASA Goddard Space Flight Center, Greenbelt, Maryland.

For the linear correlation between Dobson spectrophotometer and TOMS instrument total ozone measurements data over Athens in May 1991 the correlation coefficient is 0.96. The general remark which should be made about the intercomparison between the results from the two instruments is that there exists a very good consistency.

In order to examine the spatial distribution of total ozone in south-eastern Europe during the same time period, we compared the total ozone values over Thessaloniki (40.3°N, 22.6°E) and Sofia (42.5°N, 23.2°E) to those over Athens. The intercomparison shows that in general the total ozone concentration is gradually increased as a function of latitude as is expected from the theory.

From the above mentioned analysis we conclude that the Athens station, except for providing the regional total ozone values with enough accuracy, may also serve a valuable function in providing ground truth for satellite-based measurements and thus allow the long-term drift of satellite instrument measurements to be estimated.

11.2 Intra-annual variations of total ozone content over Athens

TOMS measurements have been broadly used for investigations related to the total ozone variability. For instance, Nimbus-7 total ozone measurements have compared to the corresponding measurements of the Dobson network. Reinsel et al. (1987) have statistically analysed total ozone data to detect trends and solar cycle relation. Varotsos (1989) analysed total ozone to suggest that when the quasi-biennial-oscillation (QBO) in the equatorial stratosphere is in its west phase the global total ozone is positively correlated with the 11-year solar cycle, while the opposite applies for the east phase of the QBO.

Varotsos et al. (1992a) analysed the 1978-1985 Nimbus-7 global data to explore the statistical significance of these data for all Umkehr layers to identify annual and semi-annual waves. It was found that the annual wave in global ozone is strong in the high and low stratosphere and weak in the middle stratosphere, whereas the semi-annual wave in global ozone is strong in the middle stratosphere. Varotsos et al. (1992b) indicated the existence of a 4-month time period wave (terannual wave) in ozone at the strato-mesosphere for both hemispheres. From the latter study three main maxima of the terannual wave in ozone were apparent: a) the tropical zone maximum centred at the middle stratosphere, and b) the two high latitude maxima near the stratopause level in both hemispheres.

Stolarski et al. (1991) analysed the reprocessed TOMS data in order to detect total ozone trends. They found that the northern mid-latitude ozone trend shows a significant seasonal cycle with a miximum in late winter (February and March). The maximum is located near 40° latitude and reaches a greater level than 0.8% per year of ozone depletion.

This section concentrates on total ozone data in the sub-tropics of the northern hemisphere and attempts to explore the statistical significance of these data to detect annual, semiannual and terannual waves in order to identify the plausible mechanism of the observed maximum ozone decrease in late winter near 40° latitude. The mean monthly total ozone values (derived from the daily total ozone measurements made by the TOMS instrument) over Athens have been Fourier analysed to deduce the amplitude α, phase φ and the percentage contribution to the total variancc v of thc first thrcc harmonics.

The harmonic analysis performed demonstrates the temporal variability of the amplitude and the percentage contribution to the total variance of the annual wave in the total ozone over Athens. The results show that the annual wave in total ozone over Athens is strong for the entire period from January 1979 to December 1991, with its maximum occurring mostly in April. It is known that the annual wave in total ozone is in phase with the strong annual wave in the low and high stratosphere (19-28 km and 43-48 km). With respect to its annual variation, the low stratosphere mainly responds to transport processes while above 35-40 km the temperature effects are dominant (Varotsos et al. 1992a). According to this suggestion, the time march of the amplitude of the annual wave in total ozone shows the time behaviour of this wave in the low and high stratosphere.

The semi-annual wave in total ozone over Athens is weak through the entire period from January 1979 to December 1991, with its first maximum occuring mostly in March and explaining up to 16% of the total variance.

It is known (Reinsel et al., 1988) that the semi-annual wave in total ozone is in phase with the semi-annual temperature wave in the middle stratosphere (33-38 km). It should be mentioned here that the equatorial semi-annual wave in ozone is generated by the negative temperature effect on ozone photochemistry associated with the semi-annual modulation of the temperature field, whereas the polar semi-annual wave in ozone is generated by photochemical production of ozone combined with the dynamical transport from the equatorial ozone production maximum (Varotsos et al., 1992a).

The terrannual wave in ozone over Athens is weak for the entire period 1979-1991 with its first maximum occuring mostly in January. An intercomparison of the semi-annual and the terannual wave indicates that the terannual wave in ozone at the same site has a significant percentage contribution to the total variance.

Recently Varotsos et al. (1992b) have suggested the existence of the terannual wave in ozone at: a) the tropical middle stratosphere with its first maximum occurring in January, and b) the high latitude stratopause with its first maximum occurring in February.

It seems that total ozone over Athens is influenced by the tropical centre of the terannual wave in ozone. This wave can be attributed to the fact that during sudden stratospheric warmings, a slight increase in ozone occurs south of the subtropics, which is probably due to the temperature decrease and which in turn leads to somewhat lower O_3 destruction rate.

From the above discussion, the main conclusion is that the heterogeneous reactions modify the normal gas-phase chemistry within the Arctic vortex leading to an acceleration of the trend. This effect is evident at mid-latitudes when transport of polar-air occurs as the vortex is distorted. Then a negative gradient in the amount of active chlorine away from the pole takes place. At mid-latitudes the sunlight available to drive the chlorine destruction processes increases (Stolarski et al., 1991).

12 Conclusions

1. It is quite clear from what has been discussed in this survey that although it is well known since long ago that the atmospheric greenhouse effect is a basic factor contributing to the earth's climate, we are still far away from a possibility to reliably assess the greenhouse contribution to the global climate change. Since more complete observations of the climate system during at least a few decades are necessary and the Global Climate Observing System (GCOS) is for the time being still in the state of development (Kondratyev, 1993) it is quite appropriate to repeat in conclusion what has been written that: "We have a long wave to go in our understanding of an ability to model the complex climate system before confidend predictions of the aspects of climate change that need to be known". Other scientists have justly cautioned in this connection: "The propagation of estimates derived from numerical climate model predictions of greenhouse-induced climate change through impact models into policy advice is a precariously uncertain process which compounds the considerable uncertainties already inherent in policy development". The high costs of implementation of "anti-greenhouse" decisions on a global scale defines a necessity of scientifically justified and carefully-weighted political decisions. Political aspects of the problem have been discussed in detail in the monograph published by Kondratyev et al. (1994).

2. There is a total ozone depletion over Greece of the order of 5% per decade.

3. The lamination effect over Greece occurs more frequently in spring and it is associated with the northern air flow.

4. The ozone depletion in the lower stratosphere over Greece is associated with the western air flow.

5. The Quasi-Biennial Oscilation and the southern oscilation signals in ozone do not influence the long-term trend in ozone Greece.

References

Angell J.K. and Korshover J., 1985, Displacement of the north circumpolar vortex during El Nino, 1963-1983, *Mon. Wea. Rev.*, 113, 1627-1630.

Angell J.K., 1992, Relation between 300 mb North Polar Vortex and Equatorial SST, QBO and Sunspot Number and the record contraction of the Vortex in 1988-89, *J. Climate*, 5, 22-29.

Basher R.E., 1982, Review of the Dobson spectrophotometer and its accuracy, World Meteorological Organization Ozone Project Report, 13, 94 pp., Geneva.

Begum D.A., 1989, Studies of the total ozone field around the sub-tropical jet stream, Ph.D. Thesis, University of Wales.

Bernard H. W., Jr., 1993, *Global Warming Unchecked: Signs to Watch For,* Indiana University Press, 181 pp..

Bhartia P.K., Klenk K.F., Wong C.K., Gordon D. and Fleig A.J., 1984, Intercomparison of the Nimbus-7 SBUV/TOMS total ozone data sets with Dobson and M83 results, *Journal of Geophysical Research*, 89, 5239-5248.

Bojkov R.D., Mateer C.L. and Hansson A.L., 1988, Comparison of Ground-Based and Total Ozonc Mapping Spectrometer Measurements Used in Assessing the Performance of the Global Ozone Observing System, *Journal of Geophysical Research*, 93, 9525-9533.

Bojkov R.D., 1986, Surface ozone during the second half of the nineteenth century, *Journal of Climate and Applied Meteorology*, 25, 343-351.

Bolin B., 1993, *In search of the "missing carbon"*, U. N. Climate Change Buil., Issue 2, p. 1-2.

Brewer A.W. and Milford J.R., 1960, The Oxford-Kew ozonesonde, *Proceedings of Royal Society*, A256, 470-495.

Bruhl C. and Crutzen P.J., 1989, On the disproportionate role of tropospheric ozone as a filter against solar UV-B radiation, Geophys. Res. Let., 16, 703-706.

Chock D.P., Kumar S. and Hermann R.W., 1982, An analysis of trends in oxidant air quality in the South Coast Air Basin of California, *Atmos. Environ.*, 16 (11), 2615-2624.

Decker C.E., Ripperton L.A., Worth J.J.B., Vukovich F.M., Bach W.D., Tommerdahl J.B., Smith F. and Wagoner D.E., 1976, Formation and transport of oxidants along Gulf Coast and in Northern U.S. Rep. EPA-450/3-76-033, U.S. Environ. Prot. Agency, Research Triangle Park, N.C.

Dobson G.M.B., 1931, A photoelectric spectrophotometer for measuring the amount of atmospheric ozone, *Proceedings of the Physical Society*, 43, 324-339.

Dobson G.M.B., 1973, The laminated structure of the ozone in the atmosphere, Quart. J. Roy. Meteor. Soc., 99, 599-607.

Evans G., Finkelstein P., Martin B., Possiel N. and Graves M., 1983, Ozone Measurements from a network of Remote Sites, *J. Air Pollut. Contr. Assoc.*, 33, 291-296.

Farman J.C, Gardiner B.G. and Shanklin J.D., 1985, Large losses of total ozone in Antarctica reveal seasonal ClO_X/NO_X interaction, *Nature*, 315, 207-210.

Finlayson-Pitts B.J. and Pitts J.N.Jr., 1986, Atmospheric Chemistry, Wiley Interscience, New York, pp. 25-32.

Fong P., 1993, *The greenhouse effect from the viewpoint of the fourth (aerosol) phase of water*, AGU 1993 Annu. Meet. EOS Suppl., p. 123.

Galbally I.E, 1968, Some measurements of Ozone variation and destruction in the Atmospheric Surface Layer, *Nature*, 218, 456-457.

Gusten H., Heinrich G., Cvitas T., Klasinc L., Ruscic B., Lalas D. and Petrakis M., 1988, Photochemical formation and transport of ozone in Athens, Greece, *Atmos. Envir.*, 22, 1855-1861.

Harrison E. F., P. Minnis, B. R. Barkstrom, V. Ramanathan, R. D. Cess and G. G. Gibson, 1990, *J. Geophys. Res.*, 95, 18687-18703.

Hughes P., 1993, Personal Travel and the Greenhouse Effect. A Strategy for Stabilizing the Atmosphere, Earthscan Publ. Ltd., London, 208 p..

Intergovernmental Panel on Climate Change, 1992, IPCC Supplement, Cambridge Univ. Press.

Komhyr W.D., 1969, Electrochemical concentration cells for gas analysis, *Ann. Geophys.*, 25, 203-210.

Kondratyev K. Ya., 1949, *Longwave Radiative Transfer in the Atmosphere*, Leningrad, Gostechizdat (in Russian).

Kondratyev K. Ya., 1956, *Radiative Heat Exchange in the Atmosphere*, Oxford, Pergamon Press.

Kondratyev K. Ya., 1969, *Radiation in the Atmosphere*, New York, Academic Press.

Kondratyev K. Ya., 1972, *Complete Atmospheric Energetics Experiment*, GARP Publ. Sec., N12, WMO, Geneva, 94 pp..

Kondratyev K. Ya., 1988, *Climate Shocks: Natural and Anthropogenic*, John Wiley and Sons, New York e. a. 296 pp..

Kondratyev K. Ya. (ed.), 1991, *Aerosols and Climate*, Leningrad, Gidrometeoizdat, 540 pp. (in Russian).

Kondratyev K. Ya., 1992, *Global Climate, Leningrad*, "Nauka", 359 pp. (in Russian).

Kondratyev K. Ya., 1993, *Complex monitoring of the Pinatubo volcanic eruptions. Studing the Earth from Space*, No. 1, 111-122, (in Russian).

Kondratyev K. Ya. and N. L. Moskalenko, 1984, *Greenhouse Effect of the Atmosphere and Climate*, Moscow, VINITI, 262 pp. (in Russian).

Kondratyev K. Ya. and H. Grassl, 1993, *Global Climate Change in the Context of Global Ecodynamics*, St. Petersburg, PROPO, 195 pp. (in Russian).

Kondratyev K. Ya. and A. P. Cracknell, 1994, *Observing Global Climate Change*, Taylor and Francis, London e. a. (in press).

Kondratyev K. Ya., L. N. Dyachenko and V. V. Kozoderov, 1988, *The Earth's Radiation Budget*, Leningrad, Gidrometeoizdat, 350 pp. (in Russian).

Kondratyev K. Ya, V. I. Danilov-Danilyan, V. K. Donchenko and K. S. Losev, 1994, *Ecology and Politics*, Nauka Publ..

Kuntasal G. and Chang T.Y., 1987, Trends and relationships of O_3, NO_X and HC in the South Coast Air Basin of California, *J. Air Pollut. Control Assoc.*, 37, 1158-1163.

Logan J.A., Prather M.J., Wofsy S.C. and McElroy M.B., 1981, Tropospheric chemistry: a global perspective, *J. Geophys. Res.*, 86, 7210-7230.

Lonneman W.A., Sella R.L. and Meeks S.A., 1986, Non-methane organic composition in the Lincoln Tunnel, *Envir. Sci. Technol.*, 20, 790-799.

Mateer C.L. and De Luisi J.J., 1981, The estimation of the vertical distribution of ozone by the short Umkehr method, Proceedings of the Quadrennial International Ozone Symposium, 1980, Ed. London J., Reidel D., Boulder, Colorado, Vol 1, 64-73.

McKenna D.S., Jones R.L., Austin J., Browell E.V., McCormick M.P., Krueger A.J. and Tuck A.F., 1989, Diagnostic studies of the Antarctic vortex during the 1987 Airborne Antarctic Ozone Experiment: Ozone miniholes, *J. Geophys. Res.*, 94, 11641-11669.

Pitt D. and S. Nilsson, 1993, *Protecting the Atmosphere. The Climate Change Convention and its Context*, Earthscan Publ. Ltd., London, 160 p..

Red Book, 1992, Ozone data for the world, Atmospheric Environment Service, Department of the Environment, Downsview-Ontario, Canada, in cooperation with the World Meteorological Organisation, 33, No. 6, November-December 1992.

Reid S.T. and Vaughan G., 1991, Lamination in ozone profiles in the lower stratosphere, *Q.J.R. Meteorol. Soc.*, 117, 825-844.

Reinsel G.C., Tiao G.S., Miller A.J., Wuebbles D.J., Connell P.S., Mateer C.L. and De Luisi J.J., 1987, Statistical analysis of total ozone and stratospheric Umkehr data for trends and solar cycle relationship, *Journal of Geophysical Research*, 92, 2201-2209.

Reinsel G.C., Tiao G.C., Ahn S.K., Pugh M., Basu S., Deluisi J.J., Mateer C.L., Miller A.J., Connell J.J. and Wuebbles D.J., 1988, An analysis of the 7-year record of SBUV satellite ozone data: Global profile features and trends in total ozone, *Journal of Geophysical Research*, 93, 1689-1703.

Report of the Fourteenth Session of the Joint Scientific Committee, 1993, Hamilton, Bermuda, 15-19 March 1993, WMO/TD-N564.

Rodhe H., 1994, *Global distributions of atmospheric sulphur compounds and their potential impact on climate, In: "EUROTRAC Symposium-94. Transport and tranformation of pollutants in the troposphere. Abstracts of lectures and posters", Garmisch-Partenkirchen, 11-14 Apri, 1994*, p. 224..

Sakellariou N., D. Asimakopoulos, C. Varotsos and O. Capsocha, 1993, Prevailing Cloud Types in Athens, *Theor. Applied Climatol.*, 48, 89-100.

Sanhueza E., Octavio K.H. and Arrocha A., 1985, Surface ozone measurements in the Venezuelan Tropical Savannah, *J. Atmos. Chem.*, 2, 377-385.

Schwartz S. E., 1993, *Comparing the global heating and cooling from fossil fuel combustion*, DOE Research Summary, N28, 4 pp..

Shreffler J.H. and Evans R.B., 1982, The surface ozone record from the regional air pollution study, *Atmos. Envir.*, 16, 1311-1321.

Stolarski R.S., Bloomfield P., McPeters R.D. and Herman J.R., 1991, Total ozone trends deduced from Nimbus-7 TOMS data, *Geophysical Research Letters*, 18, 1015-1018.

United Nations Framework Convention on Climate Change, 1993, Text. IUCC, Geneva.

Varotsos C., 1989, Comments on the connections between the 11-year solar cycle, the QBO and the total ozone, *Journal of Atmospheric and Terrestrial Physics*, 51, (5), 367-370.

Varotsos C. and Cartalis C., 1991, Re-evaluation of surface ozone over Athens, Greece for the period 1901-1940, *Atmos. Res.*, 26, 303-310.

Varotsos C. A. and D. G. Deligiorgi, 1991, Sea surface temperature and Southern Oscillation signal in the upper Stratosphere- Lower Mesosphere, *International Journal of Climatology*, 11, 77-83.

Varotsos C. and Helmis C., 1992, Deviations of the temperature models derived by remote and in-situ sensing techniques for the global middle atmosphere, *International Journal of Remote Sensing*, 13, 3127-3133.

Varotsos C. and Kondratyev K.Ya., 1994, Athens environmental dynamics: from a rural to an urban region, Optics Atmosphere Ocean, 7, 3, 3-17.

Varotsos C., Helmis C. and Cartalis C., 1992a, Annual and semiannual waves in ozone as derived from SBUV vertical global ozone profiles, *Geophysical Research Letters*, 19 (9), 925-928.

Varotsos C., Dris N. and Asimakopoulos D., 1992b, Terannual wave in the ozone and temperature in the strato-mesosphere as deduced from satellite measurements, *Journal of Climate*, 5, (2), 181-185.

Varotsos C. A., N. A. Dris, D. N. Asimakopoulos and C. Cartalis, 1992c, On the Relationship between the 10.7 cm Solar Flux, Surface Pressure and air temperature over Greece, *Theor. and Applied Climatol.*, 46, 27-32.

Viner D. and M. Hulme, 1993a, *What carbon dioxide concentrations should be used with climate change scenarios?*, U. K. Dept. of the Environ. Techn. Note N3, Norwich, 12 pp..

Viner D. and M. Hulme, 1993b, *The U. K. Met. Office high resolution GCM transient experiment*, U. K. Dept. of Environ. Techn. Note N4, Norwich, 11 pp..

Warrick R. A., E. M. Barrow and T. M. L. Wigley (Eds.), 1993, *Climate and Sea Level Change: Observations, Projections, and Implications*, Cambridge Univ. Press, 424 pp..

AMFTRAN: A New Monte Carlo Radiative Transfer Model for Calculating Air Mass Factors

L. Marquard and U. Platt

Institut für Umweltphysik, Universität Heidelberg, D-69120 Heidelberg, Germany

Abstract. The ability to properly interpret UV-visible absorptions of atmospheric chemical constituents in solar straylight is critically dependent on how well the optical path, or airmass factor, of light collected by the spectrometer is understood. Therefore a new Monte Carlo radiative transfer model (AMFTRAN) was developed. AMFTRAN is based on a Backward Monte Carlo approach, including full spherical and refractive geometry. It is able to calculate multiple scattered intensities as well as airmass factors for both satellite and ground-based observations.
AMFTRAN was successfully tested by performing a number of consistency checks. For ground-based observation the present model compares well with the spherical Monte Carlo model developed at the University of Colorado, Boulder, USA. For the satellite observation mode the model was validated through comparison with a quasi-spherical Finite Difference model, developed at the University of Bremen, Germany, within the scope of the GOME project.
AMFTRAN is currently applied to perform sensitivity studies for the airmass factor under a wide range of geophysical conditions (such as viewing geometry, refraction, surface albedo, molecular absorber profiles, aerosol loading and extinction properties).

Keywords. Radiative transfer, Backward Monte Carlo, air mass factor

1 Introduction

1.1 Definition and Implications of the Air Mass Factor

The fundamental principle underlying absorption measurements is Beer's law, wherein the attenuation of monochromatic radiation is related to the number of absorbing molecules in the optical path, so that

$$dI(s) = -I(s)\,\sigma(s)\,\rho(s)\,ds \tag{1}$$

where dI is the change of the measured flux, $I(s)$ is the incident flux, $\sigma(s)$ is the molecular cross-section of the absorbing species, $\rho(s)$ is the absorbing species number density, and ds is an element of the distance over which absorption takes

NATO ASI Series, Vol. I 53
Atmospheric Ozone Dynamics
Observations in the Mediterranean Region
Edited by Costas Varotsos

place, or optical path. In atmospheric applications, ds is often rewritten as Ldz, where dz is the vertical thickness of the optical path and L is a geometrical factor which relates the slant path length to its thickness (the 'air mass' factor). In the atmosphere, L is generally a function of altitude ($L = L(z)$), so that the absorption seen at groundlevel represents an integral over the absorbing path:

$$I(z_{min}) = I(z_{max}) \exp\left(- \int_{z_{min}}^{z_{max}} \sigma(z) \rho(z) L(z) \, dz \right) \tag{2}$$

where z_{min} is the altitude at groundlevel and z_{max} is the altitude at the top of the atmosphere.
The ability to properly interpret UV-visible absorption measurements of atmospheric chemical constituents in solar straylight is critically dependent on how well the air mass factor is understood.
The fundamental quantity measured in atmospheric absorption spectroscopy is the slant (not vertical) column abundance. Interpretation of absorption measurements in terms of the vertical column therefore requires knowledge of the air mass factor.
Use of correct air mass factors is a key requirement for the inversion of ground-based data in order to obtain vertical profiles of atmospheric absorbers.
Correct interpretation of scattered light measurements conducted by aircraft and satellite platforms is also dependent on the capability to understand the air mass factor.

1.2 Importance of Multiple Scattering

It may be presumed that multiple scattering changes the calculated air mass factors where the atmospheric density, and hence the multiple scattering, is at its largest. In the troposphere, for example, multiple Rayleigh scattering tends to increase the air mass factors when the scattering layer is above the absorbing layer, and decrease the air mass factors when the scattering layer is at or near the absorbing layer.
Multiple Mie scattering also entails implications for air mass factors, especially in the troposphere where the optical depth caused by Mie scattering often exceeds the one caused by Rayleigh scattering.

1.3 Features of the Model AMFTRAN

A Monte Carlo radiative transfer model, AMFTRAN, was developed to calculate air mass factors for both direct and scattered light absorption measurements. The model takes into consideration spherical geometry, refraction, multiple Rayleigh and Mie scattering, absorption of different atmospheric chemical constituents, aerosol loading, clouds, and surface albedo. The model can be applied to any climatic scenario. AMFTRAN is able to calculate multiple scattered intensities as well as air mass factors for both satellite and ground-based observations.

The single scattering algorithm of AMFTRAN is based on the models of *Noxon et al.* [1978] and *Solomon et al.* [1987] which have been improved by leaving out simplifications and adding new capabilities. The single scattering approach assumes that each photon reaching the detector is scattered only once before it enters the detector from its line of sight direction.
The technique used to model the multiple scattering in AMFTRAN is the backward Monte Carlo approach which has been described by *Collins et al.* [1972], *Adams and Kattawar* [1978], *Lenoble and Chen* [1992], and *Perliski and Solomon* [1993]. Compared to these models, the main improvements implemented in AMFTRAN consist of a new usage of the random generator, a new method to calculate the parameters of the spherical and refractive geometry, and a new method of calculating the direct intensities at a certain photon position. Therefore AMFTRAN is independent from the models mentioned above, also because the single scattering is modelled by a different technique.

2 Mathematical Principles of AMFTRAN

2.1 Single Scattering

In order to calculate air mass factors for a certain atmospheric species (identified by the index l in the following) it is necessary to calculate the attenuated fluxes I which are obtained if the species is included in the model atmosphere, as well as the attenuated fluxes I_l which are obtained if the species is removed from the model atmosphere. I and I_l are functions of the local solar zenith angle θ_{loc} and the altitude z, and they are given by

$$\begin{aligned} I\left(\theta_{loc}, z\right) &= \exp\left(-\int_{\vec{s}_{min}}^{\vec{s}_{max}} \left(\tau_{Rayl}\left(z\right) + \tau_{Mie}\left(z\right) + \sum_{j} \sigma_j\left(z\right)\rho_j\left(z\right)\right) d\vec{s}\right), \\ I_l\left(\theta_{loc}, z\right) &= \exp\left(-\int_{\vec{s}_{min}}^{\vec{s}_{max}} \left(\tau_{Rayl}\left(z\right) + \tau_{Mie}\left(z\right) + \sum_{j\neq l} \sigma_j\left(z\right)\rho_j\left(z\right)\right) d\vec{s}\right) \end{aligned} \tag{3}$$

where $\tau_{Rayl}(z)$ and $\tau_{Mie}(z)$ are the Rayleigh and the Mie scattering coefficents, $\sigma_j(z)$ is the absorption cross-section of the species with number j, $\rho_j(z)$ is the number density of the species with number j, and $\vec{s}_{min}$ and $\vec{s}_{max}$ are vectors which point to the beginning and end of the specified ray, respectively, both having their origin in the detector position.
The single scattering flux I^{tot} is received by a ground-based detector if the atmospheric species l is included in the model atmosphere. The single scattering flux I_l^{tot} is received if the species l is removed from the model atmosphere. These

are then given by

$$
\begin{aligned}
I^{tot} &= \int_{z_{min}}^{z_{max}} \left(p_{Rayl}(\gamma)\,\tau_{Rayl}(z) + p_{Mie}(\gamma)\,\tau_{Mie}(z)\right) I(\theta_{loc}, z)\, T(z)\, dz\,, \\
I_l^{tot} &= \int_{z_{min}}^{z_{max}} \left(p_{Rayl}(\gamma)\,\tau_{Rayl}(z) + p_{Mie}(\gamma)\,\tau_{Mie}(z)\right) I_l(\theta_{loc}, z)\, T_l(z)\, dz
\end{aligned}
\tag{4}
$$

where $p_{Rayl}(\gamma)$ and $p_{Mie}(\gamma)$ are the phase functions for Rayleigh and Mie scattering, γ is the scattering angle between the ray and the line of sight, and $T(z)$ and $T_l(z)$ are the transmissions due to absorption and scattering processes between the detector and the scattering altitude for model atmospheres including or excluding the specified atmospheric species l.
I^{tot} and I_l^{tot} are calculated by applying the ray tracing technique.

2.2 Multiple Scattering

The multiple scattering fluxes are calculated by a Backward Monte Carlo approach and added to the single scattering flux contributions given by Eqs. 4. Backward models simulate the path of light starting from the detector and ending where the light escapes from the atmosphere. This situation is equivalent to time reversal.
The simulation begins with the emergence of light from the detector up until the location of a final scattering event. In the model, the locations of the final scattering events are distributed over the optical depth of the atmosphere at specified intervals.
There is a finite probability that any scattering event could be the last one. The intensity of the first intensity history is therefore given by

$$I_1(z) = p(\theta_1)\,\sigma_{Rayl}(z)\,n_{Rayl}(z)\,T_{a,1}(z)\,T_{e,1}(z) \tag{5}$$

where $p(\theta_1)$ is the Rayleigh phase function for scattering angle θ_1, $\sigma_{Rayl}(z)$ is the Rayleigh scattering cross-section, $n_{Rayl}(z)$ is the number density of Rayleigh scatterers, $T_{a,1}(z)$ is the transmission due to absorption between the detector and the final scattering altitude, and $T_{e,1}(z)$ is the transmission due to absorption and scattering processes of the optical path between the final scattering altitude and the outer limit of the model atmosphere, i.e. the escape probability. Simulation of the second intensity history begins with the selection of a distance between the first and second scattering event. The Monte Carlo scheme to generate the corresponding probability density function is

$$\tau = -\ln(r) \tag{6}$$

where r is a random number generated with uniform probability density function over the interval [0,1]. The scattering angle which determines the direction

of travel is selected from a probability density function based on the normalized Rayleigh phase function. This distribution is generated by means of a numerical rejection scheme (see e.g. *Press et al.* [1987]).
Mie-scattering is also included in the Monte Carlo model. The Henyey-Greenstein parameterization (*Henyey and Greenstein* [1941]) is used to approximate the Mie phase function. The Monte Carlo scheme for this distribution is:

$$\theta = \arccos\left(\frac{1}{2g}\left(1+g^2-\frac{1-g^2}{\left(1-g+2gr\right)^2}\right)\right) . \tag{7}$$

The type of scattering that occurs (either Rayleigh or Mie) is selected randomly, as is the direction of the scattering in the azimuthal direction. Once the distance and direction of the previous scattering event have been selected, the location of the scattering event in Cartesian optical depth coordinates and in two spherical coordinate systems must be calculated, the first having its origin at the detector position and the second in the Earth's center. The direction cosines are given by:

$$\begin{aligned} u &= u_1 \cos\phi' + \sin\phi'\left(v_1\cos\theta' + w_1\sin\theta'\right) \\ v &= -u_1 \sin\phi' + \cos\phi'\left(v_1\cos\theta' + w_1\sin\theta'\right) \\ w &= -v_1\sin\theta' + w_1\cos\theta' \end{aligned} \tag{8}$$

where ϕ' and θ' are the values of the scattering and azimuthal angle selected in the previous simulated scattering event and:

$$\begin{aligned} u_1 &= \sin\theta\sin\phi \\ v_1 &= \sin\theta\cos\phi \\ w_1 &= \cos\theta \end{aligned} \tag{9}$$

where θ and ϕ are the angular directions selected for the current scattering event simulation.
The attenuation $T_{a,2}(z)$ due to absorption between the final and previous scattering event is then calculated, as is the escape transmission $T_{e,2}(z)$ in the direction of the sun. The intensity resulting from intensity history two is therefore:

$$I_2(z) = p(\theta_2)\,\sigma_{Rayl}(z)\,n_{Rayl}(z)\,T_{a,1}(z)\,T_{a,2}(z)\,T_{e,2}(z) . \tag{10}$$

Whether or not scattering off the ground occurs is determined by considering the coordinates of the location of a scattering event in the spherical center-of-Earth coordinate system.
The intensity history with number n including scattering off the ground is then calculated by:

$$I_n(z) = p(\theta_n)\,\sigma_{Rayl}(z)\,n_{Rayl}(z)\,a\,T_{e,n}(z)\prod_{i=1}^{n} T_{a,i}(z) \tag{11}$$

where a is the surface albedo.
The total intensity I^{tot} received by the detector is equal to the sum of the arithmetic averages of the intensity histories calculated at every interval along the detector's direction of view:

$$I^{tot} = \sum_{l} < \sum_{n} I_n (z_l) > . \tag{12}$$

3 Validation of AMFTRAN

3.1 Check of the Conservation of Energy

Every radiative transfer model must obey the laws of conservation of energy and momentum. The irradiance F is the only quantity derived from the quantity energy which is under consideration for making out an energy and momentum balance because it represents the number of photons N that cross a unit area dS in perpendicular direction per unit time dt in a beam of infinitesimal solid angle $d\omega$ in a particular wavelength interval $d\lambda$, i.e.

$$F = F_\lambda (\theta, \phi) = \frac{N}{dS\ dt\ d\omega\ d\lambda} \cos\theta \tag{13}$$

where θ and ϕ are the polar and the azimuthal angles of the detector view direction.
$F_\lambda (\theta, \phi)$ may be rewritten as $F_\lambda (\theta)$ because both Rayleigh and Mie scattering have no dependence in the azimuthal direction. The total flux I_λ leaving the atmosphere is found by integration of $F_\lambda (\theta)$ over the half sphere, i.e.

$$I_\lambda = \int_{2\pi} F_\lambda (\theta)\, d\omega = 2\pi \int_0^{\pi/2} F_\lambda (\theta) \sin\theta d\theta \tag{14}$$

where the detector is located outside the model atmosphere. This means that in order to test the energy balance for a certain radiative transfer model $F_\lambda (\theta) \sin\theta$ must be calculated at specified angles $\theta_i \in [0, \pi/2]$, $i = 1, ..., n$ and integrated numerically over θ. The model fulfills the laws of energy and momentum conservation if the result of this integration is equal to the flux entering the atmosphere divided by 2π.

This energy and momentum balance was calculated for AMFTRAN and, additionally, for GOMETRAN (*Rozanov* [1993]). The results are shown in Fig. 1. AMFTRAN fulfills the law of energy and momentum conservation extremely well, i.e. within $\pm 4 \cdot 10^{-8}$, over the spectral range from 240 nm to 790 nm. GOMETRAN conserves the total energy and momentum of the light radiated into the atmosphere within 0.8 %. The poorer performance of GOMETRAN probably results from its quasi-spherical approximation, i.e. in GOMETRAN the multiple scattering part of the radiative transfer equation is treated in plane-parallel form and only the so-called 'source term' is calculated in spherical form. AMFTRAN, however, has full spherical geometry.

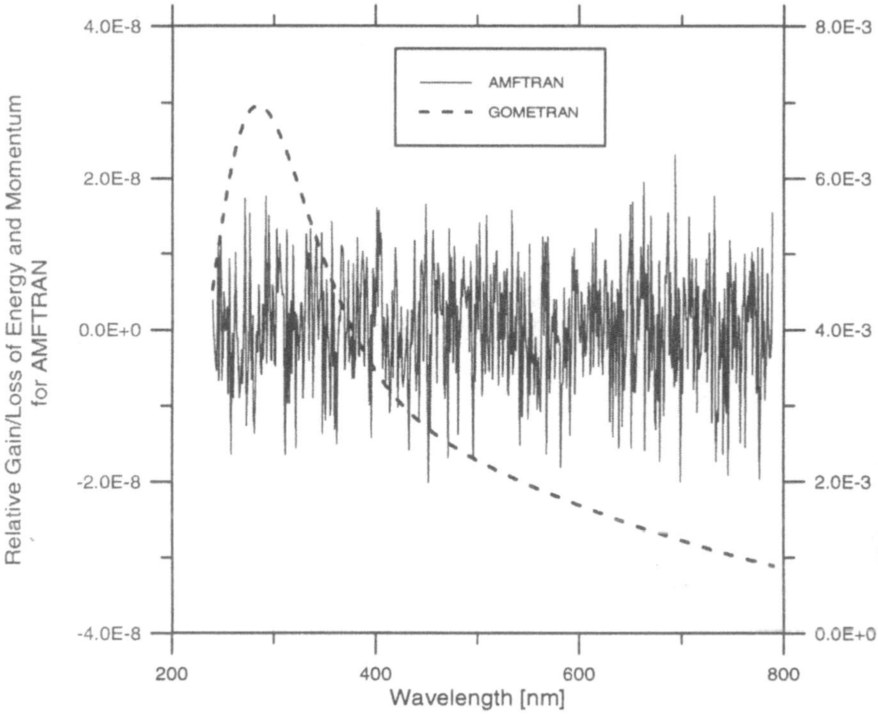

Figure 1: *Energy balance for AMFTRAN and GOMETRAN.* AMFTRAN fulfills the law of conservation of energy extremely well (within $\pm 4 \cdot 10^{-8}$) over the whole spectral range of GOME. GOMETRAN conserves the energy radiated into the atmosphere within 0.8 %.

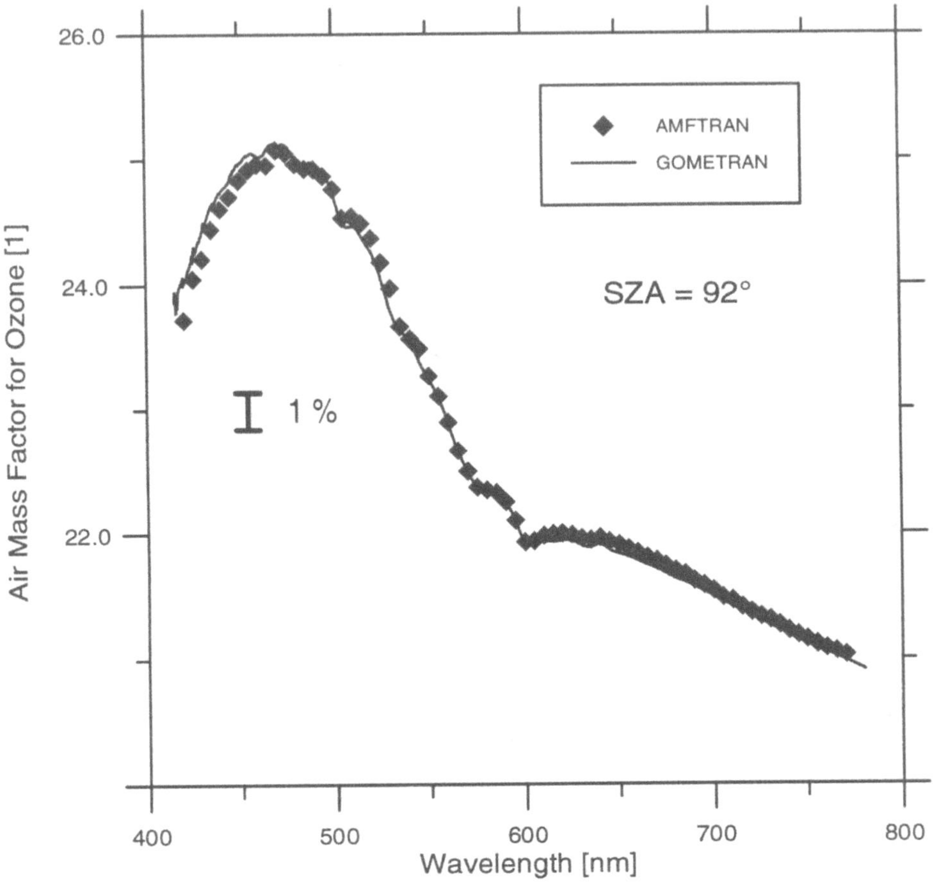

Figure 2: *Comparison AMFTRAN - GOMETRAN:* Text see Fig. 4.

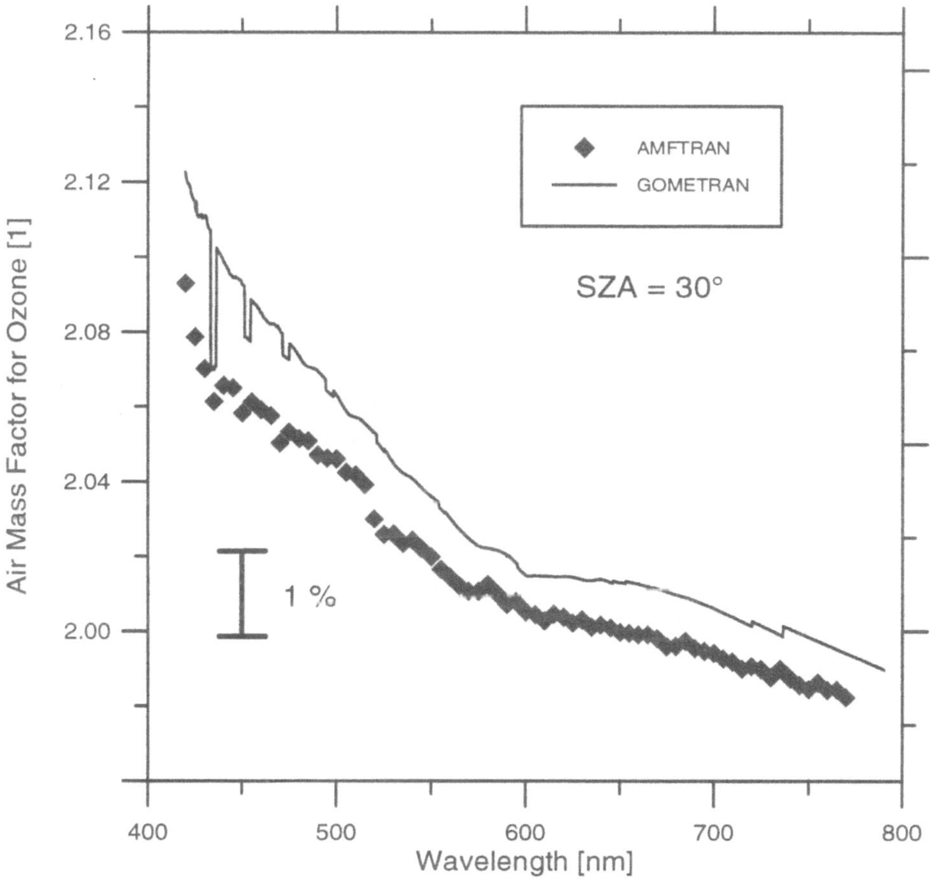

Figure 3: *Comparison AMFTRAN - GOMETRAN:* Text see Fig. 4.

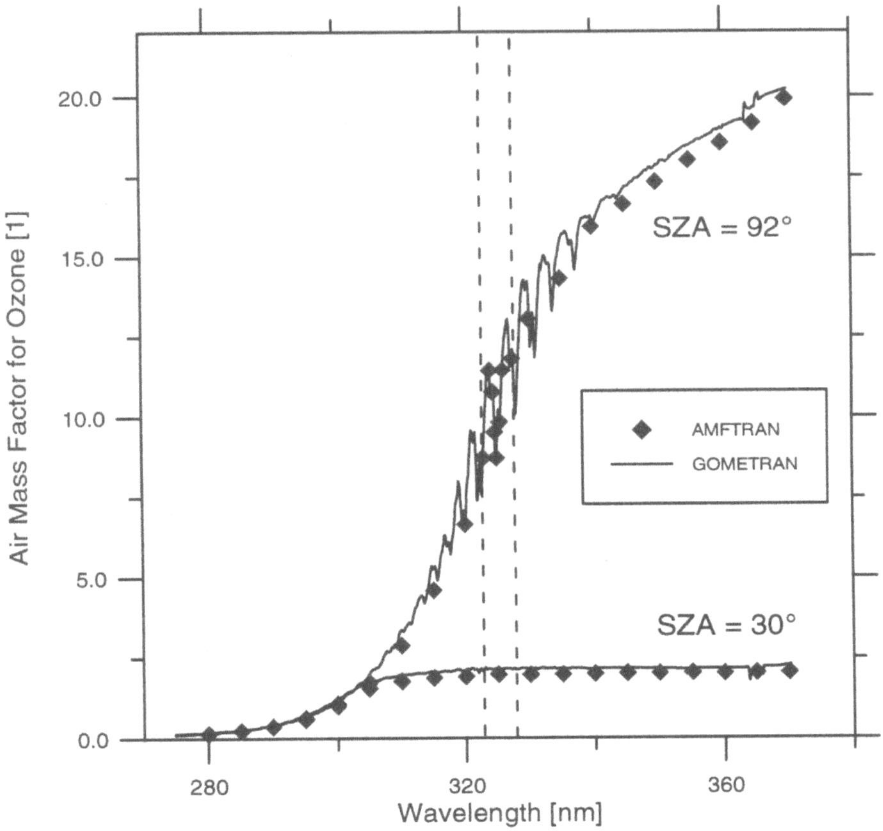

Figure 4: *Comparison AMFTRAN - GOMETRAN:* Regarding the calculation of air mass factors for ozone both models agree within 3%. Other viewing geometries and atmospheric conditions have yielded similar results, and it is expected that the differences between AMFTRAN and GOMETRAN will stay within the 5 % range for all possible scenarios.

3.2 Comparison AMFTRAN - GOMETRAN

The satellite observation mode of AMFTRAN was evaluated in comparison with GOMETRAN (*Rozanov* [1993]) within the scope of the GOME project. GOMETRAN is a quasi-spherical forward model which is based on the Finite Difference method.

Results for the comparison between both models regarding the calculation of air mass factors for ozone are shown in Figs. 2, 3, 4.
For the selected wavelength ranges the deviations between both models are less than 3%. Other viewing geometries and atmospheric conditions have yielded similar results and it is expected that the differences between AMFTRAN and GOMETRAN will stay within the 5% range for all possible scenarios.

4 References

Adams, C.N. and G.W. Kattawar, Radiative transfer in spherical shell atmospheres, I., Rayleigh scattering, *Icarus, 35*, 139, 1978.
Collins, D.G., W.G. Blättner, M.B. Wells, and H.G. Horak, Backward Monte Carlo calculations of the polarization characteristics of the radiation emerging from spherical shell atmosphere, *Appl. Opt., 11*, 2684, 1972.
Henyey, L.C., and J.L. Greenstein, Diffuse radiation in the galaxy, *Astrophys. J., 93*, 70, 1941.
Lenoble, J., and H.B. Chen, Monte Carlo study of the effects of stratospheric aerosols and clouds on zenith sky absorption measurements, Paper presented at the International Radiation Symposium, *IAMAP, Tallin Estonia*, Aug., 1992.
Noxon, J.F., R.B. Norton, and W.R. Henderson, Observation of atmospheric NO_3, *Geophys. Res. Lett., 5*, 675, 1978.
Perliski, L.M., and S. Solomon, On the evaluation of air mass factors for atmospheric near-ultraviolett and visible absorption spectroscopy, *J. Geophys. Res., 98*, 10363, 1993.
Press, W.A., B.P. Flannery, S.A. Teukolsky, and W.T. Vetterling, Numerical Recipes, *Cambridge University Press*, Cambridge, 1987.
Rozanov, V.V., A Study of methods for retrieval of atmospheric constituents: 4. Development of a forward model, *Final Report to ESA*, 1993.
Solomon, S., A.L. Schmeltekopf, and R.W. Sanders, On the interpretation of zenith sky absorption measurements, *J. Geophys. Res., 92*, 8311, 1987.

Air Pollution Models and Their Role in Environmental Policy

N. Moussiopoulos

Aristotle University, GR-54006 Thessaloniki, Greece

Abstract. Mathematical models are being increasingly used to study wind flow and air pollutant dispersion. In several cases, so-called air pollution models have been successfully used to address environmental policy issues related to air quality in urban areas. If properly validated with available measurements, air pollution model systems may be used for both the optimization of air pollution abatement strategies and short-term predictions of smog episodes in highly polluted areas.

One of the contemporary air pollution model systems is the EUMAC Zooming Model (EZM). Core models of this comprehensive model system are the nonhydrostatic mesoscale model MEMO and the photochemical dispersion model MARS. Manifold air quality studies, primarily for large urban agglomerations, have been already performed with the EZM. As examples for such studies, results are presented from the application of the EZM to Athens, Heilbronn and the Upper Rhine valley. In all these cases the results of the EZM agree fairly well with available observations.

Keywords. Urban air quality, air pollution model, mesoscale wind flow, photochemical smog, environmental policy issues.

1 Introduction

In the last decades the mechanisms of air pollutant transport and transformation in the atmosphere are the subject of intense research activities. A primary objective of these activities is to study complex source-receptor relationships, i.e. to assess the responsibility of individual source groups for the occurence of elevated air pollution levels. In the meantime most of the physical and chemical processes leading to smog situations seem to be understood. Yet, air pollution causes still serious problems in most countries of the world.

In most cases atmospheric transport phenomena may not be adequately described solely on the basis of observations. For this reason mathematical models were developed which may be used to simulate wind flow and pollutant transport and transformation in the atmosphere. In the last years such 'air pollution models' are increasingly considered to be important tools for air quality management, as they

NATO ASI Series, Vol. I 53
Atmospheric Ozone Dynamics
Observations in the Mediterranean Region
Edited by Costas Varotsos

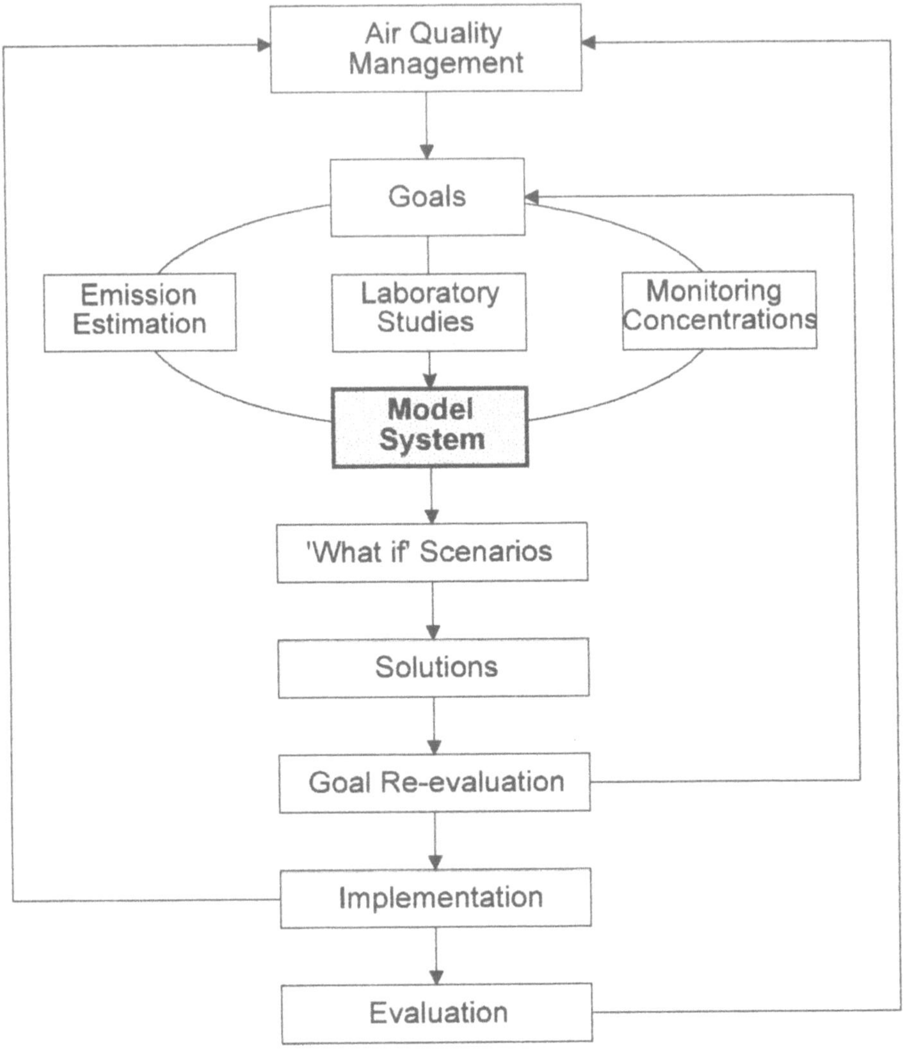

Figure 1. The role of an air pollution model system in air quality management (adapted from Longhurst et al., 1994).

may be used to quantify complex source/receptor relationships (*Figure 1*). Among their potential fields of application, the most important ones are

- the optimization of air pollution abatement strategies either locally or in a larger region;
- the short-term prediction of smog episodes in highly polluted areas.

In recent years, air pollution model development in Europe has been coordinated in the frame of EUMAC (EUMAC = European Modelling of Atmospheric Constituents). As a subproject of EUROTRAC, EUMAC's aim was the creation of a complete model family capable of both answering scientific questions and solving environmental policy problems over all relevant scales. A key role within this model family is being played by the EURAD model, which may be used to simulate continental-to-regional transport and transformation phenomena (Hass et al., 1990). The dispersion of chemically active pollutants in subregions of the EUMAC model domain can be described by the aid of the EUMAC Zooming Model (EZM; Moussiopoulos, 1995). The present paper contains a brief description of the EZM and results of selected policy-oriented applications of this model system.

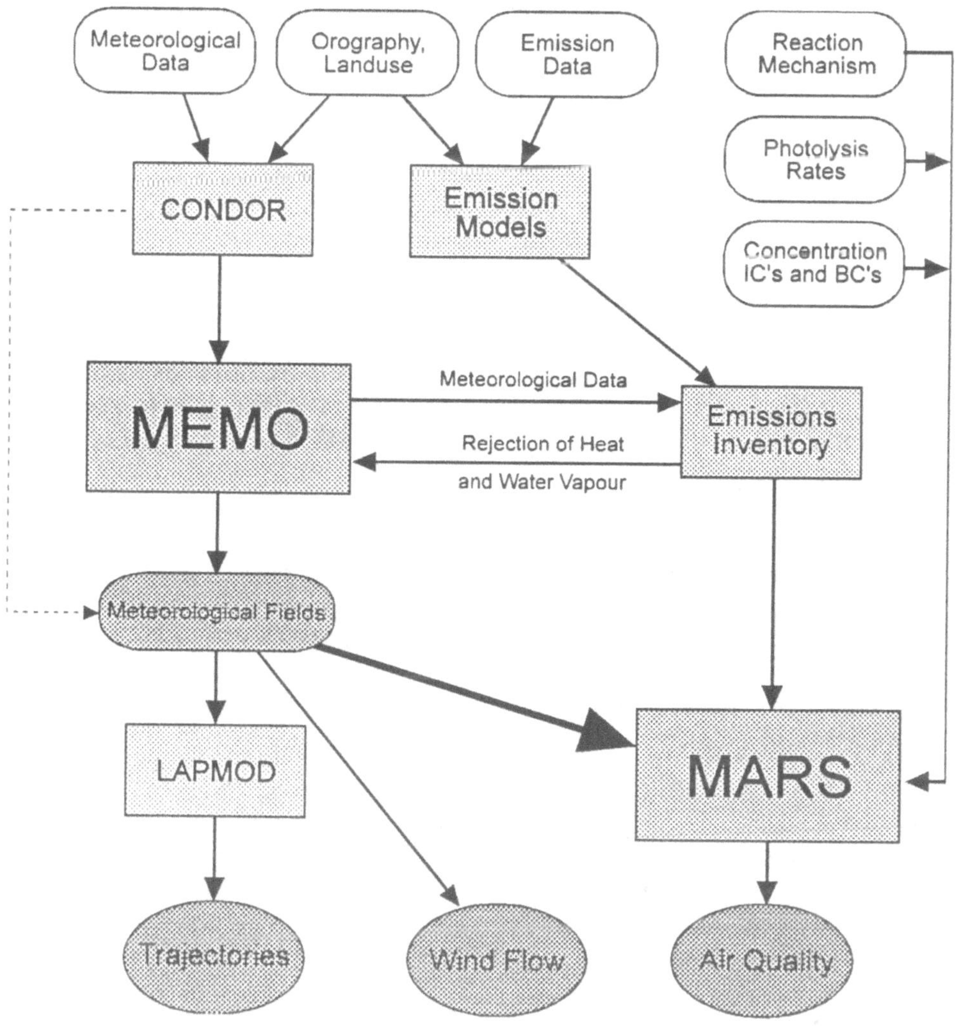

Figure 2. Structure of the EUMAC Zooming Model (EZM).

2 The EUMAC Zooming Model (EZM)

The EZM, the structure of which is illustrated in *Figure 2*, allows describing the dynamics of the atmospheric boundary layer and hence also calculating the advective and diffusive transport of air pollutants in the local-to-regional scale. It is best suited for air quality studies in urban agglomerations. Originally, the EZM was designed to be applied in conjunction with the EURAD model. In the meantime, however, a stand-alone model system evolved which may be driven solely with data derived from observations. The core models of the EZM are the nonhydrostatic prognostic wind model MEMO (Moussiopoulos et al., 1993; Kunz and Moussiopoulos, 1995) and the three-dimensional photochemical dispersion model MARS (Moussiopoulos et al., 1995b). Further constituents of the EZM are the diagnostic wind model CONDOR (Moussiopoulos et al., 1988) and the Lagrangian particle model LAPMOD (Winkler, 1994).

The decisive design criteria of the EZM have been:

- *Completeness*. The EZM incorporates a contemporary prognostic boundary layer model (MEMO) providing 'reliable' 3D meteorological fields, i.e. input which in a large number of applications was proved to agree fairly well with reality. Hence, in the case of the EZM the uncertainty in ozone predictions due to wind field input uncertainties is negligible. ('Wind field input' is only the large-scale information imposed to MEMO).
- *Consistency*. All modules of the EZM were coded in a uniform manner, especially as numerical grid, data structure and algorithm characteristics are concerned. Major features of the EZM are a high degree in modularity and ensured portability.
- *Robustness*. Algorithms of the EZM are characterized by conservativity and efficiency. In all cases estimates on the numerical error may be provided, so that inaccuracies in pollutant concentrations may be attributed to uncertainties in the input data (especially emission data) and the parameterizations used.
- *Flexibility*. The photochemical dispersion model MARS may be used in conjunction with any chemical reaction mechanism. Hence, the complexity of the approach may be adapted to the accuracy of the available input data.

3 Brief description of the individual models

3.1 The nonhydrostatic mesoscale model MEMO

MEMO is a fully vectorized nonhydrostatic mesoscale model (Moussiopoulos et al., 1993; Kunz and Moussiopoulos, 1995). Within MEMO, the conservation equations for mass, momentum and scalar quantities as potential temperature, turbulent kinetic energy and specific humidity are solved. The model uses the anelastic approximation to filter sound waves. The Boussinesq approximation may be used in the simulations, although a non-Boussinesq model version exists as well. The governing equations are solved in terrain-influenced coordinates. Non-equidistant grid spacing is allowed in all directions. The numerical solution is based on second-order discretization applied on a staggered grid. Within the transformed coordinates, Cartesian momentum components are kept but derivatives are computed from the transformed coordinates (Schumann and Volkert, 1984). Both the divergence operator and the pressure gradient contain only pure differential quotients rather than products between the latter and metric coefficients; as a consequence, conservative properties are preserved within the discrete model equations. The discrete pressure equation is solved with a direct elliptic solver in conjunction with a generalized conjugate gradient method (Flassak and Moussiopoulos, 1988).

Advective terms are treated with an explicit, monotonicity-preserving discretization scheme with only small implicit diffusion (Wortmann-Vierthaler and Moussiopoulos, 1995). Turbulent diffusion is described using K-theory. The exchange coefficients for momentum and scalars are computed with an one-equation turbulence model (conservation equation for the turbulent kinetic energy and algebraic equation for the mixing length). At roughness height similarity theory is applied, the turbulent fluxes resulting from the Businger equations. The radiative heating/cooling rate in the atmosphere is calculated with an efficient scheme based on the emissivity method for longwave radiation and an implicit multilayer method for shortwave radiation (Moussiopoulos, 1987a). The surface temperature over land is computed from the surface heat budget equation. The soil temperature is calculated by solving an one-dimensional heat conduction equation for the soil. Water temperature is kept constant during a simulation. In coastal areas a grid location is classified as 'water', if the cover rate percentage of water exceeds 50%. In the opposite case this location is treated as 'land' and the water fraction is neglected. At lateral boundaries generalized radiation conditions are implemented. The current standard version of MEMO allows performing nested grid simulations.

3.2 The photochemical dispersion model MARS

MARS is a fully vectorized photochemical dispersion model (Moussiopoulos et al., 1995b). This model uses a fully implicit selfadaptive method, i.e. an implicit algorithm characterized by a variable time step and a variable order. The overall solution procedure is controlled by the local error (the additional error of the algorithm made in a time step), which is not allowed to exceed a given tolerance.

Differently than in most of the other proposed photochemical dispersion models, MARS allows describing the combined effects of vertical diffusive transport and chemical transformation of pollutants (Moussiopoulos, 1987b). By this approach the feasible error caused by splitting the operators associated with vertical diffusion and chemistry is avoided (Graf and Moussiopoulos, 1991). Operator splitting is thus restricted to the separate treatment of advective transport. For the latter a positive definite algorithm with small implicit diffusion is utilized. Turbulent diffusion is described by the aid of an one-equation turbulence model with a suitable algebraic parameterization for the mixing length. Topography is properly considered by using terrain-influenced coordinates. Flux conservation is ensured by adopting a staggered grid. MARS may be used in conjunction with any chemical reaction mechanism.

3.3 The diagnostic wind model CONDOR

This model allows reconstructing three-dimensional wind fields over complex terrain on the basis of observations (Moussiopoulos, 1988). For this purpose, an interpolation procedure is first applied to obtain a first estimate for the wind field. Subsequently, this estimate is transduced to a mass-consistent wind field by solving an elliptic partial differential equation. A suitable parameterization allows taking into account the influence of atmospheric stability on the correction of the wind field.

3.4 The Lagrangian particle model LAPMOD

As a Lagrangian particle model, LAPMOD is suited to simulate the dispersion of point source emissions, especially in the case of buoyancy induced flow (Winkler, 1994). Plume dispersion may be described by the aid of an appropriate entrainment assumption. Dry deposition may also be taken into account, e.g. by adopting a multiple resistance model of the 'big-leaf' type.

4 EZM applications

The EZM is being utilized by several research institutions all over Europe for analyses of atmospheric transport and transformation phenomena in the local-to-regional scale. Several recent EZM applications have a strong environmental policy background. The remainder of this paper contains information on three such applications.

4.1 Athens case study

The capital of Greece is a good example for a large city continuously suffering under extremely high air pollution levels. The Athens case is very challenging also from the scientific point of view because it combines elevated anthropogenic emission levels and complex topographical and meteorological conditions. For this reason, within EUMAC the model intercomparison activity APSIS was initiated (APSIS: Athenian Photochemical Smog, Intercomparison of Simulations; Moussiopoulos, 1993).

The Athens basin is surrounded by mountains at three sides and is open to the sea to the SW. This results in a frequent occurence of sea breeze circulation events, primarily during summery anticyclonic situations. Most smog episodes in Athens take place under such meteorological conditions. The highest air pollutant concentrations in Athens occur, however, in the case of a weak synoptic forcing acting against the development of a sea breeze. Such conditions, which may easily lead to air mass stagnation above the city, were prevailing on May 25, 1990, i.e. the day selected for most of the model simulations within the APSIS activity.

For an adequate description of the mesoscale wind flow in the Greater Athens area, nested grid simulations were performed with the EZM: Results of a first application of MEMO on a coarse grid were used to formulate improved boundary conditions for the fine grid simulations. The model results support the observational evidence that the sea breeze was very weak during the whole of the considered day. The interface between the circulation cell in the Athens basin and that in the neighbouring Mesogia plain coincides with the eastern border of the basin. Correspondingly, air mass transport between the Athens basin and the surrounding areas is found to be severely restricted during that day.

Detailed comparisons of model results and observations show that the observed situation is very reliably reproduced with the EZM (Kunz and Moussiopoulos, 1995; Moussiopoulos et al., 1995). Although this finding is very encouraging, the experience gained within APSIS shows that much more detailed observational data are needed for a conclusive model evaluation.

Already in the beginning of the 90's the EZM proved useful for supporting air quality management in Athens: Results of simulations with the EZM for various emission scenarios were considered at the optimization of the air pollution abatement strategy for Athens (Moussiopoulos et al., 1992).

Recently, the EZM was applied in the frame of a study launched by the Greek Ministry of the Environment aiming to assess the environmental impact of constructing the New Athens Airport. For this purpose, nested grid simulations were performed to describe the dispersion of inert air pollutants in the Athens basin and the adjacent Mesogia plain for different meteorological conditions (Moussiopoulos et al., 1995a).

4.2 The "Heilbronn Ozone Experiment"

In view of the elevated air pollution levels in the industrialized area of Heilbronn, the "Heilbronn Ozone Experiment" was performed in the period June 23-26, 1994 with the intention to prove to what extent ozone levels may be reduced by the aid of short term local scale interventions. In order to analyse the detailed meteorological and air pollutant measurements carried out during this period, nested grid simulations were performed with the EZM for the situation prevailing during the experiment (Moussiopoulos et al., 1995c). Dispersion calculations were also performed for the hypothetical case of no emission reductions during the period of the experiment. By comparing the results of the latter calculations with those of the standard ones it is possible to quantify the effectiveness of the emission reduction interventions.

In order to take into account all relevant orographic influences on the flow field, a two-step one-way nesting technique (coarse, medium and fine grid simulations) based on the expanded radiation boundary condition was applied. The coarse grid simulations covered an area of 200×200 km^2 including most of the Black Forest and the Schwäbische Alb in the south of the domain and the Rhine valley and the Odenwald in the northwest. The high-resolution fine grid covered an area of 20×20 km^2 with a horizontal grid spacing of 500 m. It should be noted that the employed two-step nesting technique allows avoiding that the nesting factor exceeds 3.5.

For the dispersion simulations on the fine grid the concentrations at the lateral boundaries were generated by additional dispersion simulations on a coarse grid covering an area of 130×130 km^2. The large scale boundary concentrations for NO_2 and ozone on the coarse grid were derived from results of simulations carried out by TNO with the LOTOS (Builtjes, 1992).

The EZM simulation results are providing a reasonable impression of the actual observed situation. As the main exception, temperature (and therefore also, to a lower extent, the near ground wind speed) are systematically overpredicted, most probably because of the underestimation of the aerodynamic roughness of the terrain. Comparison of the predicted and measured diurnal variation of the wind speed and direction at Heilbronn on June 24, 1994 shows that both wind speed and direction are satisfactorily reproduced, the agreement between predictions and observations improving with increasing resolution.

In general, the computed concentrations are also in good agreement with observations. At most stations, deviations between predicted ozone peaks and corresponding observations do not exceed 10 ppb. The results of the simulations performed with the EZM allow estimating the changes in air pollution levels because of the implementation of short term local scale interventions. Results for the NO_2 and ozone concentrations at Heilbronn and Neckarsulm on June 24, 1994 for both scenarios, i.e. with and without the emission reductions reveal

- drastic reductions of NO_2 concentrations (by 25-40%), but
- unchanged ozone levels in the urban areas.

Specifically, marginal ozone increases can be detected which are caused by the lower ozone depletion potential, a consequence of the decrease in NO. Downwind of the urban areas the situation is reversed, but still the differences in ozone concentrations are at the most not higher than a few ppb.

As the main result of this study, short term local scale interventions prove not to decisively affect photochemical pollutant concentrations in urban agglomerations.

4.3 Wind flow and ozone formation in the Upper Rhine valley

The Upper Rhine valley is a broad valley bordered by the Black Forest to the east and by the Vosges to the west. The valley itself is a densely populated and industrialized area including the cities Karlsruhe, Strasbourg, Freiburg and Basel. The highways through the valley have a south/north orientation and represent main transportation routes for local and transit traffic. In view of the elevated air pollution levels in the Upper Rhine valley, it was of major importance to establish an emissions inventory with a high temporal and spatial resolution (Schneider et al., 1995).

The application of the EZM to the Upper Rhine valley was performed for September 16, 1992, a day in the midst of an intensive measuring campaign performed in that region with the participation of numerous scientific groups (Zimmermann, 1993). It was a summer day with surface temperatures up to 25°C and measured ozone concentration peaks exceeding 100 ppb. In view of the size of the domain and in order to perform the simulation at a reasonable computational time demand, the compact reaction mechanism KOREM was used in conjunction with the dispersion model MARS.

From the model results for the wind and ozone concentration patterns it can be followed that the valley and the surrounding mountains have a strong influence on the wind flow in the region. During the night and in the morning the wind field shows a channeled flow in the valley from south to north. As a consequence, in the interior of the valley the emissions associated with the highways lead to a total depletion of ozone in the morning. In contrast to this, elevated regions to the east and, partly, to the west of the valley retain significant amounts of ozone during the night due to the absence of strong NO sources in their vicinity. In the afternoon, however, the picture changes: Along the valley the emissions are characterized by a rather high VOC/NO_x ratio. This results in a strong ozone production in most parts of the domain. Low wind speeds in the valley also favour thc production of photooxidants.

The comparison of computed diurnal variations of ozone concentrations with available measurements at selected locations proves that in general the computed ozone concentrations are in very good agreement with the measured ones. This agreement is not only in a qualitative sense (e.g. the time when a peak occurs), but also from a quantitative point of view in spite of the fact that point measurements are compared with mean values for grid cells of the size of 9 km^2.

5 Concluding remarks

By integrating scientific results in environmental policy, air pollution models are extremely valuable tools for optimizing control strategies and for implementing effective smog alarm systems in heavily polluted areas. As one of the contemporary air pollution model systems, the EUMAC Zooming Model (EZM) has been successfully applied for air quality studies in several European areas including Athens, Heilbronn and the Upper Rhine Valley. In these applications, the results of the EZM agree fairly well with available observations.

References

Builtjes P.J.H. (1992), The LOTOS-Long Term Ozone Simulation-Project Summary Report, TNO-report IMW-R 92/240, Delft.

Flassak Th. and Moussiopoulos N. (1988), Direct solution of the Helmholtz equation using Fourier analysis on scalar and vector computers, *Environmental Software* **3**, 12-16.

Graf J. and Moussiopoulos N. (1991), Intercomparison of two models for the dispersion of chemically reacting pollutants, *Contr. Phys. Atmospheric* **64**, 13-25.

Hass H., Memmesheimer M., Geiss H., Jakobs H.J., Laube M. and Ebel A. (1990), Simulation of the Chernobyl radioactive cloud over Europe using the EURAD model, *Atmospheric Environ.* **24A**, 673-692.

Kunz R. and Moussiopoulos N. (1995), Simulations of the wind field in Athens using refined boundary conditions, *Atmospheric Environ.* **24**, 3575-3591.

Longhurst J.W.S., Lindley S.J. and Conlan D.E. (1994), Towards local air quality management in the United Kingdom, in **Pollution Control and Monitoring** (J.M. Baldasano, C.A. Brebbia, H. Power and P. Zannetti, eds), Computational Mechanics Publications, 527-532.

Moussiopoulos N. (1987a), An efficient scheme to calculate radiative transfer in mesoscale models, *Environmental Software* **2**, 172-191.

Moussiopoulos N., (1987b), A new technique for the solution of diffusion equation systems for reacting pollutants, *Bound. Layer Meteorol.* **41**, 251-264

Moussiopoulos N. (1993), Athenian photochemical smog: intercomparison of simulations (APSIS), background and objectives, *Environmental Software* **8**, 3-8.

Moussiopoulos N. (1995), The EUMAC Zooming Model, a tool for local-to-regional air quality studies, *Meteorol. Atmos. Phys.* **57**, 115-133.

Moussiopoulos N., Flassak Th. and Knittel G. (1988), A refined diagnostic wind model, *Environmental Software* **3**, 85-94.

Moussiopoulos N., Flassak Th., Sahm P. and Berlowitz D. (1993), Simulations of the wind field in Athens with the nonhydrostatic mesoscale model MEMO, *Environmental Software* **8**, 29-42.

Moussiopoulos N., Pattas K. and Samaras Z. (1992), Measures for reduction of road traffic emissions and consequences to ozone formation in Athens, in **Traffic Induced Air Pollution** (R. Pischinger, ed.), Mitteilungen des Institutes für Verbrennungskraftmaschinen und Thermodynamik, 351-362.

Moussiopoulos N., Sahm P., Gikas A., Karagiannidis K., Karatzas A. and Papalexiou S. (1995a), Analysis of air pollutant transport in the Athens basin and in the Spata area with a three-dimensional dispersion model, in **Air Pollution III, Vol. 3: Urban Pollution** (N. Moussiopoulos, H. Power and C.A. Brebbia, eds), Computational Mechanics Publications, Southampton, 141-152.

Moussiopoulos N., Sahm P. and Kessler Ch. (1995b), Numerical simulation of photochemical smog formation in Athens, Greece - A case study, *Atmospheric Environ.* **24**, 3619-3632.

Moussiopoulos N., Sahm P., Kessler Ch., Voegele T. and Schneider Ch. (1995c), Application of the EZM to the 'Heilbronn Ozone Experiment', in **Air Pollution III, Vol. 1: Air Pollution Theory and Simulation** (H. Power, N. Moussiopoulos and C.A. Brebbia, eds), Computational Mechanics Publications, Southampton, 397-404.

Schneider Ch., Kessler Ch. and Moussiopoulos N. (1995), Influence on emission input data on ozone level predictions for the Upper Rhine valley, in **Urban Air Pollution, Vol. 2** (H. Power and N. Moussiopoulos, eds), Computational Mechanics Publications, Southampton, 1-39.

Schumann U. and Volkert H. (1984), Three-dimensional mass- and momentum-consistent Helmholtz-equation in terrain-following coordinates, in **Notes on Numerical Fluid Mechanics**, Vol. 10 (Hackbusch W., ed.), Vieweg, Braunschweig, 109-133.

Winkler Ch. (1994), The Lagrangian particle model LAPMOD, in **The EUMAC Zooming Model (EZM): Model Structure and Applications** (N. Moussiopoulos, ed.), EUROTRAC Special Publication, 61-64.

Wortmann-Vierthaler M. and Moussiopoulos N. (1995), Numerical tests of a refined flux corrected transport advection scheme, *Environmental Software* **10**, 157-175.

Zimmermann H. (1993), Die gemeinsame TRACT/REKLIP-Messkampagne, in Zusammenfassungen der Beiträge zum REKLIP-Kolloquium in Basel am 4. November 1993, Basel, Switzerland.

A THERMOLUMINESCENCE DOSEMETER FOR SOLAR ULTRAVIOLET IRRADIATION

J.A. Muñoz; I. Aguirre de Cárcer; G. Lifante; F. Cussó; F. Jaque

Departamento Física de Materiales, C-IV. Universidad Autónoma de Madrid. Cantoblanco, 28049-Madrid (Spain)

Keywords: Dosemetry, ultraviolet, Solar irradiation, thermoluminescence

INTRODUCTION

Ozone depletion has received a growing attention by the general public and the scientific community because of the role of this atmospheric gas on screening harmful solar ultraviolet radiation from reaching the earth surface. Furthermore, ultraviolet dose measurements are important because of its role on the complex ozone photochemistry.

Divalent europium doped alkali halides are phosphor systems that are excited with Xe-lamp ultraviolet light [1] and have been proposed as ultraviolet dosemeters [2]. These materials show a thermoluminescence signal (TL) when excited at the upper edge of Eu^{2+} high energy absorption band in those matrix (Figure 1).

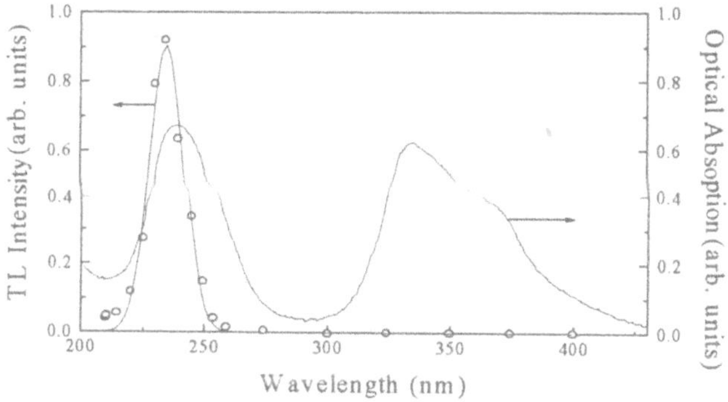

Figure 1. Absorption and thermoluminescence excitation curves for $KCl:Eu^{2+}$.

NATO ASI Series, Vol. I 53
Atmospheric Ozone Dynamics
Observations in the Mediterranean Region
Edited by Costas Varotsos

The thermoluminescence signal is linear with UV light up to 10^3 Gy (Figure 2).

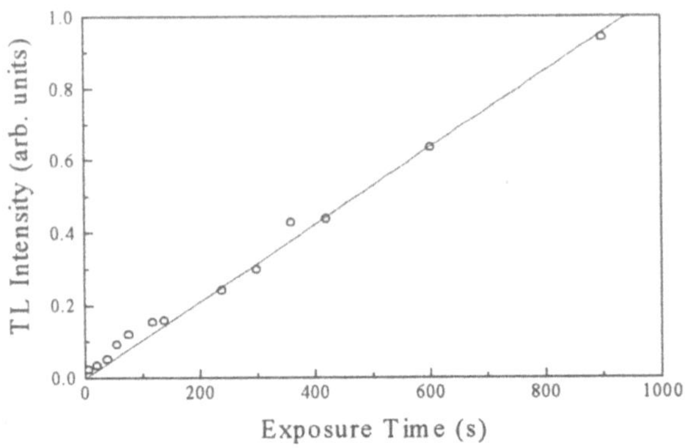

Figure 2. Dependence of the $KCl:Eu^{2+}$ TL intensity as a function of the irradiation exposure time at $\lambda = 250$ nm .

It has been reported the detection of TL signal in $NaCl:Eu^{2+}$ with unfiltered direct solar light exposure which could be related to solar ultraviolet radiance [3] . While laboratory measurements indicate that the ultraviolet excitation in the 230 - 270 nm range, is responsible for the TL signal of these materials, there is the possibility of interpreting the solar irradiation TL signal as due to the combined effect of a small residual TL excitation of the materials with a lower energy wavelength and the higher solar irradiation at $\lambda > 300$ nm. Furthermore it has been reported that $KCl:Eu^{2+}$ is photostimulated with 560 nm after UV irradiation [4] which introduces another rival explanation for the solar TL signal discussed. In this work we report the TL response of $KCl:Eu^{2+}$ to direct solar irradiance of different ultraviolet wavelength at the earth surface, in order to ascertain the origin of this solar activation.

EXPERIMENTAL PROCEDURE

$KCl:Eu^{2+}$ single crystals were grown in the Crystal Growth Laboratory of the Departamento de Física de Materiales, Universidad Autónoma de Madrid, using Czochralski method under controlled atmosphere. Doping was achieved by adding $EuCl_2$ to the melt (previously reduced). A Eu^{2+} concentration of 150 ppm was calculated by using the calibrated constants [5] associated with the optical absorption bands of Eu^{2+} ions in KCl.

Seven samples, aproximately 3mm x 2mm x 2mm size were used for the TL measurements and they were cut from the same single crystal. Before solar irradiation the samples were preheated to 500 °C to assure erasure of signals from any accidental activation of the traps. The samples were kept at dark before and after solar irradiation. TL data were taken at a heating rate of 5 °C/s with a TLDA-40 RISO system [6].

For solar irradiation the samples were exposed during 60 seconds to the light collected by a reflecting mirror telescope (focal length 999 mm, 707 cm^2 effective collection area) directed to the sun and filtered with a Jobin-Yvon monochromator (1250 rulings/mm, wavelength band-pass 2 nm) and a low pass UV filter.

RESULTS AND DISCUSSION

Solar irradiation of the samples were done at Cantoblanco, in Madrid's (Spain) suburban area (40.5° N; 3.75° W; 700 m above sea level, high visibility atmosphere) around midday on october 16 and 17, 1995 for which a 269 DU and 268 DU were reported by TOMS respectively.

Six samples were irradiated at approximately the same solar time. Sample number seven was not exposed to solar irradiation and suffered the same treatment as the others in order to control for other factors. The samples were irradiated at different wavelength within the 250 nm - 350 nm range, and their glow curves were obtained. Data shown in figure 3 correspond to the first TL peak intensity after normalization and substraction of the common background, estimated trough sample number seven TL signal.

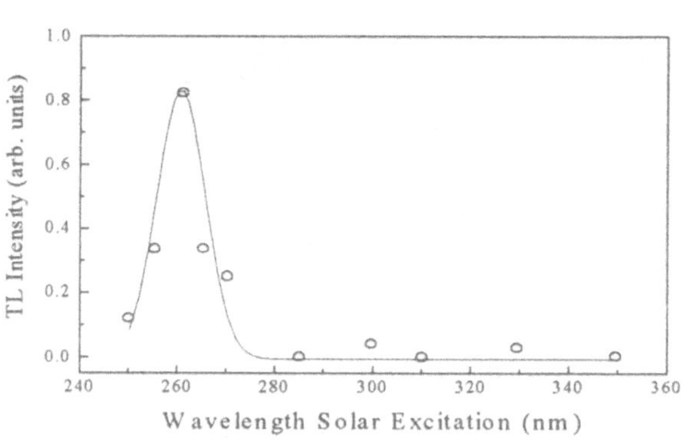

Figure 3. Thermoluminescence intensity after 2 minutes solar irradiation at different monochromator wavelengths taken at Cantoblanco (40.5°N, 3°W, 700 m above sea level) October 16th, 1995 (269 DU).

The TL excitation under direct solar irradiation (Figure 3) coincides with the excitation curve obtained with Xe-lamp in the laboratory (Figure 1) and indicates that UV-C light reaching the earth surface is responsible for the TL activation.

As an additional check similar to previous results [7] with NaCl:Eu^{2+}, the thermoluminescence signal was suppressed by a UV-cut filter, which confirms the UV-C nature of the solar irradiation TL signal, ruling out the possibility of TL signals been attributed to wavelength higher than 300 nm of the solar spectrum reaching the ground.

Solar irradiation of wavelengths shorter than 290 nm have been difficult to measure using monochromator and photomultipliers because of the contamination of the desired signal by stray light outside the spectral region of observation [8]. The stray light problem arises due to imperfections in the grating ruling and is particularly acute when, as in the solar spectrum, there is a large dynamic range of intensities.

The stray ligth problem is absent from our experiment because of the narrow TL excitation spectrum of the europium doped alkali halides (Figure 2).

This result is surprising because calculations of ozone layer depth and ozone molecule cross-sections [9] indicates that no direct solar photons in the UV-C range may reach the earth surface or they reach the ground in very small numbers.

The results of this experiment in our opinion should be interpreted as an instance of UV-C reaching the earth surface and the 260 nm clear peak above background level signal in figure 3, as the convolution of the KCl:Eu^{2+} TL excitation spectrum given in figure 1 and the dayligth clear sky UV-C irradiation spectrum at the measuring site.

From these experimental results it appears that, although small, there is a measurable fraction of UV-C irradiance that reaches the earth surface, in spite of the ozone layer (269 DU), that could play an unattended role in the biosphere dynamics.

Acknowledgements This work is partly supported by CE grant CI*-CT93-0316AR , ASINEL (Madrid, Spain) and Universidad Autónoma de Madrid.

REFERENCES

[1]. I. Aguirre de Carcer, F.Cussó and F.Jaque, Phys. Rev. B 38, 10812 (1988).

[2]. I. Aguirre de Carcer, G.Lifante, F.Cussó, F.Jaque and T. Calderón, Appl. Phys. Lett. 58, 1825 (1991).

[3]. F. Jaque ; F. Cussó ; I.Aguirre de Cárcer ; G. Lifante y T. Calderón, Health Physics 60, 579, (1991)

[4]. H.Nanto, K. Murayama, T. Usuda, F.Endo, Y.Hirai, S.Taniguchi and N. Takeuchi, J. Appl. Phys. 74 (2), 1445, (1993).

[5]. J.Hernández, W.K.Cory and J.Rubio, J.Chem.Phys. 72, 198 (1980).

[6]. L.Botter-Jensen, Nucl. Tracks Radiat. Meas. 14, 177 (1988)

[7]. E. España, T. Calderón, F.Cussó, F.Jaque, G. Lifante and P.D. Townsend, Nucl. Tracks Radiat. Meas. 20, 605 (1992).

[8]. K. Minschwaner, R.J.Thomas and D.W.Rusch, J. Geophys. Res. 100, 11157 (1995).

[9]. Molina L.T and Molina M.J, J. Geophys. Res. 91, 14501 (1986)

On the Forecasting of Intermonth Total Ozone Variations

O.Pokrovsky, I.Dalyuk

Main Geophysical Observatory, Karbyshev str.,7, St.Petersburg, Russia, 194021

Main Geophysical Observatory, Karbyshev str.,7, St.Petersburg, Russia, 194021

Abstract. An attempt has been made to investigate systematically the statistical methods of forecasting the processes which determine the environmental safety. Two statistical methods for total ozone (TO) prediction (for a month period ahead) are proposed. They are based on the Karhunen-Loeve (K-L) approximation of satellite total ozone time series for the preceding time interval. The research also includes the comparison of the two regularization types for solution of an ill-posed prediction problem. There were demonstrated that the optimal selection of K-L basic functions is a key problem determining the successfulness of the prediction. The efficiency of predictor approximation also controls the skill-score of TO prediction. Average skill-score of TO prediction is about 80%.

Keywords. Total ozone, long-term forecasting, ill-posed problem, Karhunen-Loeve expansion

1. Introduction

Depletion of stratospheric ozone caused by anthropogenic impact can modify the amount of biologically harmful ultraviolet radiation (UV-B) penetrating to the Earth's surface with potential adverse effects on the human health and also on the biosphere (including plants and microorganism in the aquatic and terrestrial ecosystems). The main factor affecting the surface UV radiation field, especially under the conditions of cloudless and weakly turbid atmosphere, is total column ozone. Although the UV radiation intensity can decrease with increasing concentration of aerosol, anthropogenic pollutants (gases) or in the case of haze, fog or cloudiness, the periods of TO negative anomalies are potentially dangerous for human health.

So several Radiological Centers intend to warn the population of these potentially dangerous days. For these purpose it is necessary to have a long-term TO prediction (for a week time interval or more). The paper Dalyuk I.V. and Pokrovsky O.M. (1994) presents the research concerning effective approximation of satellite and surface based TO time series. The statistical method proposed was demonstrated to be the premise for medium-range forecasting of

NATO ASI Series, Vol. I 53
Atmospheric Ozone Dynamics
Observations in the Mediterranean Region
Edited by Costas Varotsos

TO variations. Dalyuk I.V. and Pokrovsky O.M. (1995) gave several examples of TO predictions derived from TOMS data for spring and autumn seasons. The purpose of this study is to investigate systematically the results provided by two TO long-term forecasting techniques based on the analysis of TO time series for preceding time interval of the same length. The accuracy of TO prediction within seasons is studied here for twelve year long data.

The estimates of expected prediction errors and factors determining them are investigated also. The indicators of forecast skill are discussed.

2. Statement of Problem

The 'predictor-predictand' approach is a background of any statistical forecasting technique. Another fundamental feature characterizing the method is the utilization of training statistical sample containing the data for the preceding time period.

As the prediction problem being an ill-posed leads to the instability of calculational process, so the successfull predictor and predictand selection plays a major role. Regularization of the prediction scheme with respect to the predictor variations should be realized in the same way as in case of the remote sensing inversion problem. In particular, the optimal predictor selection problem is analogous to the problem of the optimal design of experiments (Pokrovsky, 1984). In case of the remote sensing ill-posed problem the selection of the optimal measurement scheme is connected with the realization of the minimax principle to obtain maximum of information content by minimum number of observable parameters.

In case of the prediction problem it is necessary to start from the same positions. The input data represents the time series of daily TO measurements data $\{x_j\}$. Let's formulate the classical problem: on the base of the discrete TO data set x_j over the time interval [a,b] to carry out the forecasting on the adjoined time interval $[b+1,c]$. Let's assume that both time intervals are of the same length equal to one month. Thus, having data vector $x^1 = (x_{i,\dots,n} x_n)$ we must obtain the best prediction estimates for the vector x^2. For simplicity let's assume the random variables x_t centered by their means. It is equivalent to the statement that mean values of x_t are equal to zero.

The optimal selection of a predictor and predictand is the key problem. In meteorology a specially selected subset of vector x^1 components is usually used as a predictor providing the most close correlations with values x^2. The functionals $\alpha_i = \varphi_i (x^1)^T (i = 1,\dots,p)$ of the vector x_1 are proposed to be used as a predictors - that is the principal original side of the method suggested (T - transposition sign, $\varphi_i = (\varphi_1^i,\dots,\varphi_n^i)$. As components of the vector $\varphi_i(i = 1,\dots,p)$ we propose to use the values of *i*-function of Karhunen-

Loeve (K-L) approximation set $\varphi_i(t)$ at the same time scale, i.e. $\varphi_l^i = \varphi_i(t_l)(l = 1,\ldots,n)$. Consequently, the functional is the i-coefficient of the K-L expansion $\chi^1 \approx \sum_{i=1}^{p} \alpha_i \varphi_i$. As K-L function set is the optimal basis for time series approximation from the point of view of statistical tests (Davenport and Root, 1956) so the set of functionals $\alpha_i(i = 1,\ldots,p)$ is the best set of predictors for this problem.

The optimal set of predictands $\beta_j(j = 1,\ldots,q)$, which are the coefficients of K-L expansion of the vector x^2 : $x^2 \approx \sum_{j=1}^{p} \beta_j \phi_j$ can be derived just in the same way. Here the values of K-L j -function $\phi_j(t)$ on the references corresponding to predictand data are used as components of the vector ϕ_j , i.e. $\phi_l^j = \phi_j(t_{n+1})(l = 1,\ldots,n)$. Coefficients are defined as $\beta_j = \phi_j(x^2)^T (j = 1,\ldots,q)$. Such a set of predictands is proposed because of its most compact representation of the expected x^2 vector fluctuation spectrum.

The application of the statistical forecasting technique requires the availability of multi-annual daily TO data. In this case the set of predictor coefficients $\{\alpha_j\}$ can be considered as a set of independent random values with 0-means and certain variances $\{\lambda_t^2\}$. Analogously the set of predictand coefficients $\{\beta_j\}$ can be considered as a set of independent random values with 0-means and certain variances $\{\mu_j^2\}$. Consequently, having the set $\{\alpha_j\}$ we can predict the set $\{\beta_j\}$. So by means of the last one coupled with $\{\phi_j\}$ the desired TO variations x^2 for the next month can be easily retrieved.

Let's introduce the functions $\{\chi_k\}(k = 1,\ldots,s,)$ which approximate TO variations over the interval $[a,c]$. The indicated interval includes the predictor and predictand time intervals. Thus, the problem is to approximate the joined vector $x = (x^1, x^2)$. This approximation is characterized by the set of coefficients $\{\gamma_k\}$ with variances $\{\kappa_k^2\}$.

In the frame of the problem concerning the approximation of the distributions x^1 and x^2 by K-L sets $\{\varphi_i\}$ and $\{\phi_j\}$ the selection of the values p

and q can be subjected to a 'threshold' criterion for the relative cumulative variances (RCV). The RCV of the approximation of the predictor $\sum_{i=1}^{p} \alpha_i \varphi_j$ is equal $d_{\varphi}^2(p) = \left(\sum_{i=1}^{p} \lambda_i^2\right) / \left(\sum_{i=1}^{n} \lambda_i^2\right)(p = 1,\ldots,n)$. The RCV for the predictand is analogously deduced $d_{\phi}^2(q) = \left(\sum_{j=1}^{q} \mu_j^2\right) / \left(\sum_{j=1}^{n} \mu_j^2\right)(q = 1,\ldots,n)$. The number of predictors can be defined as a minimal integer p providing $d_{\varphi}^2(p) \geq d_{\varphi}^*$ ($0 \lhd d_{\varphi}^* \lhd 1$ - the prescribed 'threshold'). The criterion for selecting q is analogous : $d_{\phi}^2(q) \geq d_{\phi}^*$.

3. Forecasting Algorithms

The vectors $\alpha = (\alpha_1,\ldots,\alpha_p)^T$ and $\beta = (\beta_1,\ldots,\beta_q)^T$ are formed from the sets of coefficients defined above. The problem is to construct an operator F providing the best statistical estimates $\hat{\beta} = F(\alpha)$ for β . In this study we confine ourselves to discussion of the class of the best linear estimates (Rao C.R., 1967). In this case it is necessary to construct a matrix F so that $\hat{\beta} = F\alpha$.

Along with the traditional statement of the prediction problem mentioned above it is possible to consider the alternative approaches. We can pose the problem of retrieval of the TO expansion coefficients $\{\gamma_k\}$ of K-L set of functions $\{\chi_k\}$ over the joined predictor + predictand time interval $[a,c]$. Furthermore, we have two possibilities. The first one is the direct estimation of the coefficients $\{\gamma_k\}$ by observational data x^1 at $[a,b]$. Since the set of 'truncated' functions $\{\chi_k\}$ defined over $[a,b]$ is not an orthonomal system , we have in order to determine $\{\gamma_k\}$ to solve the classical problem of quadratic functional minimization $\gamma = \mathbf{arg\ min}\left\| \chi_\gamma - (x^1)^T \right\|^2$. Here the vector $\gamma = (\gamma_1,\ldots,\gamma_s)^T$, matrix $\chi = (\chi_1^T|\ldots|\chi_2^T)$ is formed by column-vectors obtained from K-L expansion over discrete time lags $[a,b]$. This problem can be solved by the least squares method or its modifications .

The second approach confines the estimation of the set $\{\gamma_k\}$ by the certain set of coefficients $\{\alpha_2\}$. Thus, that is the prediction scheme $\gamma : \hat{\gamma} = G\alpha$.

The classical least squares method is associated with solving a normal equation system $\chi^T \bullet \chi \bullet \gamma = \chi^T \bullet (x^1)^T$, which in practice demonstrates instability while being solved. Two types of regularization methods are usually used to find the stable solution. The first one is coupled with the selection of the optimal number s of functions χ_k in the K-L expansion. Another one is based on the ε - parametric regularization. In this case the system of normal equations mentioned above is replaced by regularized system in the form: $(\chi^T \bullet \chi + \varepsilon\, I_s)\gamma = \chi^T (x^1)^T$ here I_s is a unit matrix of $s \times s$ dimension, ε - is a regularization parameter $(\varepsilon \rhd 0)$ (Tikhonov and Arsenin, 1974). The optimal selection of parameter ε determines the successfulness of solving the problem by this regularization method. Thus, the application of both types of regularization requires the preliminary investigations concerning selection of the optimal regularizator.

Concluding the discussion of different approaches on the prediction problem let's mention once more the existence of two different ways of forming the 'predictor-predictand' scheme: the traditional one - $\hat{\beta} = F\alpha$ and the extrapolation scheme - $\chi^T \bullet \chi \bullet \gamma = \chi^T \bullet (x^1)^T$. In both cases for obtaining a stable solution it is necessary to construct a regularizator by means of one of this two methods mentioned above. In present research the method of selection of the optimal number p of K-L expansion components $x^1 \approx \sum_{i=1}^{p} \alpha_i \varphi_i$ providing the separation of the stable part of operator F is used as a first prediction scheme. Let's use the notation *F1* for this one. To perform a second prediction scheme the ε - parametric regularization method is used. This prediction method will be designated as *F2*.

4. Examples on the Forecasting

Twelve-annual TOMS data time series were used as an input data for the forecasting experiments (1979-1990). Previous investigations revealed the fact that there are strong reasons for successful medium-range TO forecasting (up to one month ahead) based on the utilization of preceding month data. Taking into the consideration the peculiarities of inter-annual TO variations at the mid-latitudes, the special attention is paid to the transient seasons, i.e. autumn and spring, when the ozonospheric variations are the strongest. The significant increasing of stratospheric ozone due to active photochemical and dynamical processes is specific for spring and, vice versa, the most strong decreasing of

TO is observable in autumn. These two seasons appeared to be the most complicated for TO forecasting.

Experiments concerning TO forecasting were conducted for all months (1979-1990). The accuracy estimates are examined for each of the twelve years. The comparison of the prediction schemes described above was carried out as well as attempts to reveal factors detecting the efficiency of month-long TO prediction of TO anomalies, in this case TO values can exceed the limits of the TO natural variability $\left[-\sigma,\sigma\right]$.

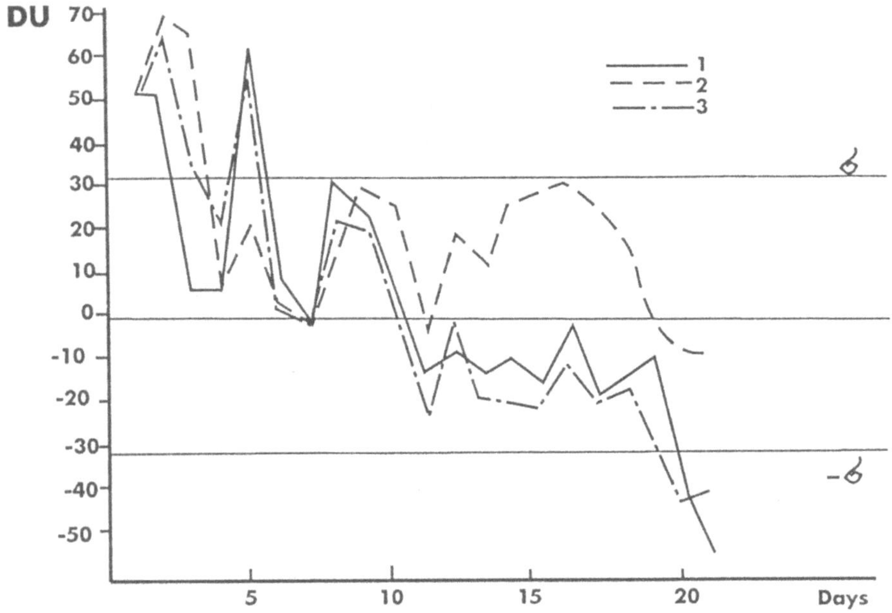

Fig.1 Forecasting of Total Ozone. Predictor - April,1979; predictand - May, 1979; *scheme* F2 (ε=3); 1- factual variations, 2 - forecast by one function, 3 - forecast by seven functions.

The efficiency of TO time series approximation by K-L expansion appeared to be one of the most significant factors determining the prediction validity. For two months long basis $\{\chi_k\}$ the level of approximation corresponding to TOMS error level of 5-6 Dobson Units (DU) is achieved by using 6-7 K-L basic functions. For example, the necessary number of basic functions is 7 for time interval 'April-May' and is 6 for 'September-October'. The forecasting experiments revealed the numerical instability, and that is usual for ill-posed problems. That's why the scheme *F2* which involves parametric regularization appeared to be rather effective. Fig.1. illustrates the example of TO prediction by the scheme *F2* for the spring of 1979. The details of inter-month TO variations with daily resolution can be retrieved by using only one function χ_1. Additionally, the duration of TO anomalies can be detected rather accurately. Another example (Autumn, 1979) (Fig.2) illustrates the influence of the regularization parameter on the prediction results. The values of the parameter ε correspond to three levels of input data error level: 1,3 and 6 DU. The calculational results demonstrate stability of the solution in the interval selected for the regularization parameter and simulate rather well the factual TO variations. It can be noted that with increasing ε, solution becomes more smoothed. In this case the optimal value of the regularization parameter is between 1 and 3. Thus, the value of ε corresponding to the error level of 6 DU could be considered rather overstated. Nevertheless, any value of parameter in the interval $1 < \varepsilon < 6$ can be used for simulating TO variations and it's principal anomalies.

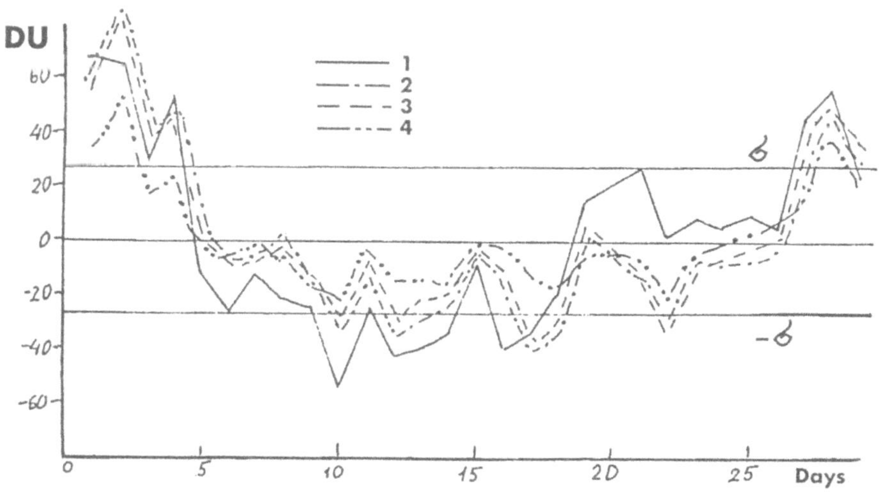

Fig.2 Total Ozone Forecasting; predictor - September, 1979; predictand - October, 1979; *scheme* F2 (six functions); 1 - factual variations; $2-\varepsilon=1$; $3-\varepsilon=3$; $4-\varepsilon=6$.

The method *F1* to an even greater degree depends on the optimal number of K-L basic functions because this aspect specifically brings out the regularization of the prediction operator (Pokrovsky,1972). The present research demonstrates that in contrast to scheme *F2*, when h the interval for selecting the parameter ε can be detected beforehand by numerical experiments, the optimal number of basic functions on the scheme *F1* depends on the certain month, peculiarities of TO variations for the preceding period and the efficiency of approximation by K-L functions. Fig.3 illustrates the application of the scheme *F1* for TO prediction at spring period of 1985. The left part of Fig.3 shows a set of predictors and the degree of their K-L approximation in relation to the number of K-L basic functions for April. The right part of the figure shows the corresponding TO prediction curves for May. The comparison of both parts reveals the dependence of prediction results on the efficiency of predictor approximation.

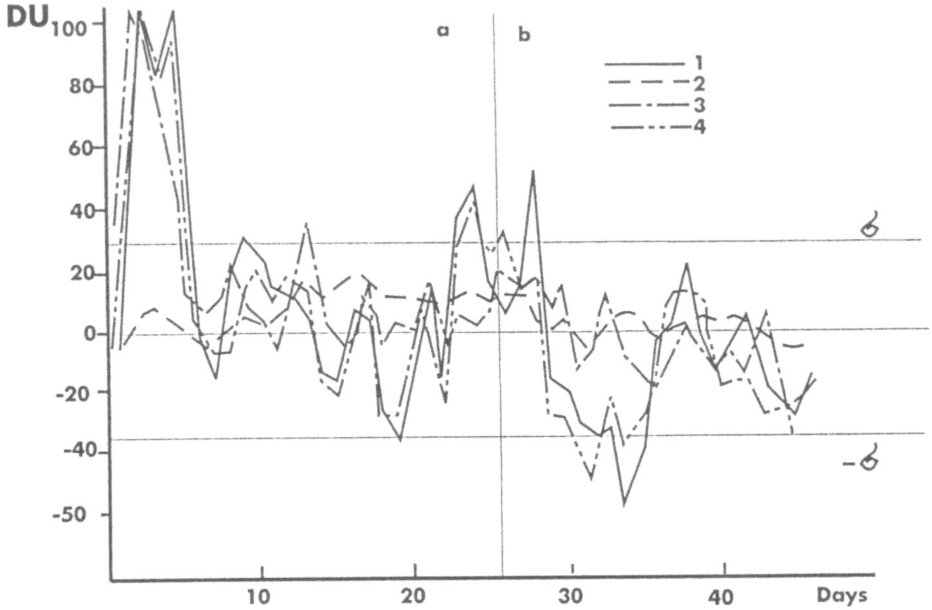

Fig.3. Approximation (a) and forecast (b) of TO variations for April (a) and May (b), 1985; *scheme F1;* 1 - factual variations; 2 - one function; 3 - three functions; 4 - six functions.

5. Comparison of methods

The uniformed data sample was used for the comparison of prediction results. Fig.4 illustrates an example of comparison of two methods, *F1* and *F2*, applied for prediction of TO variations for October, 1991. In the first half of the month the negative TO anomaly had occured, passing to the positive anomaly in the second half. The observed TO variations are simulated more successfully by the *F1* method. Moreover, the dates of crossing the level of TO variability are detected rather accurately. The result is obtained by utilizing the approximation which includes seven basic functions. The number of basic functions has been determined by analyzing the approximation validity of the predictor defined on the interval which included September. For the method *F2* the two-month-long K-L basis consisting of five functions was used. And this quantity has been determined by analyzing the relative cumulative variance (RCV) of the 'September-October' pair of months for the measurement error level of 6 DU. Applying the regularization parameter value corresponding to the measurement error level of 3 DU, the smoothed solution, which detects the negative TO anomalies, has been obtained. But the calculated curve doesn't fall outside the 1 natural variability interval when the positive TO anomaly takes place.

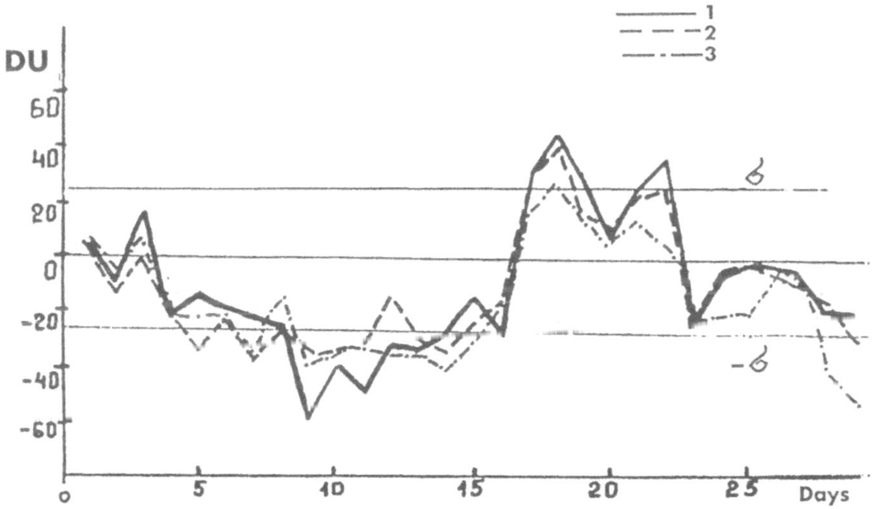

Fig.4. Comparison of two methods of TO forecast; predictor - September, 1991, predictand - October, 1991; 1 - factual variations; 2 - method F1; 3 - method F2 .

Taking into consideration the fact that the number of basic functions determines the prediction results obtained by the scheme *F1*, it is necessary to estimate the effect of this factor on the efficiency of the *F2* method. Fig.5. shows TO predicted curves for April, 1975 with different number of basic functions. The results presented reveal the fact that convergence of the solution to the real TO variations (with increasing number of basic functions) depends substantially on the choosed value of regularization parameter.

Given small value $\varepsilon = 1$ (the left part of Fig.5) the high degree of convergence of approximate solutions associated with number of basic functions increasing from 1 to 7 to the observed TO curve takes place. The factual TO variations in April can be considered anomal since diurnal TO values fall out to the limits of the natural variability interval for seventeen days, i.e. more than for half a month. The forecast involving one basic function gives only one value exterior to the interval $[-\sigma, \ \sigma]$. The prediction utilizing four basic functions has already 13 anomal values, and making use of five basic functions yields 18 values exceeding the natural variability limits. Let's note that in the latter case discrepancies between the observed and predicted TO anomal values concerned only three days.

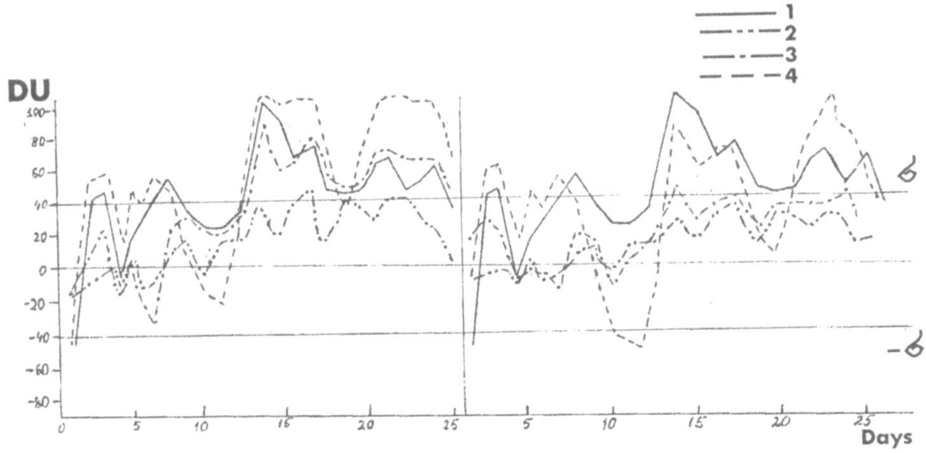

Fig.5. TO forecast, method F2; predictor - March, 1979; predictand - April, 1979; a $-\varepsilon = 1$; b $-\varepsilon = 6$; 1 - factual variations; 2 - one function; 3 - four functions; 4 - six functions.

With increasing of the regularization parameter (the right part of Fig.5.) the weaker dependence of the prediction results on the quantity of basic functions can be observed. The solution obtained by utilizing only one basic function doesn't reproduce any TO anomalies at all. The solution involving four basic functions yields only two anomal TO values. Yet, even in this case the discrepancies in the dates of anomalies (factual and predicted) concern only seven days. This can be explained by the fact that solutions obtained by making use of the higher value of the parameter α have a more smoothed character. In addition, the results considered above indirectly indicate that the random TOMS error at middle latitudes is about 1 DU. Taking into account that the total satellite measurement error is estimated to be about 5-6 DU, we can make a conclusion of a major affect of the systematic error component.

The experiments on comparison of two methods, *F1* and *F2*, show that decreasing of the regularization parameter leads to decreasing of the discrepancies between the results of two methods. The accuracy of the method *F1* appears to be better; the theoretical comparison of the two methods explains this fact [6]. However, the method *F2* has another advantage: the nonoptimal quantity of the basic functions doesn't affect the accuracy of the method so abruptly as it occurs in the case of the *F1* method.

As it was indicated above, both methods, *F1* and *F2*, appear to be effective in forecasting TO anomalies, provided that the number of basic functions has been determined successfully. Coming from this conclusion, the next section deals with aspects which determine the possibilities of selecting the optimal number of basic functions.

6. Expected prediction effectiveness

In practice it is rather significant to obtain the *a priori* estimates of the expected prediction accuracy. For each certain pair of months 'predictor-predictand' the estimates of variances for K-L time series expansion coefficients can be obtained and then transformed to the estimate variances over the time scale. For this purposes, for example, the expression for variances of regularization problem solutions can be used (Pokrovsky,1972). However, such estimates have a rather general character as they 'express' average values through all years of the statistical sample. For practical needs it is necessary to obtain estimates corresponding to the certain year. The experiments which have been carried out gave us a simple scheme of *a priori* prediction accuracy estimation for a future period.

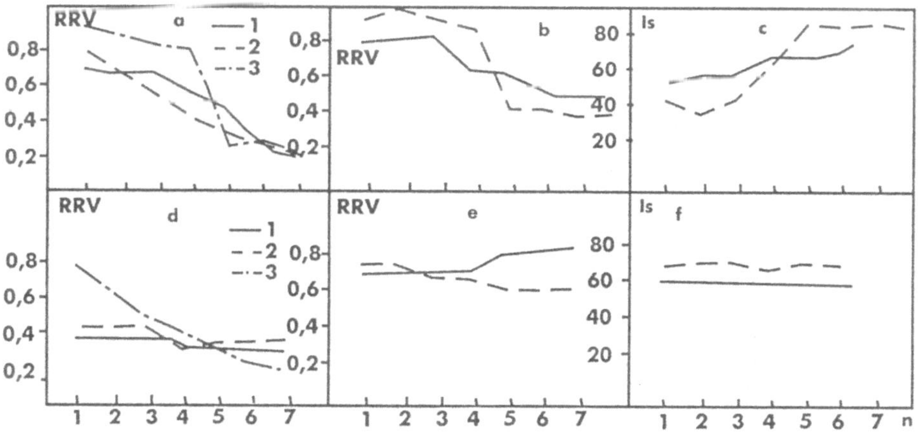

Fig.6. The relationship between RRV of the TO approximation (a,g) and forecast skill-scores, method F1; RRV (b,g); I_s - sign prediction skill-score; 1 - April-May, 1979 (a-c); 2 - September-October, 1979 (a-c); 3 - theoretical RRV distribution (a,d); 4 - April-May, 1984 (d-f); 5 - February-March, 1983 (d-f).

The prediction accuracy of the method *F1* turned out to be directly connected with the effectiveness of the approximation of the 'predictor' x_t over the interval [a,b] by K-L set of functions. As it was shown earlier (Pokrovsky, Dalyuk, 1994), the convergence (in mean over the whole sample) of the K-L expansion over the one, two and three-month interval is rather strong. The utilizing of 6-8 basic functions provides the 90% RCV level.

For estimation of the approximation residual the relative residual variance (RRV) is usually used, which is connected with the RCV by simple relation RRV = 1-RCV. The RRV for the x_t for the certain month and the RCV distribution (dependant on the number of basic functions) must be distinguished. The comparison of the RRV distribution of the current realization with averaged RRV allows to estimate *a priori* the prediction effectiveness. Fig.6. presents two types of convergence: fast and slow. In the first case (Fig.6a.) the RRV distribution (depending on the number of basic functions) for April and September, 1979 is analogous to the average (over sample) RRV. This corresponds to the also fast predictand RRV convergence (Fig.6b.). In the second case (Fig.6g.) the RRD convergence for February, 1983 and April, 1981 'strongly lacks' the average RRV distribution. With seven basic functions utilized the average RRV is equal to 0.1, and RRV for certain realizations is equal to 0.3-0.4. This corresponds to the predictand RRV curve (Fig.6d.). Thus, the rate of deviations between the average RRV distribution and the analogous characteristics of the certain realizations determine the prediction effectiveness.

7. Forecasting skill-score

According to the meteorological classification (see Kozeltseva et al, 1991) the TO forecast for a month ahead can be treated as a long-range. For estimation of the long-range statistical prediction effectiveness the skill-score index is usually used. Analogously, we used two characteristics: the first one is the TO anomalies forecast skill-score I_a, the second one is sign prediction skill-score I_s.

Let's define the first one. The prediction for each day $\hat{x}_t$ on the interval [b+1,c] is classified according to three value ranges. This value has a 0-value if it belongs to the interval of natural variability $[-\sigma, \sigma]$, it is equal to -1 if $\hat{x}_t \lhd -\sigma$ and to +1 if $\hat{x}_t \rhd \sigma$. The percent of days I_a when the coincidence (according to the ranging defined above) of the prediction $\hat{x}_t$ and factual TO variability x_t takes place, will be defined as the TO prediction anomalies skill-score. This index is rather convenient as it shows how the prediction reproduces TO natural variability as well as it's anomal values. The latter is especially important in connection with environmental problems discussed in the introduction. Table 1 contains the values of TO prediction skill-score I_a for three months: February, June, October and 12 years, 1979-1990. The adjoined columns contain the number of factual TO anomalies N_a. With increasing N_a the forecast skill-score I_a decreases reaching the level of 52% for N_a=13 in February and 55% for N_a=15 in October. Given six-seven anomal dates the index I_a reaches the level 75-80%. The index I_a is about 90-100% when N_a=1-2. The maximal number of anomalies takes place in February and October. The average value of I_a fluctuates in the interval 76-79%.

The second characteristic is a sign forecast skill-score I_s. It indicates the sign coincidence of the predicted and factual TO deviations from the multi-

annual means. The index I_s is 'milder' then I_a for it includes only sign coincidence, but not the amplitudes of x_t and $\hat{x}_t$. From the other hand, I_a registers only the signs of the values out of the $[-\sigma, \sigma]$ interval and characterizes the prediction of the anomalies particularly.
So these two characteristics are mutually complementary.

Let's note that for the months when the successful prediction is expected (Fig.6 a,b), the sign prediction skill-score I_s increases quickly with the number of basic functions (Fig. 6 c). For other months (Fig.6 d, e) the index I_s (Fig.6 f) is weakly dependent on the number of basic functions. (The index I_a has the analogous features).

Table 1. Forecast Skill-score

year	February			July			October		
	I_a	N_a	I_s	I_a	N_a	I_s	I_a	N_a	I_s
1979	96	1	87	100	1	67	84	12	90
1980	74	6	96	83	5	72	74	10	74
1981	74	6	65	72	8	78	84	15	90
1982	52	11	91	72	8	83	77	7	77
1983	52	12	74	89	4	78	94	10	100
1984	65	13	83	50	10	78	55	14	90
1985	73	4	67	83	3	67	60	15	81
1986	91	7	91	72	5	67	90	7	81
1987	83	11	87	89	2	72	90	4	87
1988	83	12	100	67	7	89	90	7	71
1989	60	13	83	78	7	83	80	8	68
1990	98	1	78	72	5	72	74	12	100
means	76	8	84	77	5	76	79	10	84

The values in Table 1 show that the sign prediction skill-score index I_s varies in a wide range from 67 to 100%. The definite dependence between the number of anomal days N_a and the index I_s can be marked. The greater values of the N_a is correspondent to the greater values of I_s . The average values of I_s vary from 76 to 84%.

Comparing the obtained values of I_s for TO with the corresponding characteristics for long term prediction of temperature and other meteorological parameters, it can be noted that the TO forecasting appears to be more effective. This fact can be explained by the stronger 'connectedness' of the processes of stratospheric dynamics determining the TO variations.

8. Conclusion

The present research is the first attempt of systematical investigation of the prediction statistical methods applied to the processes associated with environmental safety. The satellite remote sensing data are used as the database for the research. The study of the problem of forecasting TO variations (for a month period or more ahead) leads to the following conclusions:
1.The key factor determining the skill of the forecast is the optimal selection of the K-L subset of functions.
2.The analysis of RRV (Relative Residual Variance) convergence allows to estimate the prediction efficiency (particularly the potentially 'hard-predictable' periods or *vice versa*).
3.The prediction skill-score index is rather high (about 80%), so the application of this technique reavels new opportunities in TO forecasting.

REFERENCES

1. Dalyuk I.V., Pokrovsky O.M. , 1994. Comparative analysis of intra-annual total ozone time series by satellite and ground-based measurement data. - Studying of the Earth From Space,5, p.9-16 (in Russian).
2. Dalyuk I.V., PokrovskyO.M.,1995. On the forecasting of intra-monthly total ozone distribution based on TOMS data. - Studying of the Earth from Space., 3, p.25-35 (in Russian)
3. Davenport W.C., Root V.L., 1956 . Introduction in theory of random signals and noises.,N.Y., 341 p.
4. Kozeltseva V.F., Ped D.A., Sadokov V.P. 1991 , Comparison of various techniques of long-term forecasting of anomalies signs for surface temperature. - Trudy of Hydrometeorological Center, Moscow, Issue 311, p.15-27 (in Russian).
5. Pokrovsky O.M.,1972 , Comparison of statistical methods for solution of inverse problems in atmospherical optics.- Physics of Atmosphere and Ocean.,v.8, 2, p.341-348(in Russian).
6. Pokrovsky O.M., 1984 , Optimization of the meteorological remote sensing. - Leningrad,Hydrometeoizdat,264 p. (in Russian).
7. Rao C.R.,1967 , Linear statistical methods and its applications.- N.Y., 547 p.
8. Tikhonov A.N., Arsenin B.Ya. 1974Methods for solution of incorrect stated problems.-Moscow,Nauka, 224 p. (in Russian).

Analysis of Stratospheric Aerosols in the South-Eastern Mediterranean Region as deduced from SAGE I and SAGE II Measurements

A. Retalis and C. Cartalis.

Department of Applied Physics, University of Athens,
33 Ippokratous St., 106 80 Athens, Greece.

Abstract. Aerosol extinction coefficients at 1.02 ìm as deduced from measurements of the Stratospheric Aerosol and Gas Experiment (SAGE) I and II, for the periods 1979-1981 and 1985-1990, respectively, are examined for the region of South-Eastern Mediterranean, 15-30° E and 32,5-42,5° N. The most evident feature of the time series is the exponential decrease with altitude, with two peaks appearing around tropopause and 17-21 km altitude, respectively. This study also shows that the SAGE and SAGE II data at 1.02 μm are in a good agreement, for the examined region.

Keywords. Stratospheric aerosols, south-eastern Mediterranean

1 Introduction

Aerosols play a significant role in atmospheric and climatic processes. In this sense assessment of the distribution of aerosols is needed, especially in areas or for time periods, where lack of such information is recognised.

Stratospheric aerosols have been studied for decades using different and very efficient methods that employ lidar, satellite solar occultation and balloon-borne measurements of radiance and polarisation [Brogniez et al, 1992].

The Stratospheric Aerosol and Gas Experiment (SAGE) I instrument flown on the Applications Explorer Mission (AEM-B) satellite, has provided information on stratospheric aerosols at 1.02 μm and 0.45 μm wavelengths, from February 1979 to November 1981, between 70° N and 70° S, allowing a first global description of the unperturbed stratosphere and of the distribution of the Mount St. Helens volcanic material [Whitten, 1982].

The Stratospheric Aerosol and Gas Experiment (SAGE) II instrument flown on the Earth Radiation Budget Experiment (ERBE) satellite, provided stratospheric aerosol extinction profiles from October 1984 to November 1990 at four wavelengths, 1.02 μm, 0.525 μm, 0.453 μm and 0.385 μm with an altitude resolution of 1 km [WMO, 1985]. For more details regarding the SAGE II experiment see Mauldin et al. (1985).

Validations performed, in the United States and Europe [Russell and McCormick, 1989, Ackerman et al., 1989], have shown that the 1.02 μm profiles are satisfactory, from the local tropopause height up to at least 20 km above the tropopause height, i.e., 30 km in mid-latitude regions, with a precision estimated to

NATO ASI Series, Vol. I 53
Atmospheric Ozone Dynamics
Observations in the Mediterranean Region
Edited by Costas Varotsos

be around 10%. From a few kilometres above the ground up to the tropopause height the comparison between the inverted extinction profiles obtained by two different algorithms shows that the extinction data are still credible [Chu et al., 1989].

It should be mentioned that the three shorter wavelength channel aerosol data can be used with confidence only up to 15 km above tropopause, because the ozone and/or the nitrogen dioxide corrections become too large above this level. The lower "confidence" height limit of the profiles is due to Rayleigh scattering; it increases from 7 km for 0.525 μm to about 15 km for 0.385 μm. Uncertainties also increase towards the shorter wavelengths, reaching or exceeding 20% at 0.385 μm. It therefore seems appropriate to mainly use the 1.02 μm data to study the stratospheric aerosol distribution [Brogniez and Lenoble, 1991].

A lack of assessment of the stratospheric aerosol characteristics in the region of south-eastern Mediterranean, 15-30° E and 32,5-42,5° N, is noticed in the international literature. To this end, aim of this study is to assess the distribution of aerosols in the above mentioned region, based on SAGE I and SAGE II measurements.

2 SAGE II aerosol extinction profiles

Six years of SAGE II data (aerosol extinction profiles) from January 1985 to November 1990 were used in this analysis. From these profiles, the monthly peak values of aerosol extinction coefficients were estimated [Retalis, 1994], and presented in Figure 1. Bars indicate the standard deviation, representing the deviation of the individual profiles from the mean.

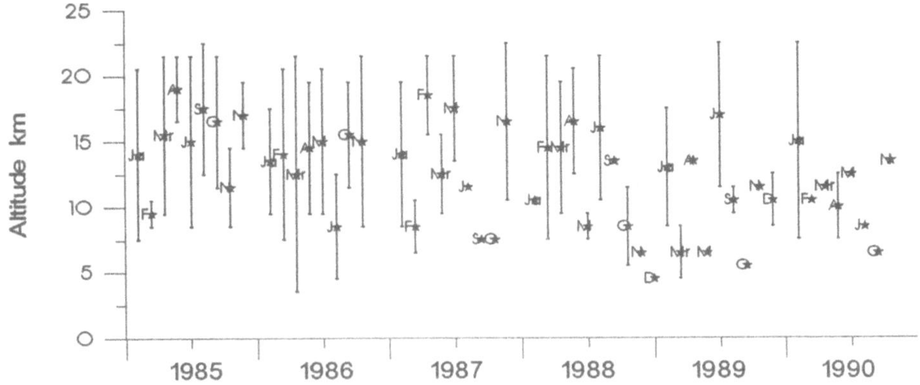

Figure 1. Monthly deviation and peaks of aerosol extinction coefficient for the period 1985-1990.

It is recognised that:

- some months (Feb. '87, Oct. '88) show a second "peak" in the upper troposphere (~ 6-10 km);
- in some cases the aerosol "peak" is not clearly detectable;
- deviations vary from 2 and 18 km with mean deviation of 10 km;
- the mean "peak" value lies within the range of 2-10 × 10^{-4} km^{-1}.

The mean seasonal aerosol extinction coefficient profiles averaged over the time period 1985-1990 are presented in Figure 2 [Retalis and Cartalis, 1995]. The following are deduced:

- a significant increase of aerosols extinction coefficients in a region of thickness of 2-4 km around tropopause, which is also seasonally variable;
- an increase of aerosol extinction coefficients in the 17-21 km layer, approximately;
- above the altitude of 30 km, aerosol extinction coefficients are non-existent;
- a trend for an exponential decrease with altitude.

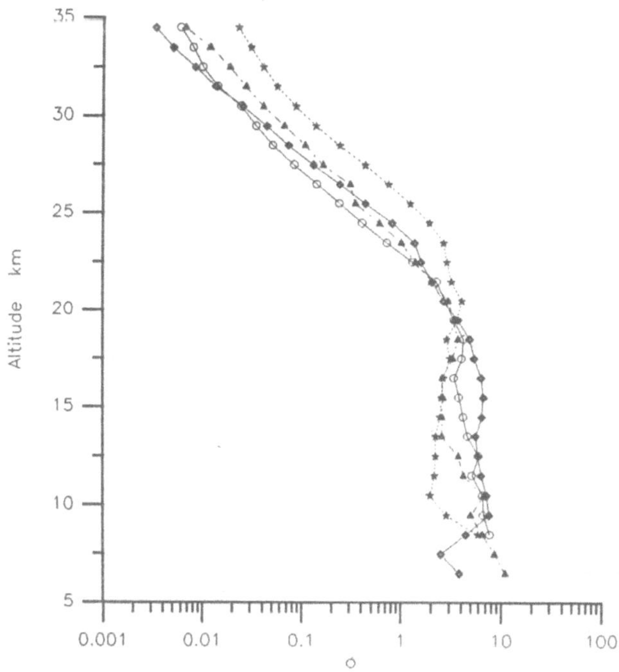

Figure 2. Seasonal SAGE II aerosol extinction coefficient (σ) profiles (10^{-4} km^{-1})
—◆— Winter, —○— Spring, - - -▲- - - Summer, ·······★······· Autumn.

At high and middle latitudes aerosols present a very homogeneous distribution within the 16-km to 21-km altitude range. Below and above this well-mixed layer the distribution is less homogeneous [Brogniez and Lenoble, 1991]. These variations are in line with the observations mentioned above for the examined period (Figure 2).

Figure 3a shows the variations of the aerosol extinction coefficient σ with time reference for two altitudes above the mean tropopause height (averaged over the six years). Figure 3a is complemented by Figure 3b, which shows the altitude fluctuations of the tropopause. Aerosol extinction coefficients decrease almost regularly at the reference altitudes, whereas slight seasonal variations are observed.

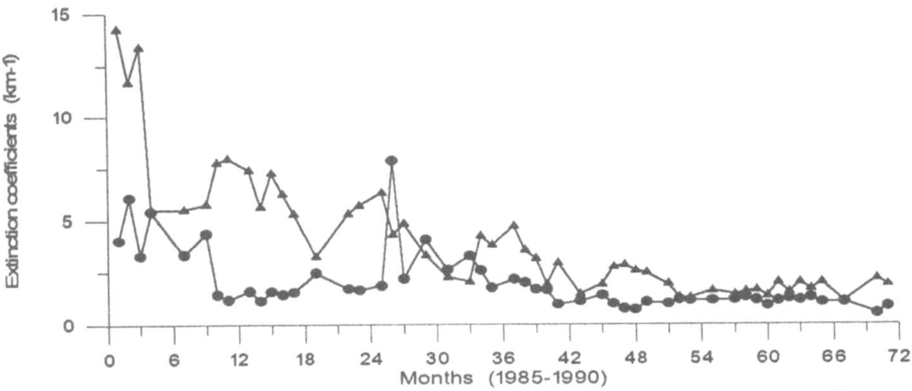

Figure 3a. Variations of the aerosol extinction coefficient σ with time reference for two altitudes (5 km circles, 10 km triangles) above the mean tropopause height

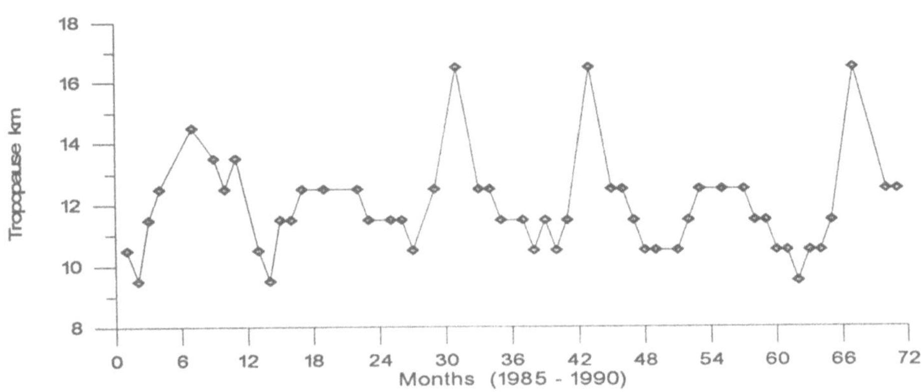

Figure 3b. Mean tropopause variation for the time period 1985-1990

From the same Figure, it yields that:

a) the aerosol extinction coefficient is about 3-7 times greater in early 1985 than in late 1990, depending on the level; Brogniez and Lenoble (1991) supported the view that this is due to the decreasing influence of the El Chichon volcano. Extinction coefficients are by the end of 1990, approximately 2×10^{-4} km^{-1} at 5 km altitude above the tropopause, which implies that the background state (1-1.5 $\times 10^{-4}$ km^{-1}) is almost recovered. This agrees well with Swissler et al. (1982);
b) the highest winter aerosol extinction coefficients at a given altitude are related to the lowest tropopause levels.

3 SAGE I and SAGE II measurements

The monthly aerosol extinction coefficient profiles for SAGE I and SAGE II measurements averaged for the periods 1979-1981 and 1985-1990 respectively, for the region of South-Eastern Mediterranean, 15-30° E and 32,5-42,5° N are given in Figures 4-12. It is clear that in the months of February and July SAGE I profiles are smoother. In addition SAGE I monthly mean maxima are greater compared to those of SAGE II, with the exception of months February and March for the regions 17 km to 21 km [Retalis and Cartalis, 1995]. The mean extinction coefficient σ is also higher in SAGE I profiles. The "aerosol layer" appears in the altitude range of 17-21 km, with the aerosol layer for SAGE I profiles initiating at lower altitudes. The trend for rapid decrease of the extinction coefficients at altitudes higher than 23 km in both cases, is recognised.

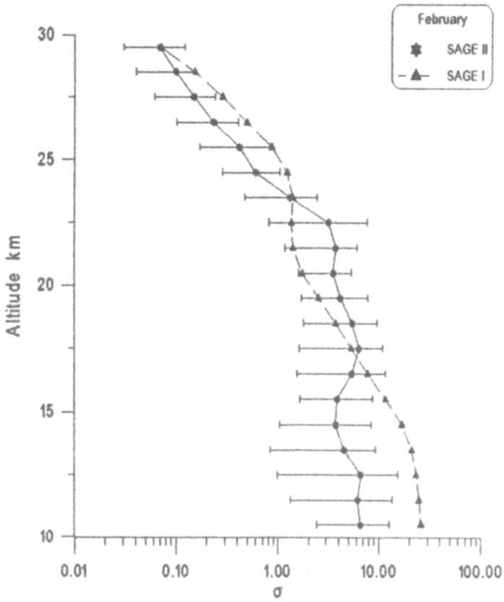

Figure 4

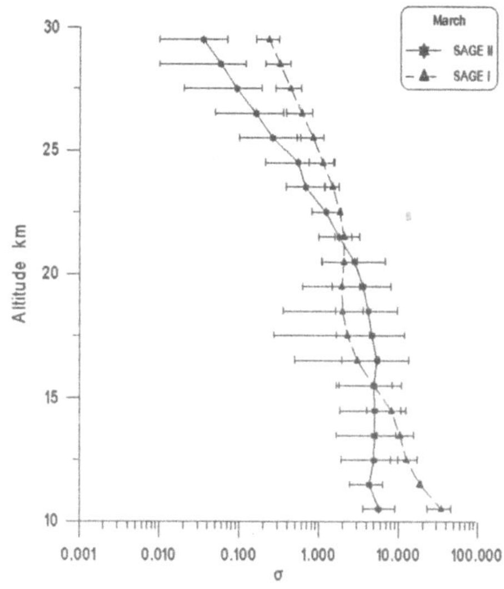

Figure 5

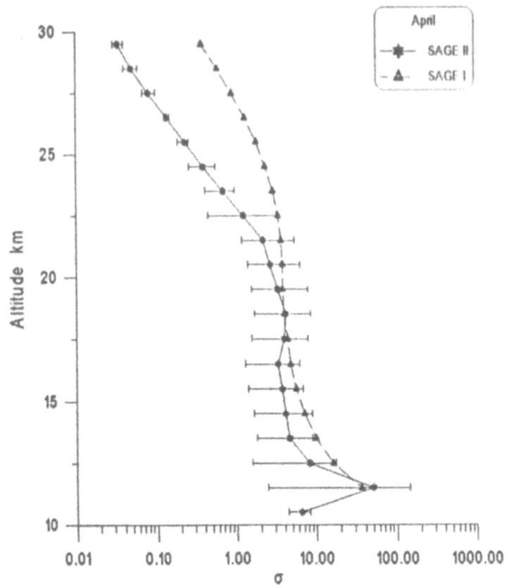

Figure 6

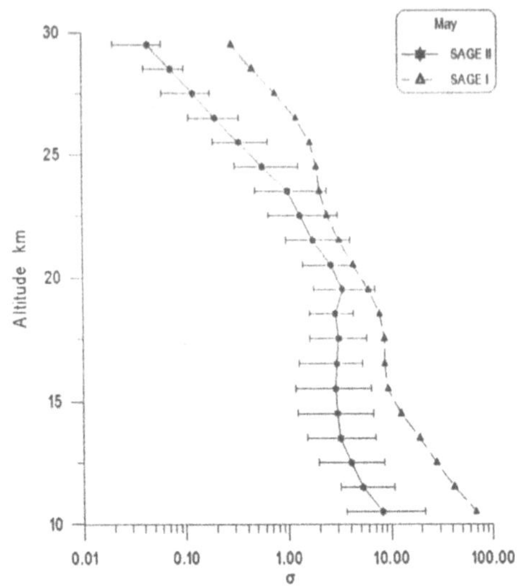

Figure 7

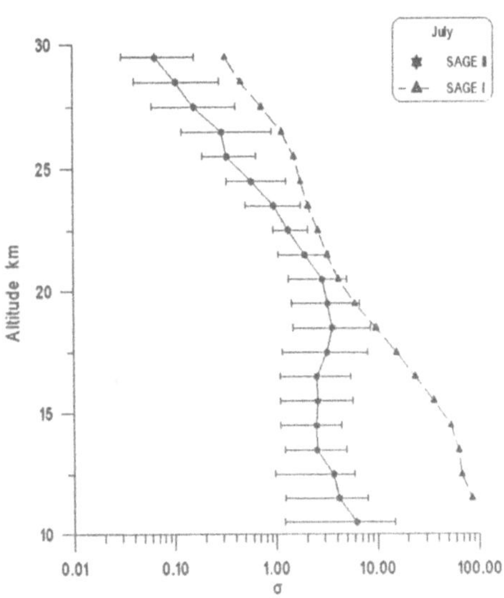

Figure 8

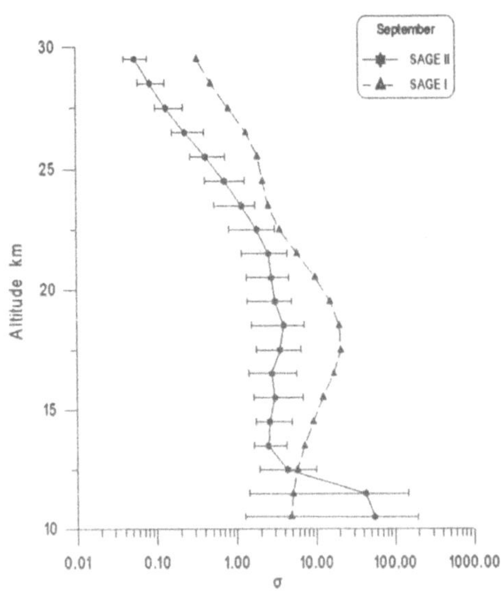

Figure 9

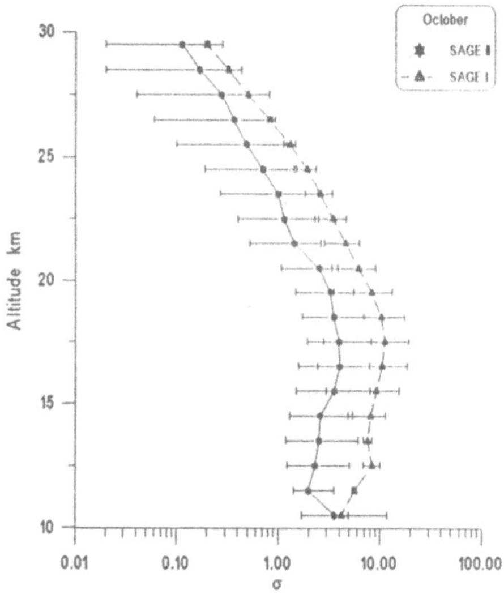

Figure 10

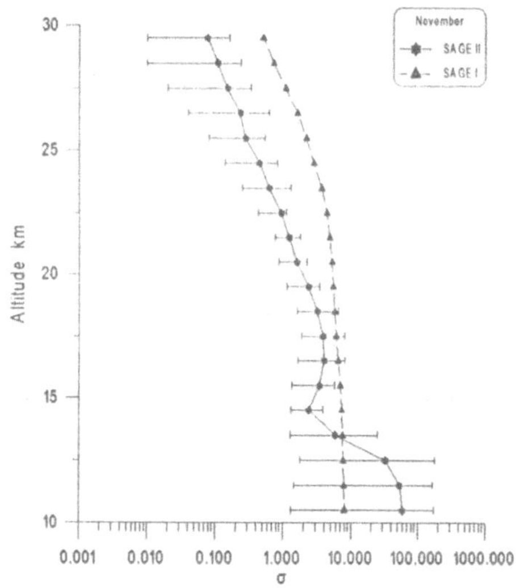

Figure 11

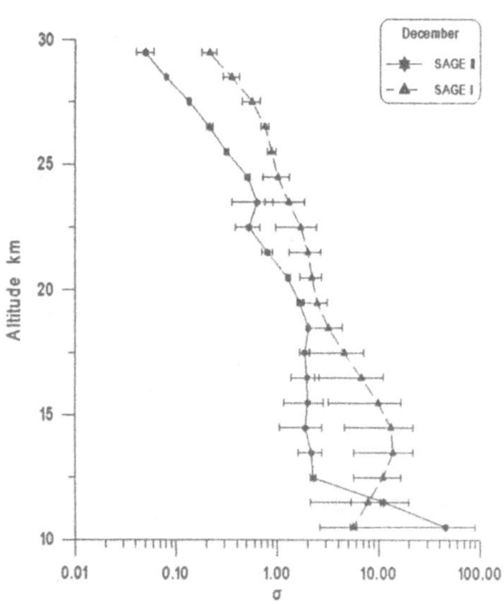

Figure 12

Figures 4-12. Monthly aerosol extinction coefficient (σ) profiles for SAGE I and SAGE II measurements averaged for the time periods 1979-1981 and 1985-1990, respectively.

4 Conclusion

Assessment of SAGE I and SAGE II aerosol extinction coefficient measurements for the period 1985-1990 in the region of South-Eastern Mediterranean, 15-30° E and 32,5-42,5° N, described herein, suggests the existence of an aerosol layer located between altitudes from 17 to 21 km. The assessment also demonstrated similar vertical distributions for SAGE I and SAGE II measurements, although aerosol extinction coefficients were found to be in general, higher for the period 1979-1981 compared to 1985-1990, and exhibited higher aerosol extinction coefficients in early 1985 compared to 1990. Finally it is deduced that the stratospheric aerosol layer does not follow the seasonal variations of the tropopause level.

The results of this paper support the use of remote sensing and atmospheric correction algorithms in the examined area and provide a comprehensive picture of the regional atmospheric characteristics. Further work should reflect an update of data, to extend the temporal coverage to 1995.

5. References

Ackerman, M., C. Brogniez, B.S. Diallo, G. Fiocco, P. Gobbi, M. Herman, M. Jager, J. Lenoble, C. Lippens, G Megie, J. Pelon, R. Reiter, and R. Santer. European validation of SAGE II profiles, *J. Geophys. Res.*, **94**, 8399-8411, 1989.

Brogniez C. and J. Lenoble. Analysis of 5-Year Aerosol Data from the Stratospheric Aerosol and Gas Experiment II, *J. Geophys. Res.*, **96**, 15,479-15,497, 1987.

Brogniez C., R. Sander, B.S. Diallo, M. Herman and J. Lenoble. Comparative Observations of Stratospheric Aerosols by Ground-Based Lidar, Balloon-Borne Polarimeter and Satellite Solar Oscillation. *J. Geophys. Res.*, **97**, 20,805-20,823, 1992.

Chue, W.P., M.P. McCormick, J. Lenoble, C. Brogniez, and P. Pruvost. SAGE II inversion algorithm, *J. Geophys. Res.*, **94**, 8339-8351, 1989.

Mauldin, L.E., III, N.H. Zaun, M.P. McCormic, J.H. Guy, and W.R. Vaughn. Stratospheric aerosol and gas experiment II instrument: A functional description, *Opt. Eng.*, **24**, 307-312, 1985.

Retalis A., Thermal and Chemical Structure of the Tropo- Stratosphere over the South-Eastern Mediterranean region as derived from Satellite observations. *MSc Thesis*, University of Dundee, Scotland, UK, 1994.

Retalis A. and C. Cartalis. On the aerosol distribution in South-Eastern Mediterranean region as deduced from satellite observations. Presented in the *20th General Assembly of the European Geophysical Society*, 3-7 April, 1995, Hamburg, Germany, 1995.

Russel, P.B., and M.P. McCormick. SAGE II data validation and initial data use: An introduction and overview, *J. Geophys. Res.*, **94**, 8335-8338, 1989.

Swissler, T.J., P. Hamill, M. Osborn, P.B. Russell, and M.P. McCormick. A comparison of lidar and balloon-borne particle counter measurements of the stratospheric aerosol 1974-1980, *J. Atmos. Sci.*, **39**, 909-916, 1982.

Whitten R.C., The Stratospheric Aerosol Layer (*Springer-Verlag Berlin Heidelberg, Germany*), 1982.

World Meteorological Organization (WMO), Atmospheric Ozone 1985, Assessment of our understanding of the processes controlling its present distribution and change. Global Ozone Research and Monitoring Project, Report No. 16, Vol. I-II-III. *World Meteorological Organization*, Geneva, Switzerland, 1985.

Atmospheric Soundings in Support of the Definition of the Tropopause Region in the South-Eastern Mediterranean Region

Retalis A., Cartalis C., Tombrou M. and Varotsos C.

University of Athens, Department of Applied Physics
33 Ippokratous St., 106 80 Athens, Greece.

Abstract. The variations of the tropopause height in the region of south-eastern Mediterranean (15-30° E and 32,5-42,5° N) as obtained from potential vorticity and potential temperature profiles and the temperature calculations obtained from the Stratospheric Aerosol and Gas Experiment II (SAGE II) instrument, flown on the Earth Radiation Budget Experiment (ERBE) satellite, are presented in an efford to explore the variation of the tropopause height in the examined area and to validate the use of satellite soundings for defining the tropopause height. It is found that the estimation of tropopause height from the use of satellite and *in situ* measurements is in good agreement (with the exception of summer months), thus validating the use of satellite data for the definition of the tropopause height.

Keywords. Tropopause, satellite soundings, potential vorticity, potential temperature, south-eastern Mediterranean

1 Introduction

Monitoring the tropopause region over a geographic area is important for a number of interrelated reasons. From a climatological point of view, it is important to investigate the behaviour of the tropopause region for several years, so that to detect any developed or under developed trends.

From a dynamic point of view, it is essential to define the tropopause region, in order to explore the stratosphere - troposphere exchange taking place over a geographic area that may be responsible for changes in the chemical composition of the atmosphere (Varotsos et al., 1992).

The study of the climatology of ozone is strongly supported by the use of satellite observations. In addition to the distribution of several chemical compounds (e.g. NO_X, aerosols), it is necessary to adequately understand the interactions between the stratosphere and the troposphere, at the tropopause level, in order to estimate the intrusion of stratospheric ozone in the tropopause.

In general, two concepts may be applied for defining the tropopause. The first concept reflects the conventional way, as the thermal structure of the atmosphere is examined with the tropopause being defined at the level where a near discontinuity of the temperature lapse rate is observed (WMO criterion). For climatological studies the criterion is satisfactory, but is inadequate when the mass exchange between the troposphere and stratosphere is to be examined. In this case the

NATO ASI Series, Vol. I 53
Atmospheric Ozone Dynamics
Observations in the Mediterranean Region
Edited by Costas Varotsos

dynamic concept is employed, namely Ertel's potential vorticity. The tropopause is defined as the region where a sharp discontinuity is observed in the values of potential vorticity, in order of one or two, with small values in the troposphere and larger in the stratosphere. Potential vorticity is a conservative quantity, implying that mass flow from the stratosphere to troposphere will result in decrease of potential vorticity through diabatic or mixing processes (Varotsos et al., 1992).

A third "concept" may be based on the use of satellite observations complemented with potential vorticity and potential temperature calculations. Aim of this study is to define - by applying all concepts - the variations of the tropopause in the region of SE Mediterranean (15-30° E and 32,5-42,5° N) in support of relevant studies or the exchange of ozone between troposphere and stratosphere.

For the purposes of the study, the potential vorticity and potential temperature profiles and the temperature profiles obtained from the Stratospheric Aerosol and Gas Experiment II (SAGE II) instrument flown on the Earth Radiation Budget Experiment (ERBE) satellite, for coinciding time periods, will be used.

2 Methodology and findings

Stratospheric Aerosol and Gas Experiment (SAGE) II is a seven-channel photometer with spectral bands at 0.385, 0.448, 0.453, 0.525, 0.600, 0.940 and 1.02 micrometers, flown on the Earth Radiation Budget (ERBE) Satellite. Solar occultation technique has been used to obtain aerosol extinction, ozone, nitrogen and water vapour profiles throughout the stratosphere and upper troposphere, over a wide range of latitudes and longitudes [Lenoble, 1989; Yue et al., 1989]. The details of SAGE II experiment are described in Mauldin et al., 1985.

Six years of SAGE II data, from January 1985 to November 1990, covering the south-eastern Mediterranean region, 15 - 30° E and 32.5 - 42.5° N, were used in the analysis. On the basis of these profiles, the tropopause heights for each month of the examining period were estimated (Retalis, 1994). In each case, the tropopause height was defined by the analytical vertical temperature profiles at the point where a significant change in the lapse rate occurred.

In Figure 1 the variation of the tropopause height for the period 1985-1990 is given. As expected the tropopause presents a winter minima and a summer maxima. Tropopause height is within the range of 9.5 to 16.5 km, with the minima occurring in February, and the maxima in July. The mean maximum tropopause height is 14.8 km, while the corresponding minimum is 10.1 km.

The annual variations of tropopause height as estimated from the potential vorticity profiles and the potential temperature profiles for the period 1978-1987, along with the previously described satellite induced temperature profiles are presented in Figure 2.

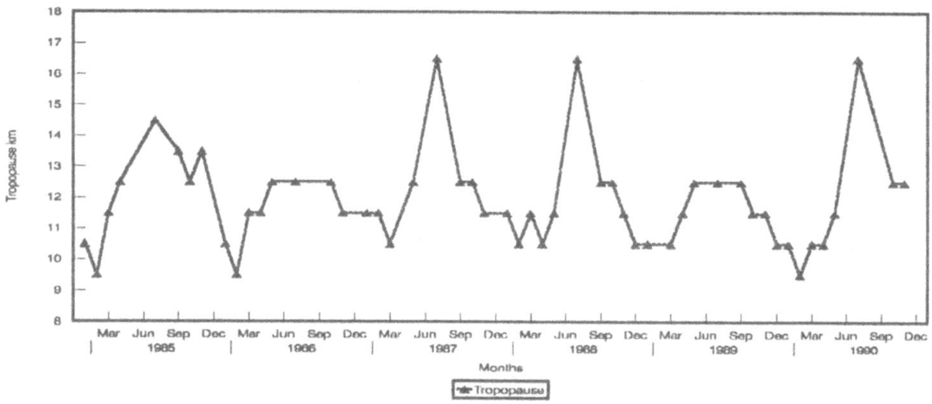

Figure 1. Variations of the tropopause height for the period 1985-1990

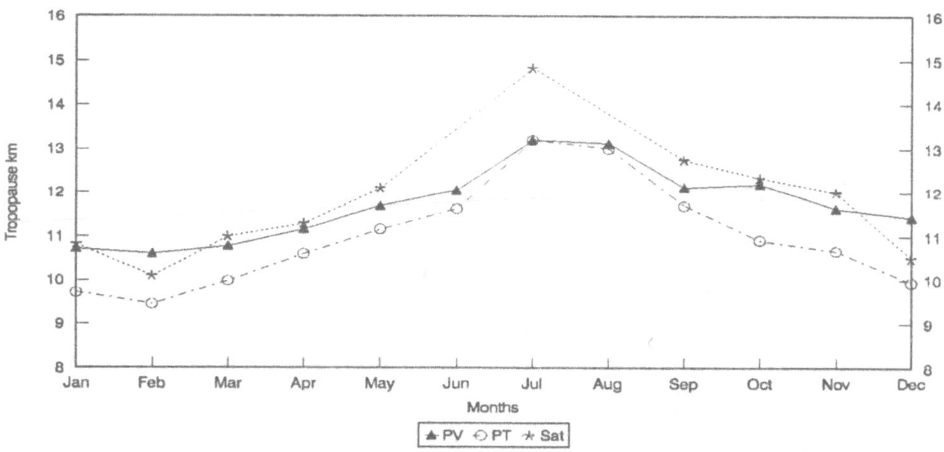

Figure 2. Variations of tropopause height from potential vorticity (PV) and potential temperature (PT) profiles for the period 1978-1987 and satellite data for the period 1985-1990

It is noticed that although all the curves follow the same course, the tropopause as estimated from satellite data is located at higher altitudes. Differences in the tropopause height as this is calculated on the basis of potential vorticity profiles and potential temperature are also recognised.

4 Conclusion

From the analysis and interpretation of the tropopause height as derived from the potential vorticity and potential temperature profiles and the temperature profiles obtained from the Stratospheric Aerosol and Gas Experiment (SAGE) II, for the area of south-eastern Mediterranean region (15-30° E and 32,5-42,5° N), it was deduced that:

- differences in the estimation of tropopause height, using different techniques, are small with the exception of summer months; thus satellite observations can be used for defining the tropopause height;
- at stations, where no measurements of total ozone were/are taken, the spatial and temporal variation of tropopause height can support the study of the dynamics of the area thus supporting the understanding of the spatial and temporal variation of total ozone;
- tropopause height is within the range of 9.5 to 16.5 km, with the minima occurring in February, and the maxima in July;
- the mean maximum tropopause height is 14.8 km, while the corresponding minimum is 10.1 km;
- no significant trend in the tropopuse height was observed for the examined area and time period.

5 References

Lenoble, J., 1989, Presentation of the European correlative experiment program for SAGE II, *J. Geophys. Res.*, Vol. 94, 8,395-8,398.

Mauldin, L.E., III, N.H. Zaun, M.P. McCormic, J.H. Guy, and W.R. Vaughn, 1985, Stratospheric aerosol and gas experiment II instrument: A functional description, *Opt. Eng.*, 24, 307-312.

Retalis A., 1994, Thermal and Chemical Structure of the Tropo- Stratosphere over the South-Eastern Mediterranean region as derived from Satellite observations. MSc Thesis, University of Dundee, Scotland, UK.

Varotsos, C., M. Tombrou, Cartalis, C. and D.N. Asimakopoulos, 1992, Using radiosonde data for the period 1978-1987 to study the tropopause region over the subtropics: an application for the south-eastern Mediterranean region. Presented in the General Assembly of EGS, Wisbanden, Germany.

World Meteorological Organisation (WMO), 1985, Atmospheric Ozone 1985, Assessment of our understanding of the processes controlling its present distribution and change. Global Ozone Research and Monitoring Project, Report No. 16, Vol. I. *World Meteorological Organisation*, Geneva, Switzerland.

Yue, G.K., M.P. McCormick, W.P. Chu, P. Wang and M.T. Osborn, 1989, Comparative studies of aerosol extinction measurements made by SAM-II and SAGE II satellite experiments, *J. Geophys. Res.*, Vol. 94, 8,412-4,424.

RETRIEVAL OF HEIGHT PROFILES OF TRACE GASES BY OPTIMAL PARAMETER ESTIMATION

E. SCHEER, [*]H. G. BOCK, [**]U. PLATT, [*]R. RANNACHER,

GRADUIERTEN KOLLEG "MODELLIERUNG & WISSENSCHAFTLICHES RECHNEN", [*]INTERDISZIPLINÄRES ZENTRUM FÜR WISSENSCHAFTLICHES RECHNEN, [**]INSTITUT FÜR UMWELTPHYSIK, UNIVERSITÄT HEIDELBERG, D-69120 HEIDELBERG, GERMANY.

Abstract. While most experiments on the subject of atmospheric physics emphasize on the detection of ozone exclusively, the GOME project was designed to monitor about a dozen additional trace gases in order to investigate the process of ozone depletion. GOME is part of the ERS2 misson and was launched in April 1995.

By the means of GOME it is possible to obtain total coverage of the earth's atmosphere for the first time within only three days in a spectral range from 240 nm to 790 nm.

Effective processing of this tremendous amount of data is only possible by developing fast retrieval algorithms. While there are several approaches for a forward model, a fast solution for the invers problem is not found yet, nevertheless important.

The classical way to solve this problem is a Gauss-Newton iteration, where the forward problem is solved for each set of parameters, while the parameters representing the height profiles of trace gas concentrations are altered for optimized output. This implies repeated calculations of the forward model and of weighting functions making this process time consuming.

This workgroup investigates a different approach. Since the forward model is based on an integro-differential equation that is solved by means of finite elements, it is possible to combine the outer parameters with the ones describing the lattice. By doing so, it is possible to solve the invers problem while iterating a slightly modified forward problem thus avoiding to solve the time consuming forward problem many times. This approach is expected to be at least two orders of magnitude less time consuming than the former.

Keywords: GOME, satellite, RTF, data retrieval, adaptive lattices

NATO ASI Series, Vol. I 53
Atmospheric Ozone Dynamics
Observations in the Mediterranean Region
Edited by Costas Varotsos

Introduction

In April 1995 the ERS2 mission was launched. Part of it is the Global Ozone Monitioring Experiment (GOME). By means of GOME it is possible to obtain total coverage of the earth's atmosphere within only three days. Effective processing of this tremendous amount of data is only possible by developing fast retrieval algorithms. This task is the subject of this work.

The Problem

Data retrieval in mathematical terms splits the problem in two parts: the forward problem and the invers or backward problem. The solution of the forward problem allows to calculate the expected result of a measurement for a given situation, i.e. all describing parameters. Typically, however, the scenery to be investigated is of the opposite character. While measured data is available, the describing parameters (i.e. trace gas concentration profiles) are to be retrieved. So to solve this task the forward problem is to be inverted. This implies to solve the forward problem first.

To do so a complete description of the forward problem has to be found. In the case of radiation transfer this is done by the Radiation Transfer Function (RTF)

$$n_\lambda \cdot \nabla \mathrm{I} = -\chi(\lambda)\rho(x) \cdot \mathrm{I} + \sigma(\lambda)\rho(x) \int \mathrm{R}(\vartheta, \vartheta') \cdot \mathrm{I} \cdot d\vartheta'$$

Although this is the most complete description many retrieval algorithms use a reduced variant of it to save computational time. Typically a modified version of Lambert-Beer's Law (LBL)

$$\mathrm{I}(\lambda) = \mathrm{I}_0 \cdot \exp(-\sigma(\lambda) \cdot \rho \cdot l)$$

is used. Since this law is applicable only for a known path of light it is neccessary to estimate a particular path of light between source and detector. Doing so is a strong restriction since in a medium as complex as the atmosphere different paths of light and multiple scattering are common (see Fig.1). To adapt calculated to measured re-sults correction factors are introduced. These fac-tors are calculated using Monte-Carlo methods.

It is obvious that for every different geometry and each set of describing parameters a special correction factor is needed leading to a large table of factors, handling each situation as a special case that was simulated at least once before.

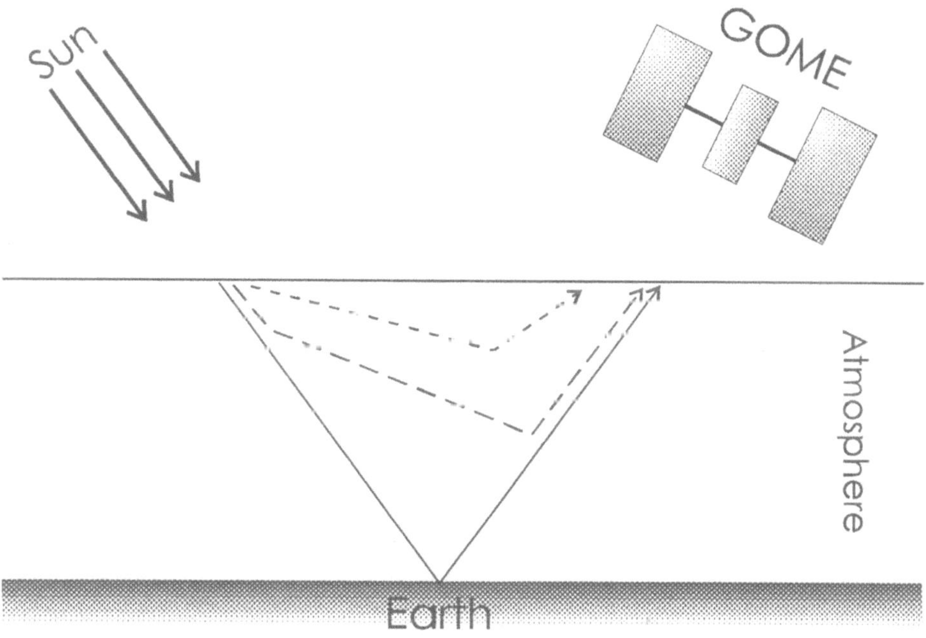

Figure 1:
This drawing sketches the situation of a reduced model as explained in the text. Missing a redistributing term the path of light has to be estimated (drawn out line). Complex phenomena like multiple scattering (dashed line) cannot be covered by these models without additional correcting mechanisms. The model also fails if the chosen path of light is not met, e.g. if a photon is scattered back in the upper atmosphere (dotted line).

The Complete Model

This work investigates a different approach. An exact solution of the forward problem is possible in a reasonable amount of time even if using the complete Radiation Transfer Function. Although an analytical solution of this equation is commonly impossible, a numerical approach is feasible. To do so an adaptive lattice is used. The advantage of adaptive lattices is due to the automatic reduction of redundant cells. If the numerical error in a cell is above a margin the cell is split into a number of smaller cells.

By applying this technique it is possible to calculate every area with the needed accuracy but without overcalculating areas of low interest. While highly dynamical areas are overlaid with a finer grid, the grid gets more and more coarse the less dynamical an area is. Notably in every area the calculation error is of the same magnitude although the size of the cells and thereby the number of calculations made in that area vary widely!

While LBL is one-dimensional, as it is along a path, the RTF is an integro-differential equation in an arbitrarily dimensional space. The equation consists of three parts, besides the transport term and the absorption term, which are common to all models, it also contains a so-called source term. This source term is the integral part of the equa-tion as it is responsible for energy redistribution between volume elements and thereby allowing multiple scattering and undistinguished paths of light.

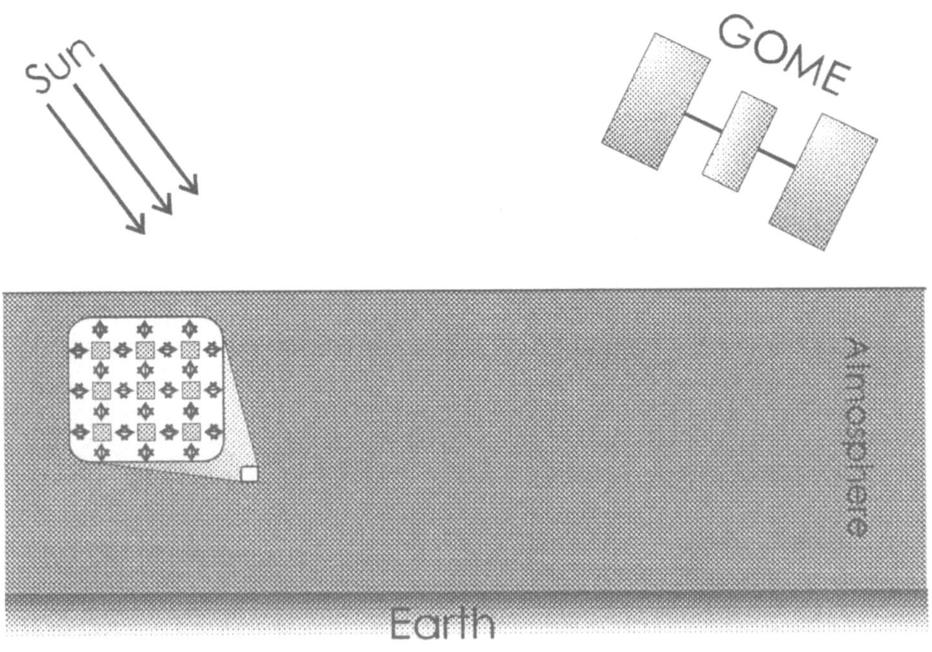

Figure 2:
This drawing sketches the situation of a model including a redistribution term using a lattice for calculation. A path of light is not drawn since no distinguished path exists. Instead a zoom shows several cells, each exchanging energy with all its neighbours. As explained in the text this model includes complex phenomena as multiple scattering, etc.

The source term is for computational reasons neglected in most models (e.g. in LBL) for it reduces the integro-differential equation to the analytically solvable form of a simple differential equation. The price to be paid for this simplification is the loss of multiple scattering and the neccessity of a known path of light, as explained above.
So where LBL only allows energy decrease along a path, in the RTF space is divided into volume elements that exchange energy in both directions with all its neighbours (see Fig.2). Obviously this covers multiple scattering and undetermined paths of light in the direct approach and without further corrections. So the RTF is the most complete description of the forward problem.

Mathematical Description

In mathematical terms the invers problem can be defined as follows.

Forward problem:

given a set of describing parameters p,
solve

$$F(x, p) = 0.$$

Invers problem:

Let μ be a measurement affected with an error ε, so

$$\mu = M(x, p) + \varepsilon,$$

where the distribution of ε shall be known.

To be found:

$$\min_{x,p} \| \mu - M(x,p) \|$$

under the condition that $F(x, p) = 0$.

A typical way to solve this problem, as it is widely used, is to estimate a starting set of parameters p, then solving the forward problem for this set. Afterwards determining $x(p)$ and using this in the solution of the invers problem. Having done that vary p to obtain a better solution.
In mathematical terms:

Problem:

Let $F_1(x, p)$ be the difference of measurement and calculation,

$F_2(x, p)$ the radiation transport function, further

$$\boldsymbol{F}(p) := F_1(x(p), p).$$

To be found:

$$\min_{p} \| F_1(x(p), p) \|$$

under the condition $F_2(x, p) = 0$.

Solution via Gauss-Newton:

p^0 :starting value,

$$p^{k+1} = p^k + t^k \cdot \Delta p^k; \quad t \in]0,1]$$

Question:

How is Δp to be chosen to gain

$$\min_{\Delta p} \left\| \boldsymbol{F}(p^k) + \frac{\partial}{\partial p} \boldsymbol{F}(p^k) \Delta p^k \right\|$$

under the above condition?

The disadvantage of this method is the immens computational time needed. Besides the calcula-tion of $x(p)$ the derivation of x in direction of p is to be calculated what brings a factorial increase in computational time.

So this way of solving the invers problem can easily exceed a factor 1000 in time relative to the forward problem.

An Improved Approach

The approach used in this work is much more efficient. Under given conditions it is possible to combine x and p to a new variable y. Also this equation is solved iteratively using Gauss-Newton. But while the former method solves the forward problem for given p first and just then varying p, this method offers to iterate both x and p in the same step. Since the number of variables in x is estimated to be about 100000 but the number of variables in p only about 50 the additional work is comparatively small.

This approach is expected to be of about a factor 25-100 less time consuming than the one sketched above.

Or in mathematical terms:

Problem:

(with the above notations)

Let $y = (x, p)$, with

y^0 :starting value,

$$y^{k+1} = y^k + t^k \cdot \Delta y^k; \quad t \in]0,1]$$

To be found:

$$\min_{\Delta y} \left\| F_1(y^k) + \frac{\partial}{\partial y} F_1(y^k) \Delta y^k \right\|$$

under the condition

$$F_2(y^k) + \frac{\partial}{\partial y} F_2(y^k) \Delta y^k = 0 \ .$$

Conclusion

As the technique of adaptive lattices is already developed and applied this work can concentrate on optimizing the invers problem.
The informatical part of this problem should be solvable by application of Successive Quadratic Programming SQP.
Again, there are general steps made but they are far from applicable to the above described special situation.

Quality Checks for Satellite and Ground-Based Total Ozone Observations

D. Spänkuch and E. Schulz

Deutscher Wetterdienst, Meteorologisches Observatorium Potsdam, Postfach 60 05 52, D-14405 Potsdam, Germany

Abstract The accuracy of total column ozone measurements, ground-based as well as satellite ones, is severely influenced by weather, in particular by the state of cloudiness at the measurement site. It is therefore of interest to find an approach to check the reliability of these measurements. Linear multiple regressions, connecting total column ozone with a variety of meteorological parameters and derived from data of a midlatidude Central European station, are shown to be a suitable quality check for total ozone measurements. Nearly all measurements with more than 20 DU deviation from the statistical estimate were identified as influenced by bad weather. The exceptions could be explained by errors in one of the predictors, the total ozone amount of the preceding day.

Keywords. Total ozone, TOMS

1 Introduction

The quality of both, ground-based as well as satellite, total-column-ozone measurements is influenced by

- the state of the instruments (degradation of optical components, etc.),
- the state of cloudiness within the field of view,
- the length of the radiation path through the atmosphere (change of wavelength pair

NATO ASI Series, Vol. I 53
Atmospheric Ozone Dynamics
Observations in the Mediterranean Region
Edited by Costas Varotsos

used from AD to CD at low sun angles in case of ground-based Dobson measurements),

- the atmospheric turbidity (less important than the other three factors, but nevertheless not to be neglected for trend analyses).

In case of ground-based measurements with Dobson spectrometers the state of cloudiness dictates the measurement approach - from direct sun - (DS), zenith blue - (ZB) to zenith cloudy - (ZB) measurements with decreasing level of accuracy. In case of TOMS satellite measurements

- a positive bias of more than 20 units (DU) is caused in regions of persistent subtropical marine stratocumulus due to lower cloud tops of these clouds than the climatological cloud top height (Thompson et al. 1993),

- a positive bias of up to 20 DU was detected at partial cloud cover, most pronounced at lower latidudes and in the case of small-scale cloud structure (Seftor et al. 1994).

Hence, quality checks of total column ozone measurements are absolutely mandatory for reliable statements of the ozone issue.

2 Approach

Total column ozone, Ω, is not an isolated atmospheric parameter but is related to other meteorological parameters as atmospheric transport processes are mainly responsible for the ozone content of the tropo - and lower and middle stratosphere (up to 35 km altidude). Consequently, there exist relatively strong statistical relations between Ω and numerous meteorological parameters. Such statistical relations have long been known (e.g. Dobson et al. 1929, Reed 1950, Normand 1953) and have recently been used successfully in short-term Ω forecasting (e.g. Burrows et al. 1994, Austin et al. 1994, Poulin and Evans 1994, Vogel et al. 1995a,b).

For the national German UV forecast (Vogel et al. 1995a,b) linear multiple regressions were derived for each month from grid point TOMS, version 7.0, and corresponding aerological data (Spänkuch and Schulz 1995). In that paper it was shown that differences between the statistically predicted Ω and Dobson measurements should not exceed ± 20 DU in maximum.

Larger differences are attributable either to unfavourable weather and, hence, doubtful quality of measurements, or to erroneous predictors (Spänkuch and Schulz 1996). This upper limit of ± 20 DU uncertainty is, hence, also valid for Ω diagnosis of ground-based measurements.

3 Quality Check for Satellite Measurements

In Spänkuch and Schulz (1995) we analyzed already large deviations (> 40 DU) between TOMS data and diagnosed total column ozone for 1979, using the multiple linear regressions between Ω and geopotential heights and thickness, derived for each month from grid point TOMS data and corresponding aerological data at Potsdam, Germany (52.23°N; 13.02°E).

This analysis is repeated here in great detail for all deviations Ω (TOMS) - Ω (estimate) larger than ± 20 DU (see Table) for two reasons. The first one is that the former analysis was based on TOMS,version 6.0 data. Meanwhile the quality of TOMS data was improved in version 7.0 with partly large corrections against version 6.0 data. The second reason is that all deviations larger than ± 20 DU are thoroughly analyzed without exception.

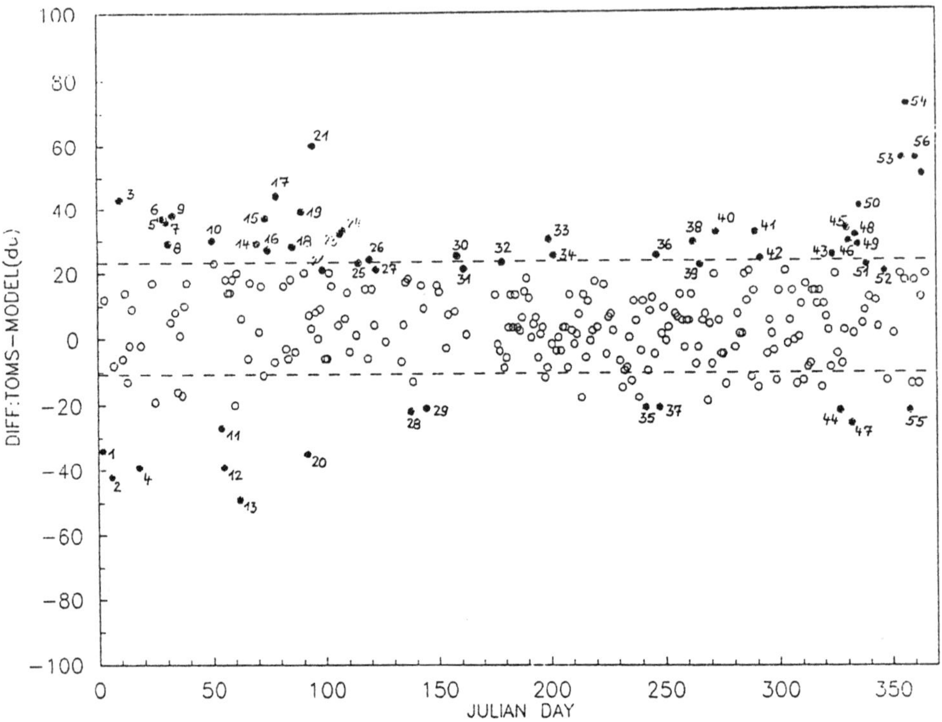

Figure Difference between TOMS grid point, version 7.0 total ozone measurements and statistical estimates, using linear multiple regressions, for Potsdam, Germany (52°23'N, 13°04'E) and 1979. For explanation of serial numbers see Table and text.

Given are in the Table the deviations (DU) and that weather information which is relevant to the accuracy of the satellite measurement, as cloudiness (in octas), cloud base height (m), precipitation, visibility (km) and wind velocity (ms^{-1} or Beaufort scale). As grid point TOMS data represent an area of about 110 x 120 km (Fleig et al. 1990) the weather information about noon (± 2 hrs) was compiled. From all days listed, there are only two days (serial numbers 46 and 55) of low cloud amount. The deviation on day 55 can be explained by the large error of $\Omega(i-1)$ of the previous day (72 DU). The estimate given in Spänkuch and Schulz (1996) would cause a resulting difference of about 20 to 40 DU. Remains day 46 with fine weather and good visibility. Evaluating the impact of an erroneous Ω(i-1) of 33 DU on the Ω forecast, the induced error would be about 15 DU, much more 'noise' than usual.

The table contains, too, the differences between TOMS, 6.0 version and the estimate for that days analyzed in Spänkuch and Schulz (1995). We note lower differences with version 7.0 than with version 6.0 data in general, indicating the right sign in the correction. The amount of the correction is, however, obviously not sufficient. There are also some cases (days 17, 56) with larger differences with version 7.0 than version 6.0 data.

4 Conclusions

The estimate of total column ozone, based on linear multiple regressions, is a suitable first quality check for ground-based as well as for satellite measurements. Deviations of the measurement from the estimate of more than about 20 to 25 Dobson units severely suggest erroneous measurements due to unfavourable weather conditions.

Acknowledgement: This study was supported by the Bundesministerium für Forschung und Technologie, Bonn, under grant 01 LO 9204/0.

Table Weather Information at Potsdam for Days in 1979 With Large Deviations Between TOMS and Diagnosed Total Column Ozone

Ser. Numb.	Jul. Day	TOMS-ESTIMATE (DU)	Weather in Potsdam
1	2	-34	6/8, 1.8 km, Beaufort 6 (bursts 8)
2	6	-42	8/8, snow, 1.4 km
3	10	43 (49)	5/8, snow, Beaufort 6
4	18	-39	7/8, snow, 3 km Beaufort 6
5	28	37	7/8, snow, 3 km
6	29	37 (67)	5/8, snow, 1.1 km
7	30	36 (65)	6/8, snow, 1.0 km
8	31	29	6/8, 500 m, snow, 1.5 to 25 km
9	33	38	7/8-8/8, snow, rain, 1.4 km, bursts 6
10	50	30	3/8 Sc, 1000 m, Cs, 6 km
11	54	-27	0 to 8/8, 600 m, snow, 1.4 km
12	55	-39	6/8, snow, 1.2 km
13	62	-49	4/8-8/8, rain, 6 km, bursts 6
14	70	29	7/8 Sc, Ac, Cs, 800 m, 7 km
15	74	37	7/8, snow, rain, 2 km, Beaufort 6 (bursts 8)
16	75	27	7/8 Sc 400 m, drizzle, 2 km
17	79	44	7/8, snow, rain, 4.5 km, bursts 6
18	86	28	7/8 Cu, Ci, Cs, 1000 m
19	90	39 (41)	8/8, snow, rain, 4 km
20	92	-35	8/8, rain, thunderstorm, bursts 6
21	95	60 (74)	7/8, rain, 1.4 km
22	99	21	8/8 Cu, Sc, 800 m, 7 km, 14 ms^{-1}

Table Weather Information at Potsdam for Days in 1979 With Large Deviations Between TOMS and Diagnosed Total Column Ozone

Ser. Numb.	Jul. Day	TOMS-ESTIMATE (DU)	Weather in Potsdam
23	107	32 (58)	7/8, St 500 m, snow, 3.5 km
24	108	33	2/8-7/8 Cu, Sc 800 m, 15 km
25	115	23	6/8 Cu 400 m, rain
26	120	24 (49)	8/8; 3/8 Cu, rain
27	123	21	7/8 Cb cal 750 m, rain
28	138	-22	4/8 Cu 1000 m, rain, 10 km
29	145	-21	6/8 Cu 800 m, rain
30	159	25	7/8 Cu 400 m, rain
31	162	21	8/8 Sc 1800 m, rain
32	180	23	5/8 Cu med 700 m, 14 km
33	200	35	6/8-8/8 Cu con, Sc 600 m, rain
34	202	25	5/8-7/8 Cb, Cu, Sc 600 m, rain
35	242	-21	4/8 Cu 800 m, 9 km
36	247	25	8/8 Cu 600 m, rain
37	248	-21	4/8 Cu 800 m, 35 km
38	264	29	7/8 Cu med 900 m, (rain), 7 km
39	266	22	3/8 Cu med 700 m, 10 km
40	273	32	5/8 Cu con 800 m, 0.5 to 14 km
41	290	32	8/8 Sc 400-800 m, 0.04 to 8 km
42	292	24	5/8 Cu med 700 m, 11 ms^{-1}
43	324	25 (41)	8/8 500 m, rain
44	327	-22	5/8 Cu med 700 m, 4.5 km
45	330	33	8/8, rain, 5 km, bursts 8
46	331	29	1/8 Cu 600 m, 28 km
47	332	-26	8/8, drizzle, 0.6 km

Table Weather Information at Potsdam for Days in 1979 With Large Deviations Between TOMS and Diagnosed Total Column Ozone

Ser. Numb.	Jul. Day	TOMS-ESTIMATE (DU)	Weather in Potsdam
48	334	31	6/8 Ac, (rain), 13 km
49	335	28	7/8 350 m, rain, 4.5 km
50	336	40 (41)	8/8, drizzle, bursts 6
51	339	22	6/8 2500 m, (rain), 10 km, 19 ms^{-1}
52	354	19 (42)	7/8-8/8 Sc str op, fog, rain
53	355	55 (63)	7/8-8/8 Sc str op 900-1200 m
54	357	72 (64)	7/8-8/8 Sc str op 600-1000 m
55	358	-22	2/8 Ac tr
56	361	55 (79)	Cs, Ci fib, 2.5 km, bursts 13 ms^{-1}
57	364	60 (44)	7/8, snow, rain, 5 km

The numbers in brackets in the 3rd column are related to TOMS, version 6.0. Weather information includes cloud amount in octas, cloud base height (m), precipation, visibility (km), wind velocity (ms^{-1} or Beaufort scale).

References

Austin J., Barwell B. R., Cox S. J., Hughes P. A., Knight J. R., Ross G., Sinclair P. and Webb A. R. (1994) The diagnosis and forecast of clear-sky ultraviolet levels of the Earth's surface. *Meteorol. Appl. 1,* 321-336.

Burrows W. R., Vallè M., Kerr J. B., Wilson L. J. and Tarasick D. W. (1994) The Canadian operational procedure for forecasting total ozone and UV radiation. *Meteorol. Appl. 1,* 247-265.

Dobson G. M. B., Harrison D. N. and Lawrence J. (1929) Measurements of the amount of ozone in the earth's atmosphere and its relation to other geophysical conditions. part III., *Proc. Roy. Soc. A 122,* 456-486.

Fleig A. J., Mc Peters R. D., Bhartia P. K., Schlesinger, Cebula R. P., Klenk K. F., Taylor S. L. and Heath D. F. (1990) Nimbus 7 solar backscatter ultraviolet (SBUV) ozone products user's guide. *NASA Ref. Publ. 1234,* 19-25.

Normand C. (1953) Atmospheric ozone and the upper air conditions. *Q. J. R. Meteorol. Soc., 79,* 39-50.

Poulin L. and Evans W. F. J. (1994) METOZ: Total ozone from meteorological parameters. *Atmos. Ocean, 32,* 285-297.

Reed R. J. (1950) The role of vertical motions in ozone-weather relationships. *J. Meteorol. 7,* 263-267.

Seftor C. J., Taylor S. L., Wellemeyer C. G. and Mc Peters R. D. (1994) Effect of partially clouded scenes on the determination of ozone, Proc. Quadr. Ozone Symp., Charlottesville, V. A., *NASA Conf. Publ. 3266,* 919-922.

Spänkuch D. and Schulz E. (1995) Diagnosing and forecasting total column ozone by statistical relations. *J. Geophys. Res., 100,* D9, 18,873-18,885.

Spänkuch D. and Schulz E. (1996) Short-term total-column-ozone-forecast errors induced by forecast errors of the statistical predictors. *Atmos. Environment* (submitted for publication)

Thompson A. M., Mc Namara D. P., Pickering K. E. and Mc Peters R. D. (1993) Effect of marine stratocumulus on TOMS ozone. *J. Geophys. Res., 98,* 23,051-23,057.

Vogel G., Spänkuch D., Schulz E., Feister U. and Döhler W. (1995a) Regional short-term forecast of total column ozone. *Atmos. Environment 29,* 1155-1163.

Vogel G., Schubert U., Kirchner R. and Günther A. (1995b) Methodische Grundlagen der operationellen UV-Vorhersage des Deutschen Wetterdienstes für Mitteleuropa. *Ann. Meteorol. 31,* 346-347.

Airline Crew, Cosmic Radiation and Cancer. Status of the Current Norwegian Study and the Proposed European Study

Ulf Tveten[1], Tor Haldorsen[2] and Jon Reitan[3]

[1] Institute for Energy Technology, Kjeller, Norway
[2] Cancer Registry of Norway, Oslo, Norway
[3] Norwegian Radiation Protection Authority, Østerås, Norway

Abstract. Airline personnel are exposed to cosmic radiation to a varying degree, depending primarily upon flight altitude and the type of routes flown. The number of flight hours flown per year has been compiled on a person-by person basis for all pilots certified for commercial aviation in Norway since 1947. To a large extent the files also contain various information on type of aircraft flown, and employing air company for these pilots, and this information will be used to estimate the individual radiation histories. In the project described in the following, the intention is to link estimated radiation exposure histories of individual airline pilots with data from the Cancer Registry of Norway to reveal possible relationships between dose and health effects. There are firm plans for expanding this project to include a number of European countries. At present an intent to participate has been expressed from Denmark, Finland, Germany, Greece, Italy, Netherlands and Norway. Strong interest has also been expressed from France and Belgium.

Keywords. Cosmic radiation, aviation, pilots, cabincrew, cancer, mortality, working environment

1 Introduction

Airline crew is probably the only relatively large population group where occupational exposure to ultraviolet radiation and ozone may have a health impact. The intensity of ultraviolet radiation may be more intense at cruise altitudes, and the concentration of ozone in the air in the cockpit (and cabin) at cruise altitudes may be higher than in air at the ground.

Although the epidemiological study which will be described in this paper is concerned neither with ozone nor ultraviolet radiation, but with cosmic radiation, there is a certain logical link between the problem areas. The cohort is the same, and the logical structure of studies on these problem areas will have much in common. To compile a complete cohort is a very work-demanding part of an epidemiological study, and once it has been performed for e.g. a project concerned with cosmic radiation, it is perfectly possible to use the cohort for other purposes, like studying the impact of excess exposure to ultraviolet radiation and exposure to higher-than-normal concentrations of ozone. Both for ultraviolet radiation and ozone concentration it is possible that there is a dependence on cruise altitude, somewhat like what is the case for cosmic radia-

NATO ASI Series, Vol. I 53
Atmospheric Ozone Dynamics
Observations in the Mediterranean Region
Edited by Costas Varotsos

tion, so that even some of the exposure-related parameters collected in a cosmic-radiation-related study may be relevant to these other types of studies, although the exposure pattern in connection with ultraviolet radiation will obviously be more comples, as the exposure will depend both on the time of day, geographical position and direction of the airplane relative to the position of the sun.

The study which will be described in this paper, however, is related to cosmic radiation, and the main purpose is to use the cohort to obtain new information on the relationship between low doses of ionizing radiation and health impacts. Many attempts have been made to determine the relationship between low doses of ionizing radiation and incidence of cancer with epidemiological methods, none with definite success. There have been several reasons for the limited success, e.g. that in many studies individual dose estimates have been lacking or uncertain.

By studying airline pilots and crew some of these inherent weaknesses are less serious. It is possible to obtain rather good dose estimates, and as a population group pilots and crew are quite homogenous, socially, economically and regarding general health status.

The present project is based upon a concept conceived in 1981, further developed in 1982, and subsequently submitted to the Norwegian Research Council. Although the proposed project won the approval of the Civil Aviation Administration and was granted the necessary permissions from the Cancer Registry and the Data Inspectorate, adequate funding was not obtained, and the project was dormant until 1992, when a renewed and extended project proposal was submitted to the Commission of the European Communities' Radiation Protection Programme jointly by Norway (project leader), Germany and Netherlands. The project was approved by the RPP, as a feasibility study, which should include airline pilots and cabin crew licensed in Norway.

The feasibility study was initiated in the late autumn of 1992. Almost a full year was spent informing the involved parties (airline companies, pilots and cabin crew associations), signing agreements, as well as obtaining the necessary permissions from the various relevant authorities. Throughout 1994/95 data from the files of the licensing department of the Norwegian Civil Aviation Administration has been compiled in a computer data base, and this information has been checked against and supplemented with information obtained from the files of the Medical Department of the Civil Aviation Administration. Further information may be obtained from the Institute of Aviation Medicine, which contains flight information on pilots in the military, as well as from the military careers of pilots subsequently employed in civil aviation.

Procedures for estimating individual radiation exposures histories from flight information (flight hours, aircraft, airline company information etc.) is under development. This is done in close cooperation with the Norwegian Airline Pilot's Association.

2 Cohort of pilots and cabin crew

Civil and military pilots have to renew their licenses at least every year. Data relating to license renewals are in Norway stored at the Civil Aviation Administration (pilots and cabin crew in civil aviation) and the Institute of Aviation Medicine (military pilots). Cabin crew renew their licenses less frequently; at the present time every fifth year, earlier every second year.

The cohort consists at present of all pilots and all cabin crew members who have had licenses for commercial air traffic in Norway since 1947. Pilots with a purely military career may be added, but a final decision has not been taken regarding this aspect. Military pilots in general operate at rather low altitudes, have fewer flights hours per year than commercial pilots, and also differ somewhat from commercial pilots in social parameters.

3 Data from the Civil Aviation Administration

The Civil Aviation Administration gave valuable assistance in locating occupational files. Some of the files of deceased and/or retired persons had been removed from the "daily access" storage, and had been placed in boxes in the basement. The immense importance of making sure that the material is complete should be underlined, and should be kept on top of the mind when the project will now hopefully be expanded to include a number of European countries. If the files that had been moved also had been destroyed, the project would not have been possible, unless complete files could have been found elsewhere.

There is good reason to believe that we have succeeded in including in the data base almost all airline pilots and cabin crew members active in Norway since 1947. The data base that has been compiled contains ca. 6,810 persons (ca. 3745 pilots and ca. 3065 cabin crew), about 9,320 licenses; covers about 20,600 permissions to fly specified types of aircraft, and about 80,280 license renewals have been examined and entered into the files generated by this project.

The files of the Civil Aviation Administration were not stored electronically, except for the years from and including 1987. The older files consisted of individual folders, one for each pilot, containing a rather large number of sheets of paper of varying content. The flight hour information was marked down on specific forms, but some of the other relevant information (e.g. the airline company) was much more difficult to find, and required looking through most of the miscellaneous papers contained in the folder.

The data base has been constructed in ACCESS, a standard and presently much used commercial data base program in Windows. The data base constructed contains a large number of useful functions, and entering of data in the base is simple, and designed to be performed by e.g. an assistant/secretary. There are searching and sorting and screening options.

In Norway every person is identified by a unique identification number. This system was introduced in 1964. For a number of persons in the file (about 1,700) the person-number was missing for various reasons. Fortunately it was relatively easy to find the numbers of most of these persons from the Central Person Register. About 100 - 200 persons, however, are more problematic. Some of these are women (mostly cabin crew) who have changed name through marriage. Some others are foreigners. This may often be easy to see from the name. Persons who died before 1960 were never assigned person-numbers and must be identified by name and birth date alone. Apart from the extra working time required to find missing person-numbers or identify persons, these problems are not serious, and it is expected that only a very small portion of the cohort will remain unidentified.

The data collected for each person is the birth date, person-number, an internal identification number (within the airline company), name, license type, class and restrictions, date when the license was issued, date for each renewal, and the accumulated flight hours at the time of each renewal. Flight hours as captain and co-pilot are given separately. In addition are given for some dates (when there has been a change in the aircraft types for which the license is valid) code identifications of aircraft types. There is also information on the airline company in which the person is working, but unfortunately this information is rather incomplete.

The number of identification codes for aircraft types in these files is surprisingly large; more than 400. It is not less surprising that there are more than 200 airline company identification codes, even after obvious misspellings etc. have been sorted out. The Norwegian Airline Pilot's Association and SAS Pilot's association have given assistance in identifying the meaning of these codes within the context of the study, where interest is obviously focused on the information these codes give which is relevant to radiation exposure.

4 Individual radiation exposure histories for pilots

The unique feature of this study is the possibility of a much better estimation of individual exposure histories than what is the case in most occupational studies going rather far back in time. From the registers of the Civil Aviation Administration and the Institute for Aviation Medicine the number of flight hours between license renewals have been stored, as well as the types of aircraft.

The flight hours per year alone is not sufficient to give a dependable measure of the exposure to cosmic radiation. It must be combined with information which indicates the types of routes flown and at which altitude. The flight time given in the registers is reckoned from the time the plane leaves the gate at departure until it is parked at the gate at arrival. This means that part of the registered flight hours is taxing on the ground as well as ascent to and descent from cruising altitude. The ascent and descent times are not too different for different flights, but the fraction of the flight hours spent at cruising altitudes may differ considerably. For example, on a Europe - USA flight about 70 - 80% of the flight time could be at altitudes interesting from a radiation exposure point of view, while for a short-haul flight (Oslo - Gothenburg) the plane will often not reach these altitudes at all.

The most exact manner in which to determine the exposure history of individual pilots would be to use their personal information, from their logbooks. There are several difficulties with this approach; primarily that it will be impossibly time-consuming, secondly that it may be difficult or impossible to acquire the information for deceased pilots, and finally that log-books may not exist, since most pilots in recent times do not write log-books.

An alternative method has to be chosen. One can not expect to find a method that will be correct for all pilots. However, various types of information are available that may be used to estimate the radiation exposure, like typical career patterns, relevant characteristics of aircraft types, and airline schedules of the air companies in the different years. The Norwegian Airline Pilot's Association has provided very helpful advice and guidance regarding this type of information, and are helping in choosing schemes that are as sensible and realistic as possible. In the present project it has been chosen to focus on the aircraft type and air company for each individual pilot and each individual year of his/her career.

One other tool will be employed within the project in this connection. Dr. Wallace Friedberg at the U.S. Federal Aviation Administration has developed a computer program, CARI-3N, which will calculate the cosmic radiation along any flight route at any calendar time in the past (taking historical variations due to fluctuations in solar activity into account). The latest version of this program, which has also very kindly been somewhat modified according the specific wishes from this project, is operational in Norway and is used in the project.

5 Cosmic radiation

The primary cosmic radiation entering the upper layers of the earth's atmosphere consists mainly of protons. Subsequent reactions with the atoms in the atmosphere change the characteristics of cosmic radiation, and at sea level the major component is gamma radiation. At typical flight altitudes of modern aviation, the proportion of the cosmic radiation which is neutrons or heavy particles is significant, contributing roughly 50% of the dose equivalent. The biological significance (effectiveness in causing biological damage) of this type of radiation as compared to that of gamma radiation and other "low-LET"-radiation is, however, not well known. It is possible that the present study may also shed some light on this problem.

The table below gives the equivalent doserate from cosmic radiation for some altitudes, at geographical position 60^0N and 10^0E and a heliocentric potential of 500 MV, which is an "average" value:

Altitude (feet above sea level)	Radiation doserate (μSv/h)
Sea level	0.03
20,000	0.73
25,000	1.51
30,000	2.80
32,000	3.46
35,000	4.62
38,000	5.92
42,000	7.67
50,000	11.2

Another aspect of some concern is the possible significance of so-called solar flares. These are episodes of intense solar activity of quite short duration; hours to perhaps a day or so. The information reviewed so far indicate that solar flares will not contribute any significant portion of a pilots or cabin crew members life-time radiation dose.

6 Statistical analysis

Pilots and cabin crew enter the study population at the date of their first license and are followed until the end of 1993 or until the date of death/emigration. The cohorts will be linked to the Cancer Registry by the personal identification number, or sex, name and date of birth for persons deceased before 1960. The incidence of different types of cancer in the cohort will be compared to the incidence in the general population by sex, age and period. However, particular emphasis will be put on the planned internal analysis of the cohort, in which the incidence of cancer according to the individual estimates of radiation exposure and other factors will be studied. The analyses will be done by descriptive epidemiological methods and modern methods of multivariate analysis (i. e. Poisson regression and Cox regression).

7 Statistical power of the study of the Norwegian cohort

The Norwegian pilot cohort will consist of at least 3400 persons, and the cumulative observation time for the cohort has been estimated to approximately 65,000 person-years. Pilots with an exclusively military career may come in addition, if it is decided to include the group. The cohort of cabin crew members will be somewhat smaller. The exposure profiles of the pilots will differ. Some members of the cohort will have little or no occupational radiation exposure, while others, who for extended time periods have been engaged in long distance flying, will have been exposed to a considerably larger extent.

Based upon estimates of the distribution of exposure profiles and the individual observation time, it is possible to calculate the statistical power of the study. These calculations will give the probabilities of obtaining significant results from the study, assuming certain hypothetical values of the excess risk connected to radiation doses. A preliminary calculation of statistical power has been carried out, and the conclusion is that the probability of finding significant results from the Norwegian study population alone is rather small, especially for some types of cancer. It seems reasonable to believe, however, that significant results can be obtained from a study population which is larger, and it does not seem to have to be extremely much larger.

8 Proposed adjoint studies

Among the many research groups in a number of countries to which contact has been established during the planning period for an extension of the project were also some groups interested, not directly in cosmic radiation, but in problem areas related to other aspects of the working conditions of pilots.

A group at Athens University had initiated work on the influence of ultraviolet radiation in the cockpit, and measurements during flight conditions had also been carried out. The variation pattern for this type of radiation is quite different from that of cosmic radiation, as it will not only be dependent upon flight altitude, but also to a very strong degree upon the time of the day and the plane orientation relative to the position of the sun. Ultraviolet radiation differs from cosmic radiation also in that it may be feasible to protect pilots against possible damage by relatively simple means. It might, however, be interesting to go back and see if there is indeed an excess of melanoma in the cohort. Information on this aspect is available at Andreas Sygros Hospital in Athens.

The second ajoint problem area concerns chromosome aberrations in pilots. Some measurements of chromosome aberration in groups of pilots with varying exposure to cosmic radiation have been performed at the University of Rome "Tor Vergata". It has been proposed to incorporate a continuation of this work in the project as an adjoint study. Some interest in this problem areas has also been expressed from the Danish partner.

Unfortunately, signals from the Commission indicate that it would be wise not to include these ajoint studies in a project proposal to the Commission.

9 Planned expansion of the project

The present phase of the project is intended as a feasibility study and it has shown that the approach adopted is sound and feasible, and that a reasonable extension of the cohort may indeed produce very valuable information about the health impacts of ionizing radiation.

Initiatives for expanding the project were taken in late 1994 and early 1995. A project proposal was submitted to the Fourth Framework Programme of the Radiation Protection Programme of the European Union in the Spring of 1995, but this proposal did not obtain the necessary rating for obtaining funding. Since then several additional countries have expressed a strong interest in joining the project. At present the committed partners of the project are: Denmark, Finland, Germany, Greece, Italy, Netherlands and Norway. Strong interest in the project has also been expressed by potential partners from Belgium and France. A new and further improved project proposal will be submitted in the end of February 1996.

Regional Distribution of Total Column Ozone Changes in Central Europe

G. Vogel, U. Schubert and D. Spänkuch

Deutscher Wetterdienst, Meteorologisches Observatorium Potsdam, Postfach 60 05 52, D-14405 Potsdam, Germany

Abstract The availability of Total-Ozone-Mapping-Spectrometer (TOMS) data since the end of 1978 gives the opportunity of investigating regional patterns of total-column-ozone (Ω)-changes in great detail. Charts of Ω differences between the first, and the second (and last) eight-year-period were produced for Central Europe and every decade of the year. The paper discusses the regional decadal distribution of Ω in spring and early summer and its change from the first to the second eight-year-period. The decline of Ω from the first to the second period is twice as much in the North of the coastal area of Germany than in southern Germany. Described, too, is the regional distribution of the UV index for the second decades from May to August.

Keywords. Total ozone, UV-B radiation

1 Introduction

The growing public concern about the decrease in stratospheric ozone and in total column ozone and, hence, in harmful UV-B radiation at the surface of our planet has required more and more detailed information on this issue. It is not sufficient for practical applications to know gross features as e.g. zonal means of the geographic distribution of variations and/or trends in total column ozone and UV-B radiation. People learn environmental benefits and drawbacks from experience on local and regional scale, and global patterns and issues have to be translated into these scales to be transparent for everybo-

NATO ASI Series, Vol. I 53
Atmospheric Ozone Dynamics
Observations in the Mediterranean Region
Edited by Costas Varotsos

dy. In Enke et al. (this issue) we show that trends of total column ozone in Central Europe are significantly dependent on weather patterns. Bojkov et al. (1995a) show the regional distribution of the 1994-1995 ozone anomaly during the northern hemisphere winter and spring on the basis of groundbased total ozone observations. In this contribution we give a detailed picture of changes in total column ozone and resulting potential UV-B radiation in Central Europe for spring and early summer comparing TOMS, version 6.0 data of the two 8 year-periods 1979 to 1986 and 1987 to 1994. The limitation in geographic area and time is not substantial but the result of limited space and capacity. The procedure is applicable to any geographic region and any time interval.

2 Input data and their handling

Daily TOMS, version 6.0 data for Central Europe were aggregated to form decadal means for every decade of the years. Then, decade means for the whole and the two eight-year periods were calculated as well as the difference between both to get information on the regional distribution of Ω changes throughout the year.

3 Climatology of total column ozone and its recent change

To understand the following discussion some introductional remarks are useful. Changes in total column ozone are not simply transferable to changes in UV impacts on biosphere and human society due to different spectral response functions of the processes of interest and due to the weight of astronomical (solar zenith angle) and of meteorological factors (cloudiness, turbidity) in particular on the locally available UV radiation.

As a rule of thumb, it is often quoted that a 1% change in total column ozone is equivalent to a

2% change in UV-B radiation (United Nations Environmental Program, 1989; German Bundestag, 1989, van der Laun et al., 1991). However, the shorter the wavelength, the stronger is the ozone-induced change of UV radiation (Bais et al. 1993, Wang and Lenoble 1994, Bojkov et al. 1995b).

Figs. 1 and 2 give the decadal means of total column ozone for April, May and June of the two eight-year-periods 1979 to 1986 (Fig. 1) and 1987 to 1994 (Fig. 2) in Central Europa. The political boundaries indicated are for better orientation. The colour scale is the same for both time periods with red tones for high (> 400 DU), yellow and green tones for medium (340-390 DU) and blue tones for low ozone amounts (< 340 DU). Both figures indicate the seasonal course as well as the latitudinal effect: The decline in ozone from April to June, and lower ozone amounts, up to about 15 DU, in the South than in the North.

The changes in ozone between both periods for the same decades as in the preceding figures are shown in Fig. 3. Red tones mean ozone increase, yellow, green and blue tones mean decrease in ozone during the last years. Although there are intervals and areas of temporarily more ozone in the last years, particularly in the first June and second May decade, due to the synoptic patterns dominant during that time, the overall picture is the decline in ozone within the last years. This decline is particularly strong in the last April and the first May decade with up to 50 DU and more in some regions. The dependence of ozone amount on weather patterns is reflected in the distinct geographic structure of ozone change. During the time period discussed the decline of ozone amount in the North is twice as much as in the South. This difference in ozone amount changes within such a rather small regional scale is remarkable and stresses the need for continuing monitoring and analysis to meet practical requirements.

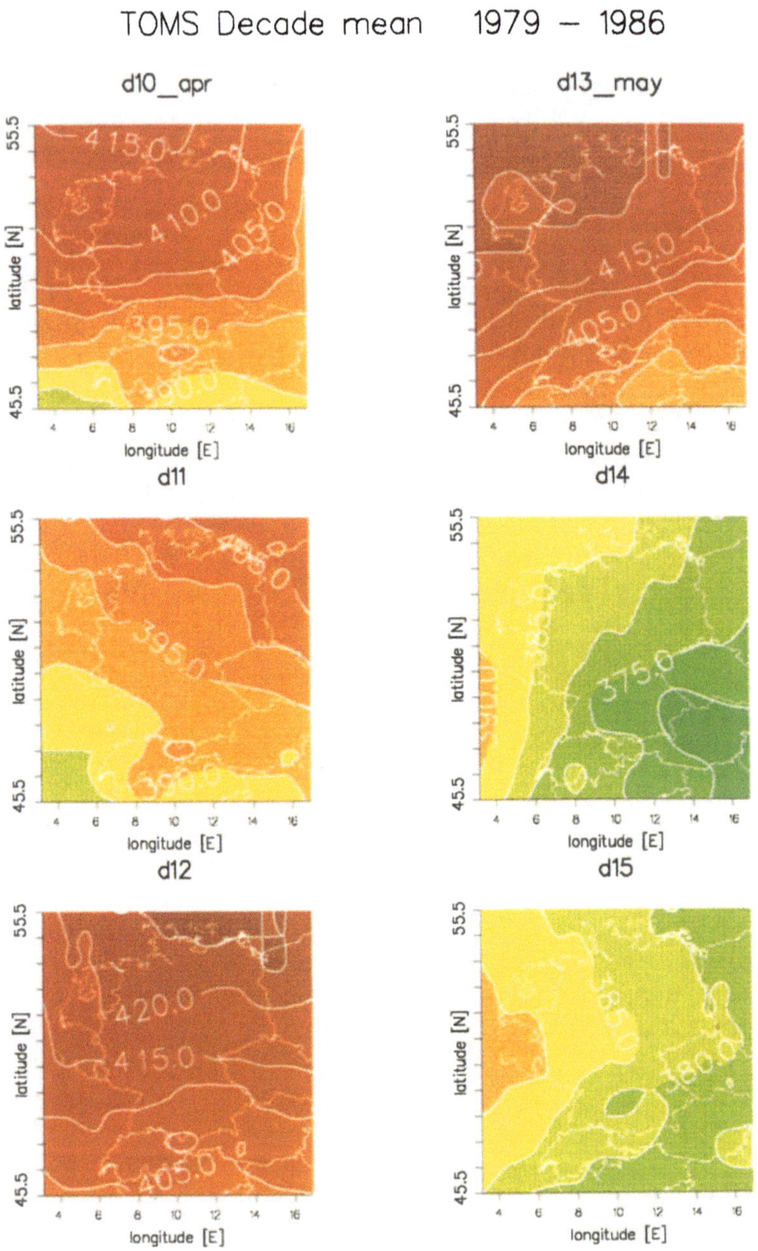

Figure 1a TOMS total ozone decadal means
1979-1986
left: April decades, right: May decades, di-ith decade of the year
Colour scale see Fig 1b

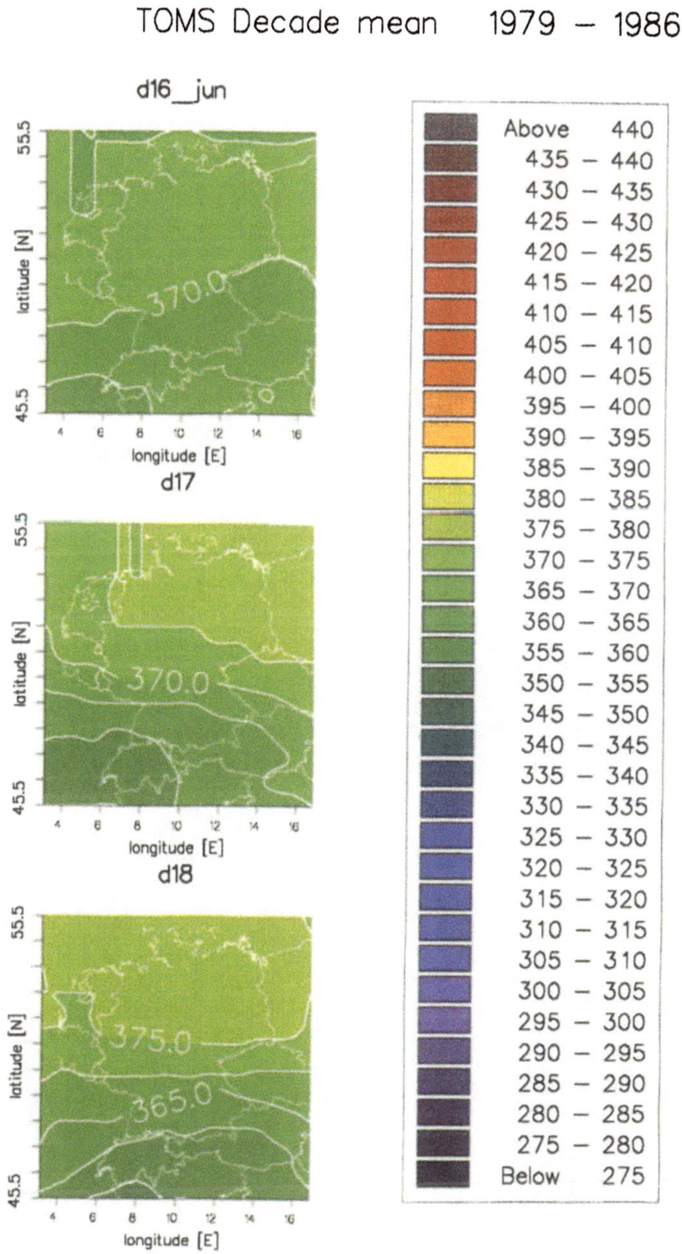

Figure 1b TOMS total ozone decadal means
1979-1986
June decades

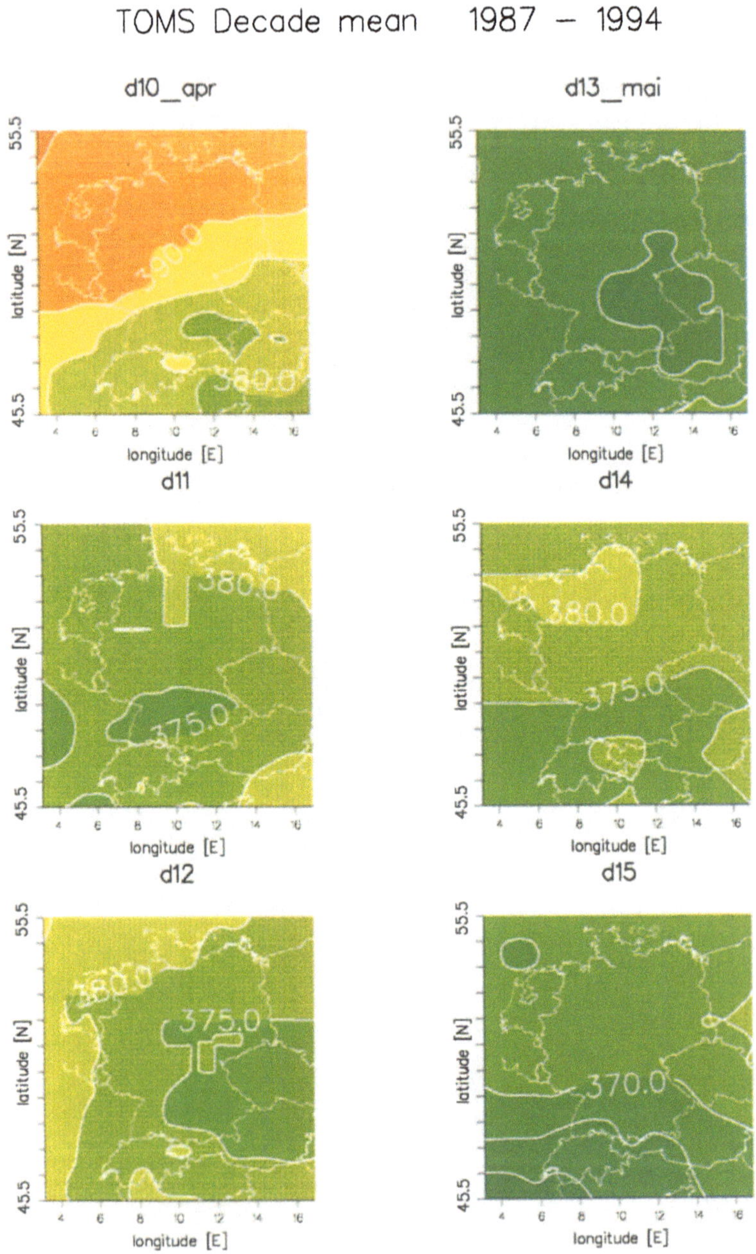

Figure 2a TOMS total ozone decadal means 1987-1994
left: April decades, right: May decades, di-ith decade of the year
Colour scale see Fig 2b

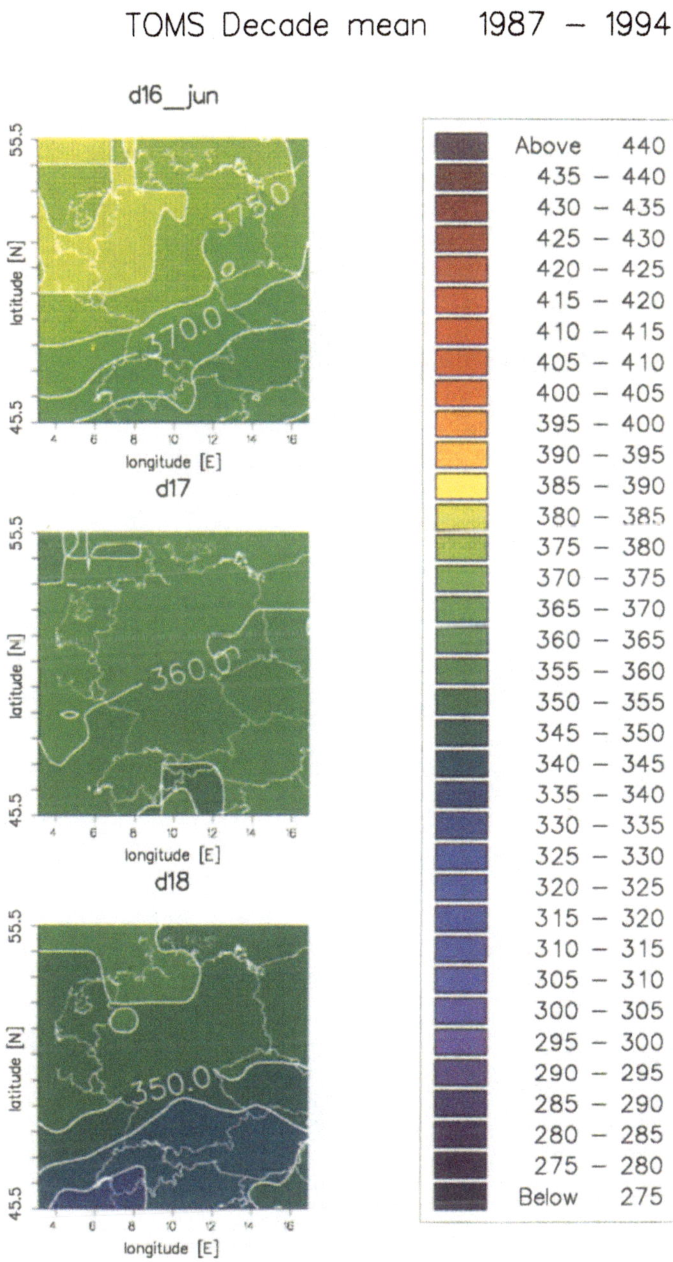

Figure 2b TOMS total ozone decadal means
1987-1994
June decades

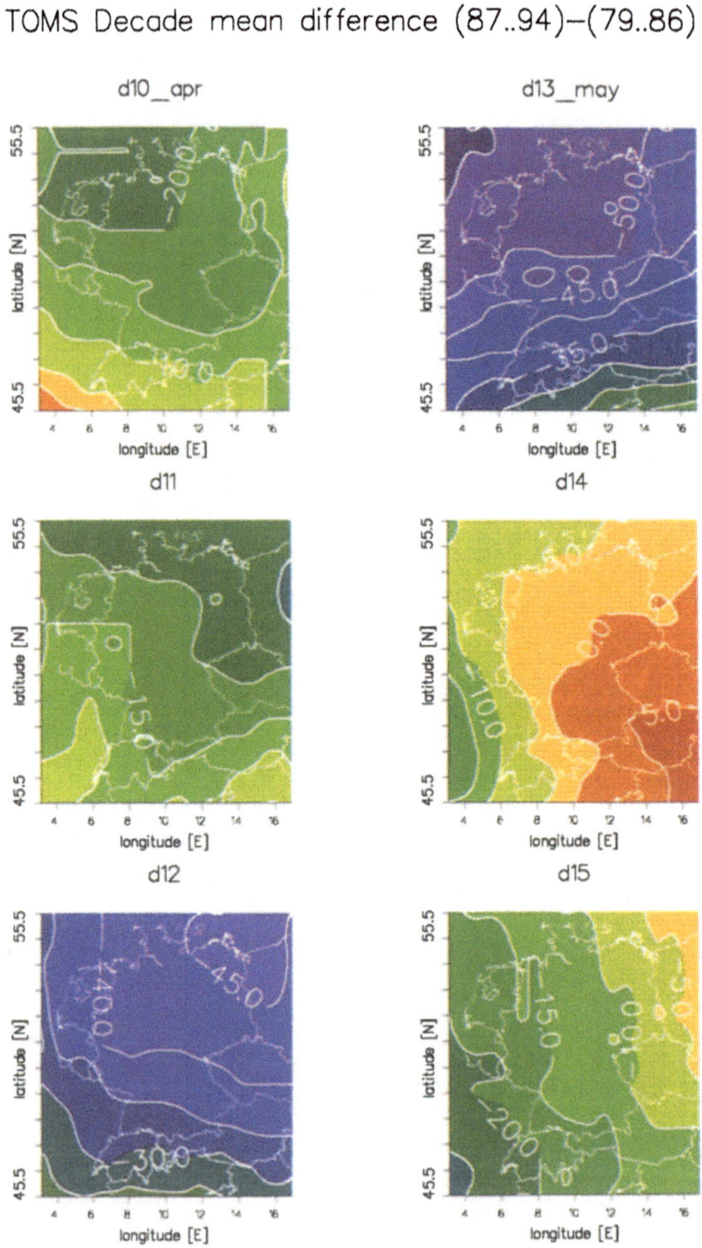

Figure 3a Decadal differences in total ozone 1987/94-1979/86
left: April decades, right: May decades, di-ith decade of the year
Colour scale see Fig 3b

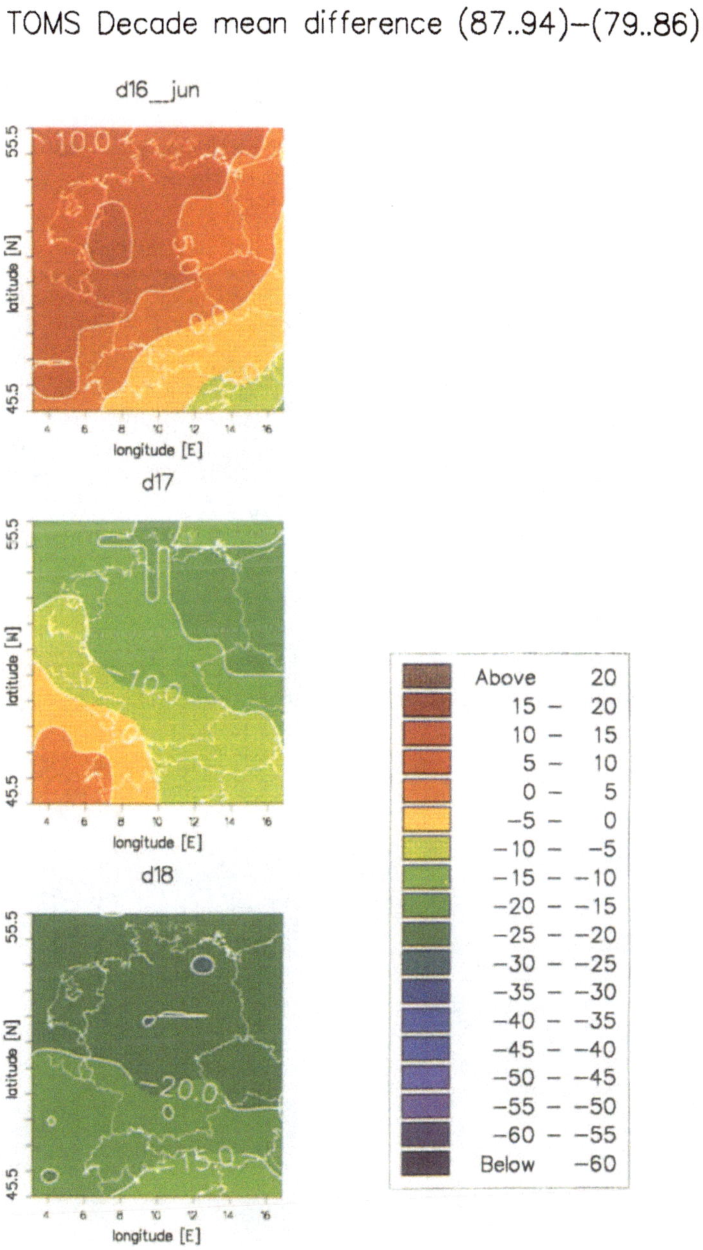

Figure 3b Decadal differences in total ozone
1987/94-1979/86
June decades

4 Climatology of the UV Index

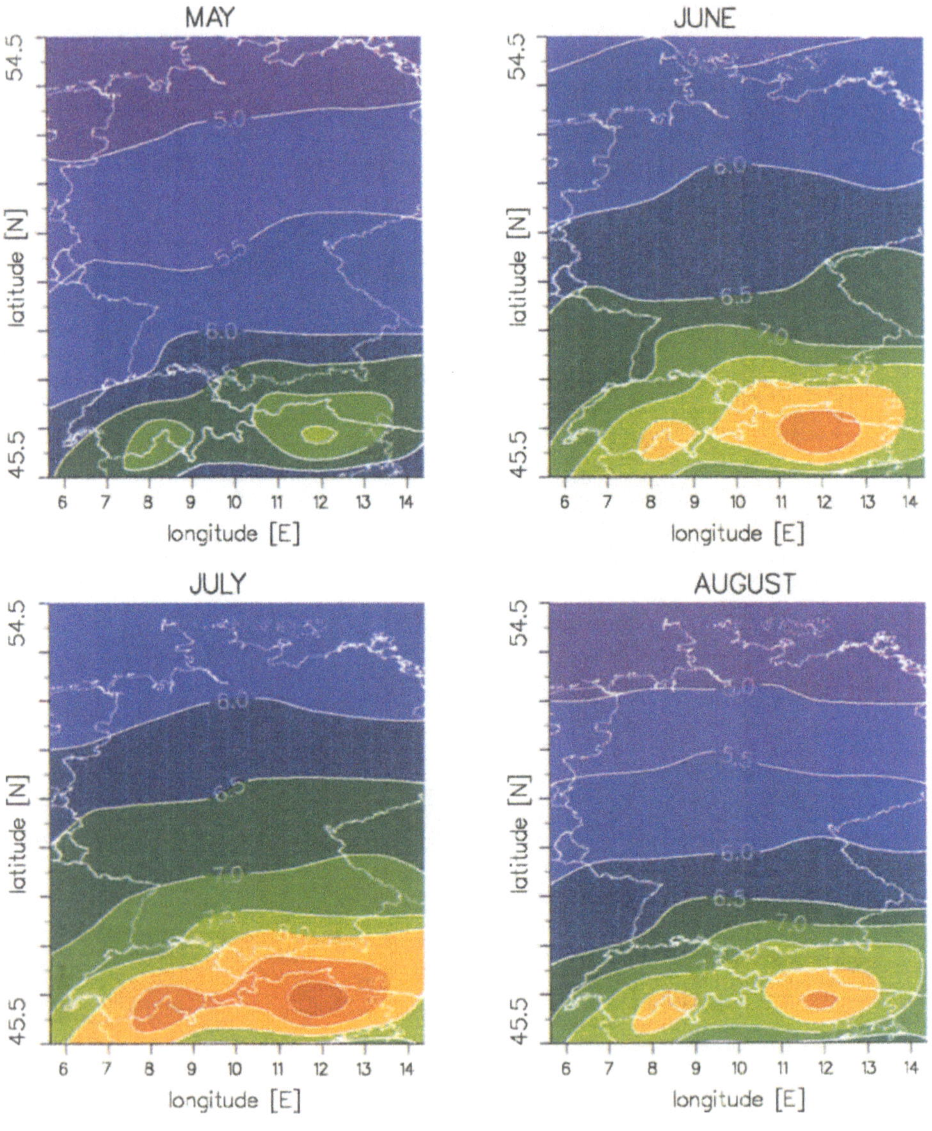

Figure 4 Regional distribution of the potential UV index 1987 to 1994 for the second decades from May to August for clear sky

It is a straightforward task to derive the climatology of the potential UV radiation, i.e. of the UV radiation at cloudless sky, from the Ω climatological charts. This is done in Figure 4 for the months May to August and the period 1987 to 1994 using LOWTRAN 7. The UV radiation is given by the UV-index

Table UV Index, Behaviour and Self Protection Time of Untanned Skin

UV Index at Midday	UV Irradiance	Recommendation	Skin Type I	Skin Type II	Skin Type II	Skin Type IV
			Time to sunburn for untanned skin [min]			
above 9.0	extremly high	non sun bathe, activities in open air only with severe protection against sun irradiance, even for face, neck, arms and legs. Keep in shadow	about 10	about 15	less than 25	about 30
7.0 to 8.9	high	avoid sun bathe, prefer shadow, protect against sun radiation by sunglasses, lotion, clothes	15-20	about 20	25-30	about 35
4.0-6.9	moderate	avoid sun bathes at midday	20-25	25-35	35-45	40-60
below 4.0	low	none	< 30	< 40	< 60	< 75

Skin types I to IV: extremely sensitive, sensitive, moderate sensitive, least sensitive

$$UV = 40 \int_{290}^{400} \frac{\partial G(\lambda)}{\partial \lambda} \; S_{er}(\lambda) \, d\lambda$$

with λ wavelength in nm, G total solar radiation (Wm^{-2}) and S_{ser} erythem impact function (Mc Kinlay and Diffey 1987a,b). There is generally an increase in the UV index of about 2 from the coastal areas in the North to the mountain regions in the South and of about 1 for June/July against May/-August due to the higher sun declination in June and July than in May and August. The maximum UV index in the Alps is the result of high elevation and the albedo effect of the snow cover.

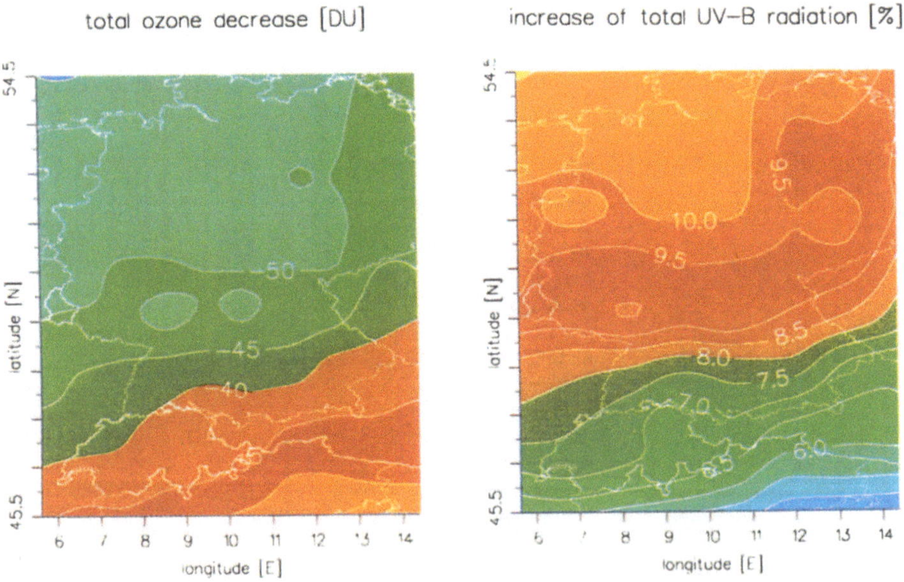

Figure 5 Difference in total ozone (DU) during the first May decade between the time periods 1987/94 and 1979/86 (left) and resulting increase in potential UVB radiation in per cent (right).

The relation between UV index and the time of occurence of sunburn effects is given in the Table for untanned skin and the four skin classes (Staiger 1995, personal communication). Moderate UV irradiance in the North with steady increase to high UV irradiance in the South is observed during these months.

Figure 5, right shows the percentage increase of UV-B radiation in the second 8-year-period (1987-1994), compared to the first 8-year-period (1979-1986) during the first May decade, due to the difference in Ω in both periods (shown left). The 35 to 50 DU decrease in Ω results in 6 to 10% increase in potential UV-B radiation in May in mid latidudes.

5 Conclusions

The regional distribution of total ozone amount as well as the resulting potential UV-B radiation and their changes in space and time are easily studied by using long-term high quality satellite data. Only a small part of all data aggregated and analyzed at the Potsdam Observatory are presented. Within the period discussed we found two times more reduction in ozone (50 DU) within the last 8 years in Northern than in Southern Germany. The decrease of 35 to 50 DU means an increase in potential UB-B radiation of 6 to 10%. The differences within a rather small regional scale are remarkable and need continuing monitoring.

Acknowledgement: The study was in part supported by the Bundesministerium für Forschung und Technologie, Bonn, under grant 01 LO 9204/0.

References

Bais A. F., Zerefos C. S., Melet C., Ziomas I. C. and Tourpali K. (1993) Spectral measurements of solar UVB radiation and its relations to total ozone, SO_2, and clouds, *J. Geophys. Res., 98,* D3, 5199-5240.

Bojkov R. D., Fioletov V. E., Balis D. S., Zerefos C. S., Kadygrova T. V. and Shalamjansky A. M. (1995a)) Further ozone decline during the northern hemisphere winter-spring of 1994-1995 and the new record low ozone over Siberia. *Geophys. Res. Lett.* (in print)

Bojkov R. D., Fioletov V. E. and Diaz S. B. (1995b)) The relationship between solar UV irradiance and total ozone from observations over southern Argentine.*Geophys. Res. Lett. 22,* 1249-1252.

Enke W., Vogel G. and Spänkuch D. (1996) Weather-pattern dependent total column ozone trends in Central Europe *(this issue)*

German Bundestag (ed.) (1989) Protecting the Earth's Atmosphere, an International Challenge, Bonn, p. 310 ff.

Mc Kinlay A. F. and Diffey B. L. (1987a)) A reference spectrum for ultraviolet induced erythema in human skin. In Poschler W. R. and Bosnajokovic F. M. Human Exposure to Ultraviolet Radiation: Risks and Regulations. Elsevier, Amsterdam

Mc Kinlay A. F. and Diffey B. L. (1987b)) A reference action spectrum for ultraviolet induced erythema in human skin. *Commission Internationale de l' Eclairage (CIE)-Journal 6,* 17-22.

UNEP (1989) "Environmental Effecte Panel Report", United Nations Environmental Programme, Nairobi, Kenya.

van der Laun J. C., Terini M. and Worrest R. C. (1991) In UNEP-WMO, Environment Effects of Ozone Depletion, 1991 Update. Nairobi, Kenya.

Wang P. and Lenoble J. (1994) Comparison between measurements and modeling of UV-B irradiance for clear sky: A case study, *Appl. Opt.* *33*, 3964-3971.

Determination of Cloud Properties and Cloud Type from DOAS-Measurements

T. Wagner, T. Senne, F. Erle, C. Otten, J. Stutz, K. Pfeilsticker and U. Platt

Institut für Umweltphysik, University of Heidelberg, INF 366,
D-69120 Heidelberg, Germany

Abstract. Zenith scattered DOAS-measurements (Differential Optical Absorption Spectroscopy) have been developed into a powerful tool for the detection of a variety of atmospheric trace gases such as O_3, NO_2, NO_3, OClO, BrO and the tropospheric species O_4 and H_2O [Solomon et al. 1987]. Up to now it was generally assumed, that tropospheric clouds have only a very small effect on these measurements [Noxon et al. 1979, Sarkissian 1991]. In recent investigations [Roozendael et al. 1994, Erle et al. 1995] it was found that the influence of clouds cannot be neglected. This finding is not only important for the correction of atmospheric measurements but also provides valuable information about cloud properties and their potential influence on the global climate.
Here examples for the influence of clouds on quantities measured by the DOAS-technique, like optical density of absorption features, light intensity, and colour-index (intensity-ratio of wavelenghts 682 nm and 388 nm) are presented.
The data were taken independently with different spectrometers operated at Kiruna (northern Sweden), airborne from the Transall over northern Europe and shipbased on a cruise from Germany to South Africa.

1. Influence of Clouds on Atmospheric DOAS-Measurements

A spectrometer analysing zenith-scattered light observes an intensity-integral over the height above the spectrograph. For clear sky conditions this height-dependent intensity is a product of the intensity of the direct light from the sun and of the scattering probability which is proportional to the atmospheric pressure (Fig.1a). In the presence of a cloud layer an important change occurs. On the one hand direct light from the sun does not even reach the airmasses below the cloud and on the other hand the scattering probability, in the height of the cloud is largely enhanced due to additional Mie-scattering (Fig.1b). In addition inside clouds multiple Mie-scattering leads to enhanced light paths, which cause increased trace gas

NATO ASI Series, Vol. I 53
Atmospheric Ozone Dynamics
Observations in the Mediterranean Region
Edited by Costas Varotsos

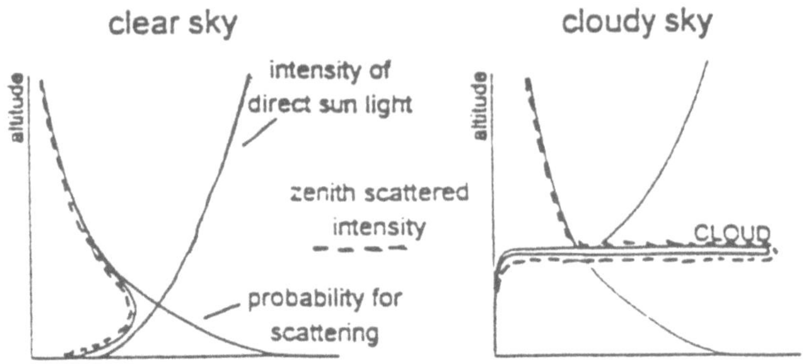

Figure1: Light intensities in dependence of the height (dashed lines) which originates the spectrograph from the zenith for clear (left) and cloudy sky (right).

absorptions. Therefore DOAS-spectra for clear and cloudy sky show several important differences:
The top of Fig.2 shows two DOAS-spectra measured at the same solar zenith angle of 85.5°. While spectrum **B** was recorded under clear sky conditions spectrum **A** was measured while clouds were present. It is quite remarkable that the intensity of spectrum **A** is larger than of spectrum **B**. In the center of Fig.2 the ratio of both spectra is shown. It can be noticed that in the short wavelength region the intensity-ratio is around unity, while for the larger wavelengths the ratio increases to > 3. When clouds appear in the instruments field of view the light detected by the spectrograph originates not only from Rayleigh but also from Mie-scattering. The much weaker wavelength dependence of this Mie-scattering increases the intensities of long wavelengths relatively to the shorter and thus enhances also the colour-index ($I_{682\ nm} / I_{388\ nm}$) .
In the bottom of Fig.2 the high-pass filtered spectrum of the intensity-ratio is shown. Strong absorption bands of O_4 and H_2O are clearly visible. In particular in the case of O_4 these absorptions cannot be explained with natural changes of the atmospheric content. Therefore they are likely caused by multiple scattering inside clouds.
Fig.3 shows the diurnal variation of these three quantities for Feb. 4, 1994. On that day a high correlation between the O_4-absorptions, the light intensity and the colour-index is found as illustrated in Fig.2. This case of simultaneous increase in the presence of clouds is frequently observed and should in the following be called **Case 1**.
Clouds do not only enhance trace gas absorptions. They also change the geometry of the light path through the atmosphere and thus lead to a decrease of the photon path length through the troposphere. Therefore thin high clouds may cause even a decrease of **tropo**spheric absorptions. An example for such a case is shown in Fig.4. In the following it will be called **Case 2**.

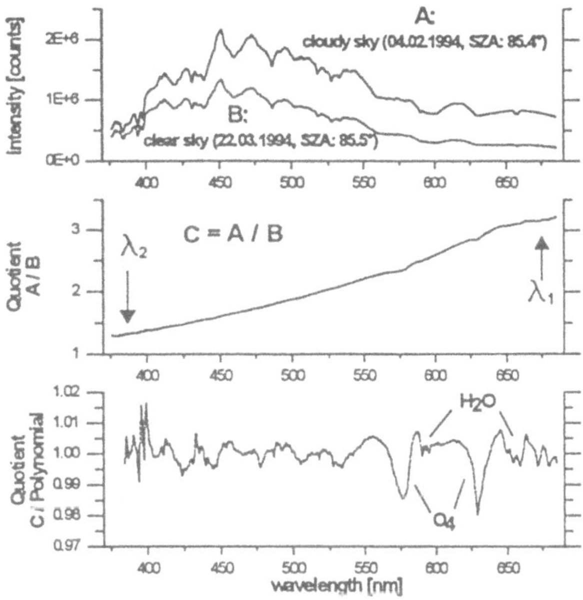

Figure 2: top: DOAS-vis-spectra for cloudy and clear sky at the same SZA.
center: The ratio of both spectra shows the large intensity strongly varriing with wavelength.
bottom: Spectrum of Fig. 2 center after passing a high-pass filter.

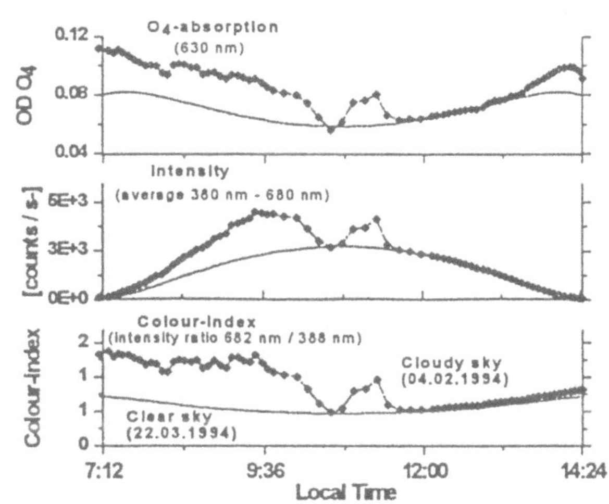

Figure 3: Example for absorption enhancement due to clouds: Diurnal variations of O_4-absorptions (top), light intensity (center) and colour-index (bottom) for cloudy sky (drawn lines) and clear sky (dotted lines).

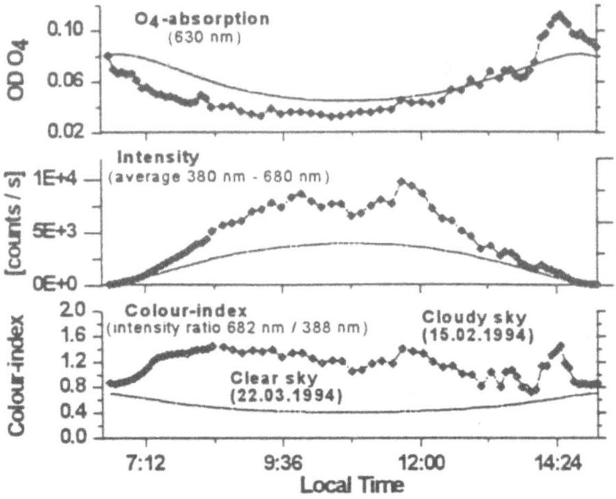

Figure 4: Example for absorption decrease due to clouds: Diurnal variations of O_4-absorptions (top), light intensity (center) and colour-index (bottom) for cloudy sky (drawn lines) and clear sky (dotted lines).

When calculating optical path length enhancements from the differences of trace gas absorptions between clear sky and cloudy sky spectra always this absorption-decreasing effect of clouds must be taken into account.

Finally thick clouds enhance the trace gas absorptions and the colour-index but decrease the measured light intensity because of large Mie-extinction inside the cloud. This **Case 3** was very rare in the Arctic winter. One example is shown in Fig.7. Fig.5 illustrates how clouds principly influence the absorption of tracegases.

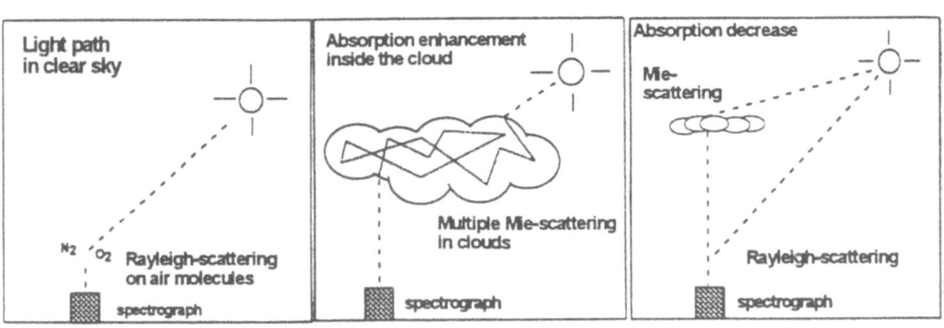

Figure 5: Lightpath through the troposphere under clear and cloudy sky conditions.

2. Determination of Cloud Height

In the case of thick clouds the absorption enhancements of the different trace gases can be determined with a good signal to noise ratio. Then it is possible to calculate an average cloud height because of the different concentration profiles of several trace gases (Fig.7). On March 5 1994, when a warm front passed Kiruna large absorption enhancements were measured for all tropospheric absorbers (Fig.6). For an average cloud height of about 4-6 km the calculated light path enhancements for the different trace gases agree within their errors (Fig.8). This average cloud height is in good agreement with cloud observations from satellite which give an cloud extention from 500 m to 9 km height for that day (Eberhard Reimer, FU Berlin).

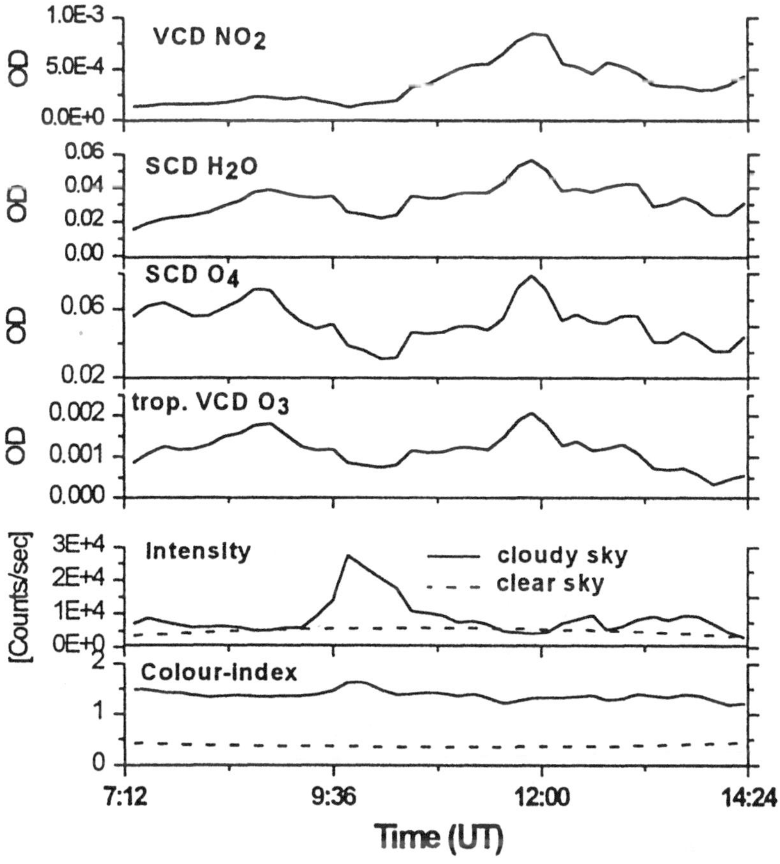

Figure 6: Diurnal variations of different trace gas absorptions, light intensity and colour-index for March, 5 1994 when a warm front passed Kiruna. The dashed vertical line shows the time for which an average cloud height was determined (Fig.8). For this situation the measured light intensity is smaller than for clear sky (Case 3) while the colour-index is still enhanced. The fact that the thickness of clouds hardly affects the colour-index but strongly affects the intensity could be used to estimate the thicknes of clouds.

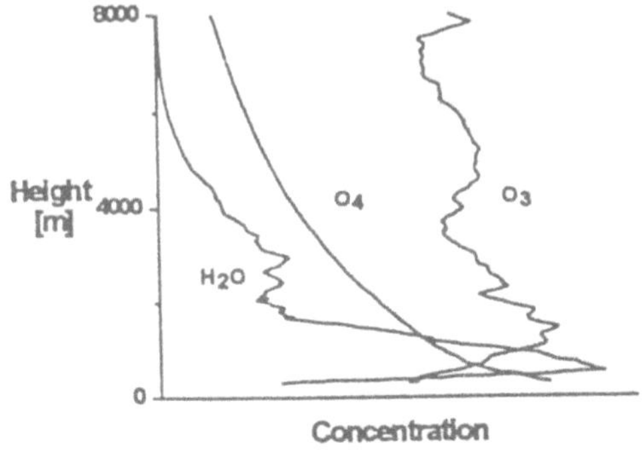

Figure 7: Typical tracegas profiles of O_3, O_4 and H_2O over Kiruna in the winter (derived from O_3-sondes).

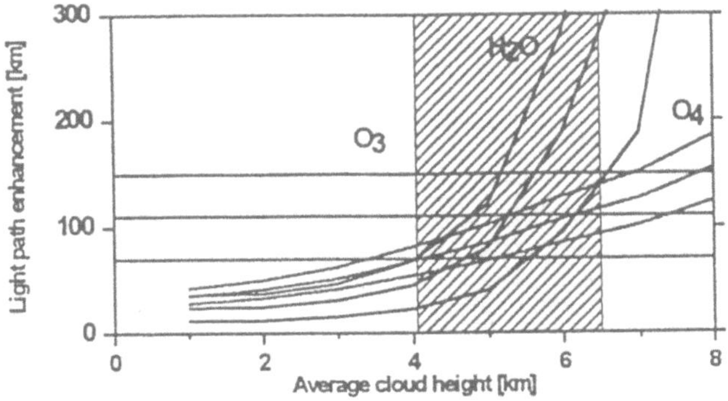

Figure 8: Determined light path enhancements for the different tracegases in dependence of the average cloud height. For an height of about 4-6 km the calculated tropospheric light path enhancements agree within their errors.

3. Frequency Distribution of Tropospheric Light Path Enhancements In the Arctic Climate

Within the precission of our study we assume that the O_3-concentration is nearly constant throughout the troposphere (Fig.7). Therefore O_3-measurements are well suited to calculate tropospheric light paths, because clouds may enhance the O_3-absorptions nearly independently of their height. For the time from Feb 19 to March 27, 1994 about 1300 single O_3-measurements are used to calculate a frequency distribution of tropospheric light path enhancements for the arctic climate in the winter (Fig.9). An average light path enhancement was found to be about 20 km. This distribution is related to mostly stratiform clouds. Large values (greater than 100 km) as detected on March, 5 (Fig.6) belong to air from mid-lattitudes and are very rare (<1%). The different width of the distribution for the several SZA-ranges reflects the larger statistical error in the calculation for higher SZA.

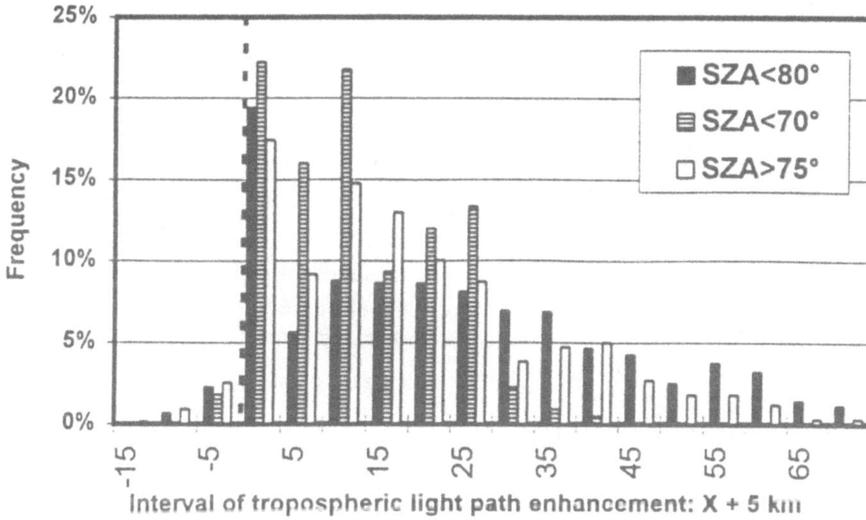

Figure 9: Frequency distribution of tropospheric light path enhancements caused by clouds in the arctic winter. The different width of the distribution for the several SZA-ranges reflects the bigger statistical error in the calculation for higher SZA. The calculations were performed for a time from Feb. 19 until March, 27 including all measurements within the chosen SZA-ranges. The maximum of the distribution and the average value of about 20 km are nearly the same for the different SZA-ranges. This distribution is related to mostly stratiform clouds. The rare cases with tropospheric light path enhancements greater than 100 km (<1%) when airmasses from mid-lattitudes appeared (Figures 6,8) are not shown in this distribution.

4. Determination of Cloud Type

DOAS-measurements in the visible region allow the discrimination between the cloud type C_L (Cu, Cb, Sc, St) and type C_M (Ac, As, Ns) or C_H (Ci, Cs, Cc). Furthermore it is possible to detect from DOAS spectra whether the type C_L in the instrumental field of view is located higher or lower than 1 km.
The spectra were recorded on a cruise of the German research vessel „Polarstern“ from Germany to South Africa in October 1993. Additionally weather observations were provided from the DWD (Deutscher Wetterdienst) for that cruise and data from the cloud-height laser (AWI). For the detection of cloud effects it was made use of the optical density of the O_4-species at 477 nm. By comparing the optical densities between clear sky days and cloudy sky days three different cases can be observed.

Observed phenomena in the DOAS-straylight spectra	*Cloud Type*
a) O_4 optical density enhance-ments from which photon path lenght enhancement up to about 300 km can be determined.	clouds of the group C_L (the effect in the O_4 was observed on all 29 days when these clouds occurred)
b) O_4 optical density reductions for all SZA's	clouds of the groups C_M and C_H (the effect was observed on all 12 days when these clouds occurred)
c) O_4 optical densities as during clear sky but characteristic reductions between a SZA of 82° to 84° to higher SZA.	Clouds of class C_L with bottoms > 1 km (14 days)

Table 1: Cloud Types detected by DOAS. (C_L, C_M, C_H clouds in low, middle and high height)

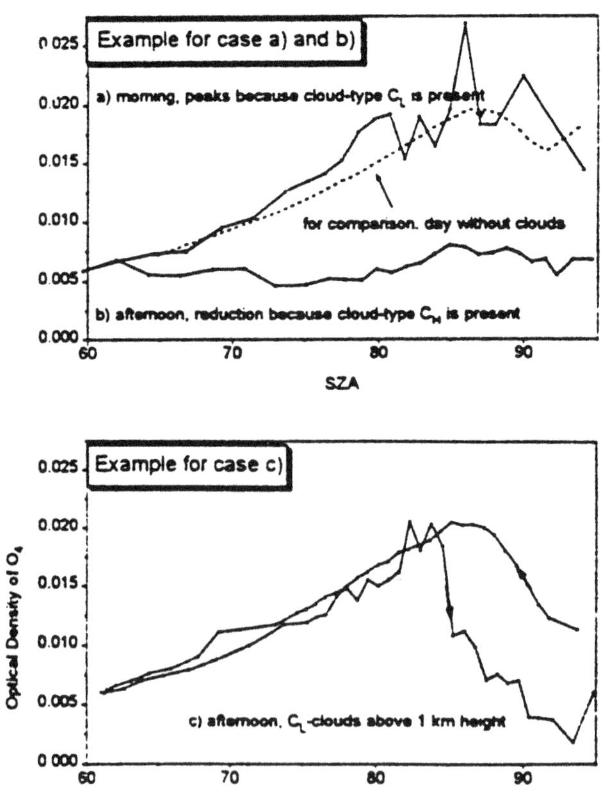

Figure 10: Examples for cloud detection with the SZA-dependence of the optical density of O_4.

5. Conclusion

Clouds influence DOAS-measurements in several ways:

a) For zenith-scatterd sunlight clouds cause large changes of intensity and colour-index. Especially the colour-index can be used as an indicator for the appearance of clouds because even thin clouds lead to significant changes.

b) The light intensities in the presence of clouds vary widely with the cloud thicknes. Thus this quantity could be used together with the colour-index to estimate the thicknes of clouds.

c) Because of multiple Mie-scattering inside clouds the absorptions of tropospheric species are enhanced compared to clear sky conditions. For most clouds which appeared over Kiruna (northern Sweden) several trace gas absorptions were increased simultaneously with the light intensity and the colour-index.

d) Clouds do not only enhance the photon path lengths but they principly change the geometry of the light path, which leads to an absorbtion decrease in the troposphere compared to the clear sky absorptions. Normally this decrease is smaller than the absorption enhancement inside clouds and thus overall an absorption increase is typically found. But the absorption-decreasing effect of clouds has always to be taken into consideration when calculating light path enhancements.

e) Because of the different tropospheric trace gas concentration profiles it is possible to calculate an average cloud height from the absorption enhancements of several species.

f) From O_3-measurements a nearly independent of cloud height tropospheric light path enhancement can be determined because of the relatively constant tropospheric O_3-concentration. For tropospheric clouds in arctic winter a frequency distribution of light path enhancements shows an average value of about 20 km, representing mostly stratiform clouds.

g) Some special cases of cloud-influence on the diurnal variations of O_4-absorptions can be significantly assigned to different types of clouds. Thus DOAS-measurements could be used for automatic cloud observations.

h) For some rare cases in the arctic climate and more cases in the tropics and mid-lattitudes tropospheric light path enhancements > 100 km were measured.

6. ACKNOWLEDGEMENTS

The authors thank E. Reimer from the FU Berlin for his friendly interpretation of satelite data in reference to cloud properties. We also like to thank Gert König-Langlo from AWI / Bremerhaven for his help on cloud questions.

7. References

Erle et al. 1995 — On the Influence of tropospheric clouds on zenith-scattered-light-measurements of stratospheric species, *Accepted by Geophys. Res. Lett (22.08.1995)*

Noxon et al. 1979 — Stratospheric NO_2, Observational method and behavior at mid-altitude, *J.Geophys. Res, 84, 5047-5065, 1979*

Pfeilsticker et al. 1995 — 'Can enhanced tropospheric photon path lengths explain the anomalous cloud absorption phenomenon?', *Poster at the Third European Symposium on Polar Stratospheric Ozone, Schliersee, Bavaria, Germany, 18.-22. September 1995.*

Roozendael 1994 — Ground-based visible measurements at the Jungfraujoch station since 1990, *J. Quant. Spectrosc. Radiat. Transfer, 52, 3/4, 231-240, 1994*

Sarkissian et al. 1991 — Identification of polar stratospheric clouds from the ground by visible spectrometry *Geophys. Res. Lett. 18, 4, 779-782, 1991*

Solomon et al. 1987 — On the interpretation of zenith sky absorption measurements, *J.Geophys. Res., 92, 8311-8319, 1987*

The ASI Series Books Published as a Result of Activities of the Special Programme on Global Environmental Change

This book contains the proceedings of a NATO Advanced Research Workshop held within the activities of the NATO Special Programme on Global Environmental Change, which started in 1991 under the auspices of the NATO Science Committee.

The volumes published as a result of the activities of the Special Programme are:

Vol. 1: **Global Environmental Change.**
Edited by R. W. Corell and P. A. Anderson. 1991.
Vol. 2: **The Last Deglaciation: Absolute and Radiocarbon Chronologies.**
Edited by E. Bard and W. S. Broecker. 1992.
Vol. 3: **Start of a Glacial.** Edited by G. J. Kukla and E. Went. 1992.
Vol. 4: **Interactions of C, N, P and S Biogeochemical Cycles and Global Change.**
Edited by R. Wollast, F. T. Mackenzie and L. Chou. 1993.
Vol. 5: **Energy and Water Cycles in the Climate System.**
Edited by E. Raschke and D. Jacob. 1993.
Vol. 6: **Prediction of Interannual Climate Variations.**
Edited by J. Shukla. 1993.
Vol. 7: **The Tropospheric Chemistry of Ozone in the Polar Regions.**
Edited by H. Niki and K. H. Becker. 1993.
Vol. 8: **The Role of the Stratosphere in Global Change.**
Edited by M.-L. Chanin. 1993.
Vol. 9: **High Spectral Resolution Infrared Remote Sensing for Earth's Weather and Climate Studies.**
Edited by A. Chedin, M.T. Chahine and N.A. Scott. 1993.
Vol. 10: **Towards a Model of Ocean Biogeochemical Processes.**
Edited by G. T. Evans and M. J. R. Fasham. 1993.
Vol. 11: **Modelling Oceanic Climate Interactions.**
Edited by J. Willebrand and D.L.T. Anderson. 1993.
Vol. 12: **Ice in the Climate System.** Edited by W. Richard Peltier. 1993.
Vol. 13: **Atmospheric Methane: Sources, Sinks, and Role in Global Change.**
Edited by M. A. K. Khalil. 1993.
Vol. 14: **The Role of Regional Organizations in the Context of Climate Change.**
Edited by M. H. Glantz. 1993.
Vol. 15: **The Global Carbon Cycle.**
Edited by M. Heimann. 1993.
Vol. 16: **Interacting Stresses on Plants in a Changing Climate.**
Edited by M. B. Jackson and C. R. Black. 1993.
Vol. 17: **Carbon Cycling in the Glacial Ocean: Constraints on the Ocean's Role in Global Change.**
Edited by R. Zahn, T. F. Pedersen, M. A. Kaminski and L. Labeyrie. 1994.
Vol. 18: **Stratospheric Ozone Depletion/UV-B Radiation in the Biosphere.**
Edited by R. H. Biggs and M. E. B. Joyner. 1994.
Vol. 19: **Data Assimilation: Tools for Modelling the Ocean in a Global Change Perspective.**
Edited by P. O. Brasseur and J. Nihoul. 1994.

Vol. 20: Biodiversity, Temperate Ecosystems, and Global Change.
Edited by T. J. B. Boyle and C. E. B. Boyle. 1994.
Vol. 21: Low-Temperature Chemistry of the Atmosphere.
Edited by G. K. Moortgat, A. J. Barnes, G. Le Bras and J. R. Sodeau. 1994.
Vol. 22: Long-Term Climatic Variations – Data and Modelling.
Edited by J.-C. Duplessy and M.-T. Spyridakis. 1994.
Vol. 23: Soil Responses to Climate Change.
Edited by M. D. A. Rounsevell and P. J. Loveland. 1994.
Vol. 24: Remote Sensing and Global Climate Change.
Edited by R. A. Vaughan and A. P. Cracknell. 1994.
Vol. 25: The Solar Engine and Its Influence on Terrestrial Atmosphere and Climate.
Edited by E. Nesme-Ribes. 1994.
Vol. 26: Global Precipitations and Climate Change.
Edited by M. Desbois and F. Désalmand. 1994.
Vol. 27: Cenozoic Plants and Climates of the Arctic.
Edited by M. C. Boulter and H. C. Fisher. 1994.
Vol. 28: Evaluating and Monitoring the Health of Large-Scale Ecosystems.
Edited by D. J. Rapport, C. L. Gaudet and P. Calow. 1995.
Vol. 29: Global Environmental Change Science: Education and Training.
Edited by D. J. Waddington. 1995.
Vol. 30: Ice Core Studies of Global Biogeochemical Cycles.
Edited by R. J. Delmas. 1995.
Vol. 31: The Role of Water and the Hydrological Cycle in Global Change.
Edited by H. R. Oliver and S. A. Oliver. 1995.
Vol. 32: Atmospheric Ozone as a Climate Gas.
Edited by W.-C. Wang and I. S. A. Isaksen. 1995.
Vol. 33: Carbon Sequestration in the Biosphere.
Edited by M. A. Beran. 1995.
Vol. 34: Climate Sensitivity to Radiative Perturbations: Physical Mechanisms and Their Validation.
Edited by H. Le Treut. 1996
Vol. 35: Clouds, Chemistry and Climate.
Edited by P. J. Crutzen and V. Ramanathan. 1996.
Vol. 36: Diachronic Climatic Impacts on Water Resources.
Edited by A. N. Angelakis and A. S. Issar. 1996.
Vol. 37: Climate Change and World Food Security.
Edited by T. E. Downing. 1996.
Vol. 38: Evaluation of Soil Organic Matter Models.
Edited by D. S. Powlson, P. Smith and J. U. Smith. 1996.
Vol. 39: Microbiology of Atmospheric Trace Gases.
Edited by J. C. Murrell and D. P. Kelly. 1996.
Vol. 40: Forest Ecosystems, Forest Management and the Global Carbon Cycle.
Edited by M. J. Apps and D. T. Price. 1996.
Vol. 41: Climate Variations and Forcing Mechanisms of the Last 2000 Years.
Edited by P. D. Jones, R. S. Bradley and J. Jouzel. 1996.
Vol. 42: The Mount Pinatubo Eruption: Effects on the Atmosphere and Climate.
Edited by G. Fiocco, D. Fuà and G. Visconti. 1996.
Vol. 43: Chemical Exchange Between the Atmosphere and Polar Snow.
Edited by E. W. Wolff and R. C. Bales. 1996.

Vol. 44: **Decadal Climate Variability: Dynamics and Predictability.**
Edited by D. L.T. Anderson and J. Willebrand. 1996.

Vol. 45: **Radiation and Water in the Climate System: Remote Measurements.**
Edited by E. Raschke. 1996.

Vol. 46: **Land Surface Processes in Hydrology: Trials and Tribulations of Modeling and Measuring.** Edited by S. Sorooshian, H. V. Gupta, and J. C. Rodda. 1997.

Vol. 47: **Past and Future Rapid Environmental Changes: The Spatial and Evolutionary Responses of Terrestrial Biota.**
Edited by B. Huntley, W. Cramer, A. V. Morgan, H. C. Prentice, and J. R. M. Allen. 1997.

Vol. 48: **The Mathematics of Models for Climatology and Environment.**
Edited by J. I. Díaz. 1997.

Vol. 49: **Third Millennium BC Climate Change and Old World Collapse.**
Edited by H. N. Dalfes, G. Kukla, and R. Weiss. 1997.

Vol. 50: **Gravity Wave Processes: Their Parameterization in Global Climate Models.**
Edited by K. Hamilton. 1997.

Vol. 51: **Sediment Records of Biomass Burning and Global Change.**
Edited by J. S. Clark, H. Cachier, J. G. Goldammer, and Brian Stocks. 1997.

Vol. 52: **Solar Ultraviolet Radiation. Modelling, Measurements and Effects.**
Edited by C. S. Zerefos and A. F. Bais. 1997.

Vol. 53: **Atmospheric Ozone Dynamics. Observations in the Mediterranean Region.**
Edited by C. Varotsos. 1997.

SPRINGER NATURE

GPSR Compliance

The European Union's (EU) General Product Safety Regulation (GPSR) is a set of rules that requires consumer products to be safe and our obligations to ensure this.

If you have any concerns about our products, you can contact us on ProductSafety@springernature.com

In case Publisher is established outside the EU, the EU authorized representative is:

Springer Nature Customer Service Center GmbH
Europaplatz 3
69115 Heidelberg, Germany

FSC
www.fsc.org
MIX
Papier | Fördert
gute Waldnutzung
FSC® C083411

Kolibri 360 GmbH
Ferdinand-Jühlke-Straße 7
99095 Erfurt, Deutschland
produktsicherheit@kolibri360.de